W0043860

IMMUNOBIOLOGY OF PROTEINS AND PEPTIDES VII

Unwanted Immune Responses

ADVANCES IN EXPERIMENTAL MEDICINE AND BIOLOGY

Editorial Board:

NATHAN BACK, *State University of New York at Buffalo*

IRUN R. COHEN, *The Weizmann Institute of Science*

DAVID KRITCHEVSKY, *Wistar Institute*

ABEL LAJTHA, *N. S. Kline Institute for Psychiatric Research*

RODOLFO PAOLETTI, *University of Milan*

Recent Volumes in this Series

A Continuation Order Plan is available for this series. A continuation order will bring delivery of each new volume immediately upon publication. Volumes are billed only upon actual shipment. For further information please contact the publisher.

IMMUNOBIOLOGY OF PROTEINS AND PEPTIDES VII

Unwanted Immune Responses

Edited by

M. Zouhair Atassi
Baylor College of Medicine
Houston, Texas

SPRINGER SCIENCE+BUSINESS MEDIA, LLC

Library of Congress Cataloging-in-Publication Data

International Symposium on Immunobiology of Proteins and Peptides (7th
  : 1992 : Edmonton, Alta.)
    Immunobiology of proteins and peptides VII : unwanted immune
responses / edited by M. Zouhair Atassi.
        p.   cm. -- (Advances in experimental medicine and biology :
  v.347)
    "Proceedings of the Seventh International Symposium on
Immunobiology of Proteins and Peptides, held October 1-6, 1992, in
Edmonton, Alberta, Canada"--t.p. verso.
    Includes bibliographical references and index.

    ISBN 978-1-4613-6030-8   ISBN 978-1-4615-2427-4 (eBook)
    DOI 10.1007/978-1-4615-2427-4

    1. Autoimmunity--Congresses. 2. Allergy--Congresses.
  3. Proteins--Immunology--Congresses. 4. Peptides--Immunology-
  -Congresses.   I. Atassi, M. Z. II. Title. III. Series.
  QR188.3.I58 1992
  616.97--dc20   94-7976
                                                               CIP

Proceedings of the Seventh International Symposium on Immunobiology of Proteins and Peptides, held
October 1-6, 1992, in Edmonton, Alberta, Canada

SCIENTIFIC COUNCIL OF THE SYMPOSIUM

M.Z. Atassi, *President*
Howard L. Bachrach
Alec Sehon
Constantin A. Bona
Colin R. Young
John J. Marchalonis
Garvin Bixler
Nickolas J. Calvanico, *Secretary*

SYMPOSIUM SPONSORS

The following organizations were *major sponsors* of the symposium:

Applied Biosystems
Connaught Laboratories
Schering Plough Research
Upjohn
ImmuLogic

The following organizations were *contributors* to the symposium:

Abbott Laboratories
Amgen
Amicon
Fisher Scientific
Mr. Leonard L. Fisk
Pierce
Pharmacia
Smithkline Beecham Pharmaceuticals
Steragene
Sandoz

PREFACE

The articles in this volume represent papers delivered by invited speakers at the 7th International Symposium on the Immunobiology of Proteins and Peptides. In addition, a few of the abstracts submitted by participants were scheduled for minisymposia and some of the authors, whose presentations were judged by the Scientific Council to be of high quality, were invited to submit papers for publication in this volume.

This symposium was established in 1976 for the purpose of bringing together, once every two or three years, active investigators in the forefront of contemporary immunology, to present their findings and discuss their significance in the light of current concepts and to identify important new directions of investigation. The founding of the symposium was stimulated by the achievement of major breakthroughs in the understanding of the immune recognition of proteins and peptides. We believed that these breakthroughs will lead to the creation of a new generation of peptide reagents which should have enormous potential in biological, therapeutic, and basic applications. This anticipated explosion has in fact since occurred and many applications of these peptides are now being realized.

The seventh symposium focused on immune responses that have undesirable effects on the host, hence we named them *unwanted immune responses*. Two major aspects of unwanted immune responses were discussed at the symposium: Allergy and Autoimmunity. We believed that these immune reactions may be controlled or manipulated by similar strategies. This volume presents papers on the molecular nature of allergens, IgE-mediated responses and the IgE receptors. The papers on autoimmunity deal with autoimmune antibody and T-cell recognition of self antigens in selected well-characterized autoimmune diseases (e.g., rheumatoid arthritis, myasthenia gravis, systemic lupus erythematosus and autoimmune, type 1 diabetes). Stress proteins, naturally occurring autoantibodies and regulatory autoantibodies are now known to play important roles in autoimmunity. Papers by leading investigators present recent advances in this field. Manipulation of autoimmune responses by synthetic pathogenic autodeterminants provides promising strategies for control of autoimmune diseases whose antigens have been well characterized. The feasibility of this exciting direction is demonstrated here with autoimmune diseases, but should in principle be applicable to allergic responses as well. These and other important questions related to molecular and cellular characterization of allergic and autoimmune responses are discussed in this volume.

Finally, I should like to express, on behalf of this organization, our gratitude to our sponsors whose generous support made this conference possible.

M. Zouhair Atassi

CONTENTS

IS THERE A LINK BETWEEN THE NATURE OF AGENTS THAT TRIGGER MAST CELLS AND THE INDUCTION OF IMMUNOGLOBULIN (IG)E SYNTHESIS?

Birgit A. Helm

Krebs Institute for Biomolecular Research
Department of Molecular Biology and Biotechnology
The University of Sheffield
PO Box 594, Sheffield S10 2UH, G.B.

THE ALLERGIC RESPONSE

What is allergy? Usually the statement "*I am allergic*" suggests that one responds to the contact with a seemingly harmless extraneous entity in an adverse and exaggerated fashion. The term was originally coined by Pirquet in 1906, meaning "altered reactivity", it was used to characterise the change in reaction which has occurred following exposure to an antigen. Present day immunologists attribute the manifestations of allergy to a condition where the immune system replies to the initial encounter with a foreign substance by producing an antibody of the immunoglobulin (Ig)E isotype, and the role of IgE antibodies in mediating the clinical symptoms of both immediate and delayed hypersensitivity reactions has been extensively documented since their discovery as a separate Ig class in 1966 [1].

IgE antibodies are made up of two light chains (κ or λ isotype) which combine with two ε-chains to form the variable region involved in antigen recognition. The ε-chain is a five domain glycoprotein which forms a covalent homodimer via two cystine residues linking opposite $C\varepsilon2$ domains. In normal individuals, the plasma levels of this antibody are lower than those of any other class; their synthesis is thought to be tightly controlled by a network of B- and T-lymphocytes [2], which is probably a reflection of the involvement of this

Immunobiology of Proteins and Peptides VII
Edited by M.Z. Atassi, Plenum Press, New York, 1994

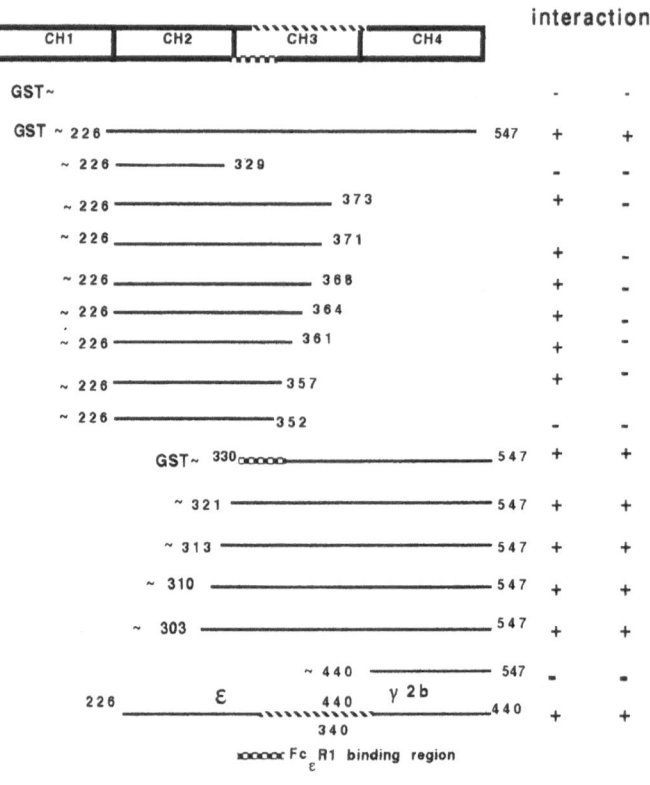

Mapping of receptor binding regions in human IgE

Figure 1. Recombinant ε-chain fragments were expressed in *E. coli* as gluthathione S-transferase fusion proteins and isolated by standard procedures. To measure rFcε/FcεR1 interaction, the transfected RBL-2H3 cell-line was incubated for 24h at 37°C with 10^{-6} M dexamethazone to induce receptor expression[18], after washing with buffer, cells were incubated with 200 fold molar excess rFcε peptides for 45 min at 20°C before addtion of ^{125}I-labelled hIgE V_{NP}. Cells were washed, lysed, and radioactivity was counted.

FcεR2/CD23 interaction was measured by FACScan of 8866 lymphoblastoid leukaemia cells bearing FcεR2[8].

antibody isotype in the activation of the release of cytotoxic mediators from effector cells that bear the class-specific Fc-receptors for this Ig [3]. IgE has not only the lowest levels but also the shortest serum half-life of all classes of immunoglobulins. The majority of IgE molecules are found bound to cells expressing high- (FcεR1) and low-affinity receptors (FcεR2) [4].

These receptors are present on leukocytes in the blood and the cells lining the mucosal membranes of the gut and the respiratory organs. Employing a series of overlapping ε-chain derived fragments shown in Figure 1 we found that the site(s) in IgE that interact with both receptors are located in the Cε3 domain [5], an observation also made by others [6,7]. The region(s) in IgE that interact with FcεR2 are close to, but not identical to the site(s) that engage the mast cell and basophil receptor [5,7,8]. There is evidence that IgE bends out of

2

plane at the interface between the Cε2 and Cε3 domain with both Fab arms and the C-terminal end of the Fc segment sticking out, away from the membrane [9].

Cross-linking by cognate antigen/allergen of cellular receptors occupied by IgE results in the release of the pharmacologically active agents that cause the clinical symptoms of allergy; but there is also compelling evidence that these mechanisms will confer protective immunity to parasites and some viral infections [3,10]. This beneficial effect is however generally considered to be of limited value in populations that live in an environment where parasitic infestations are rare, and the synthesis of an antibody of the IgE isotype is usually regarded as an unwanted immune response. Allergic reactions are rarely fatal, but with more than 20% of the human population being afflicted by some type of allergy and evidence that the incidence of the disorder is increasing world-wide, the development of effective therapeutic interventions in atopic disease is of considerable medical, commercial and social importance and an improved understanding of the molecular mechanisms involved in the genesis of the allergic response may expedite such a quest.

FACTORS INFLUENCING THE ANTIGEN/ALLERGEN SPECIFIC IMMUNE RESPONSE

At birth, IgE levels are not measurable or exceedingly low and the mechanisms that give rise to an IgE response in certain individuals as the result of the encounter with a particular antigen are at present unresolved and the subject of intense investigations in many laboratories. Several diverse factors seem to contribute towards the development of allergic disease:

1. Individual predisposition

When exposed to e.g. airborne antigens, about 20-30% of the population develop IgE antibodies. Individual variations in mucosal permeability, which may influence antigen presentation, may account for some of the observed differences in response to allergens. In addition, familial tendencies in the development of atopic disease have been observed, nearly 50% of individuals who have two allergic parents develop allergies, compared to 15% in the normal population, although no simple pattern of Mendelian inheritance could be established. There is evidence for MHC-gene control, and it has been suggested that a recessive allele determines high IgE levels, but other genes may also influence the trait, and linkage studies suggest that a gene present on chromosome 11 controls the development of atopic disease [11].

2. Mediator release by effector cells

The symptoms of immediate hypersentivity are caused by the secretion of preformed and newly synthesized mediators from mast cells and basophils, which express high-affinity IgE receptors. Mast cell-mediated cytotoxicity resembles natural cytotoxic activity since the release of mast cell-derived chemotactic factor of anaphylaxis (ECF-A), tumor necrosis (TNF) and related cytotoxic factors can be demonstrated from cells of basophil/mast cell lineage. The chemotactic effect of these agents for the recruitment of monocytes, neutrophils and eosinophils is well known and infiltration of e.g. lung tissue by these cells has been observed in asthmatic patients in the late phase [12]. In addition, eosinophil-mediated cytotoxicity is dependent on mast cell-mediators, and in tissues affected by immediate and delayed type hypersensitivity reactions, eosinophils, mast cells/basophils are frequently found in close association. Another relevant observation in this context is that IgE-activated eosinophils generate mediators such as eosinophil peroxidase and oxygen metabolites which can induce mast cell secretion through a non-immunological stimulus [13]. There is also compelling evidence that mediators released by mast cells and basophils induce the selective activation and expression of $Fc\varepsilon R2$ on inflammatory cell sub-populations [3] which are involved in the late phase allergic reaction. An initial inhibition of mast cell and basophil activation is therefore highly desirable in order to reduce the clinical symptoms associated with both immediate and delayed hypersentivity.

Empirical observations suggest that the manifestation of allergic disease is closely linked to the serum concentration of IgE. The precise nature and sequence of events that determine class switching is still unresolved. Isotype specific synthesis is known to be influenced by cytokines [14], and the role of cytokines produced by T-cells in the regulation of IgE synthesis by B-cells has been recognised for some time [15]. In the mouse[16], two types of helper T cells have been distinguished based on their pattern of cytokine production: TH_1 cells produce IL-2, interferon- (IFN)γ, IL-3, and granulocyte macrophage colony stimulating factor (GM-CSF), while TH_2 cells produce IL-3, GM-CSF, IL-4, IL-5, and IL-10. Of particular interest is IL-4 which is known to promote $Fc\varepsilon R2$ expression on inflammatory cells, it is essential for the induction of IgE transcription by B-cells, while other TH_2 cytokines (Il-5) augment IL-4 induced IgE production, or suppress the negative effect of IFN-γ on IgE synthesis (IL-10). It is important to note in this context that the profile of cytokines secreted from activated mast cells/basophils which include IL-3, IL-4, IL-5, IL-6 and GM-CSF, resembles that of TH_2 cells [17]. Mast/basophil exocytosis may therefore play an important role in the differentiation of TH lymphocytes and the regulation of IgE production by B-cells.

3. The nature of the antigen

The encounter with a number of antigens seems to give rise with preference to the synthesis of an IgE antibody, suggesting that the isotype response may be related to the nature of the antigen. Most allergens are (glyco)proteins in the molecular weight range between 5-75 kD. At present, it is not known why certain seemingly innocuous and diverse substances like pollen grains, mould spores, house dust mite or cockroach emanations,

4

animal dander or insect venoms, latex proteins or industrial pollutants elicit an allergic response in some subjects but not in others. Similarly, the production of high levels of IgE in response to parasitic infestations is a well recognised feature of the immune response.

Employing protein chemistry, structural studies and recombinant DNA technology, the possibility that a variety of antigens share common epitopes that give preferentially rise to an IgE response has been investigated, but despite intensive efforts and considerably increased information regarding the molecular structure of many common allergens no unifying feature has been proposed.

Our own studies, however, suggest an alternative explanation for the selective isotype induction observed for the above described antigens. Recently, observations from my laboratory have led to the development of a hypothesis that may help to explain the preferential selection of an IgE isotype commonly observed following exposure to certain antigens.

MANY ANTIGENS WHICH GIVE RISE TO AN IGE RESPONSE WILL TRIGGER MAST CELL SECRETION IN THE ABSENCE OF IGE SENSITIZATION

These findings became apparent when we tested a secretor variant of the rat basophilic leukaemia (RBL-3H3) cell-line (an accepted model system for the study of mucosal mast cell function), which had been transfected with the α-chain chain of the human high-affinity receptor complex [18]. Since human IgE only recognises primate receptors and no permanent cell-line is available that binds human IgE with high-affinity and responds to an immunological stimulus with mediator secretion, we chose this cell-line as host for α-chain gene transfection on the basis of earlier observations which had shown that γ-chains of rodent origin can facilitate the expression of human a chains in e.g. COS7 cells [19]. The high sequence homology between rat and human α chain in the transmembrane domain suggested that the transfected human α-chain should form a functional complex with the subunits of the rodent receptor and that sensitization with hIgE should activate the host cell's signal transducing machinery following an antigenic stimulus. Experimental details regarding the engineering and biochemical characterisation of the transfected cell-line are shown in Figure 2.

When the transfected cells were sensitized with the serum from a bee venom phospholipase A_2 sensitive individual (EC), mediator release could be demonstrated following challenge with purified bee venom phospholipase A_2 (mellitin free). A typical bell-shaped dose response pattern was observed when the sensitized cells were challenged with increasing doses of antigen. Control experiments, where non-sensitized cells had been incubated with the same concentration of antigen in the absence of sensitizing serum also showed degranulation of mediators, but this time mediator release increased in response to antigen concentration (see Figure 3).

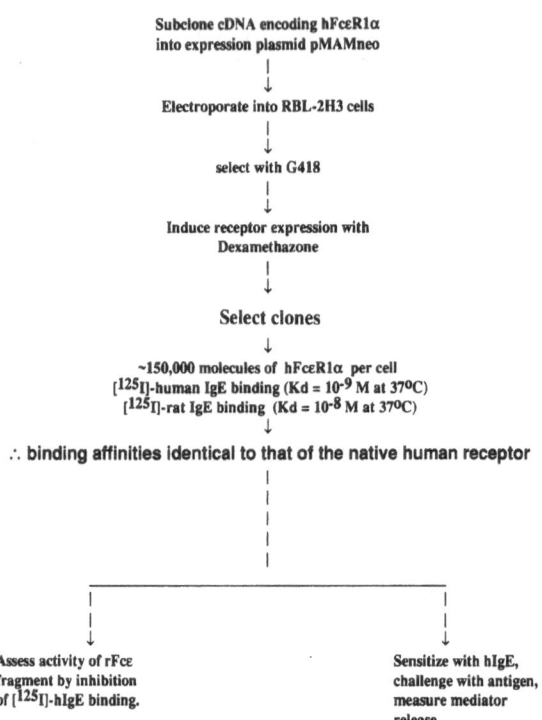

Subclone cDNA encoding hFcɛR1α
into expression plasmid pMAMneo
↓
Electroporate into RBL-2H3 cells
↓
select with G418
↓
Induce receptor expression with
Dexamethazone
↓
Select clones
↓
~150,000 molecules of hFcɛR1α per cell
[^{125}I]-human IgE binding (Kd = 10^{-9} M at 37ºC)
[^{125}I]-rat IgE binding (Kd = 10^{-8} M at 37ºC)
↓
∴ binding affinities identical to that of the native human receptor

Assess activity of rFcɛ
fragment by inhibition
of [^{125}I]-hIgE binding.

Sensitize with hIgE,
challenge with antigen,
measure mediator
release.

Figure 2. Expression of the human α-chain of the high-affinity receptor complex for immunoglobulin E in rat basophilic leukaemia cells.

The engineered cell-line binds human IgE with high-affinity. It was employed to map the FcɛR1 binding site in hIgE, using a family of overlapping IgE derived peptides shown in Figure 1. Transfected cells respond to a human IgE-mediated antigenic stimulus with mediator secretion as shown in Figure 3.

Employing this cell-line, we were able to demonstrate that even in the absence of IgE several well defined allergens, (which in susceptible individuals give rise to an IgE response following the initial encounter) such as bee and vespid proteins, phospholipases, proteases from house dust mites and fungal spores, lectins present in pollen and grain, latex-associated products and spermicides, or aspirin based drugs, can trigger the release of substantial levels of mediators of the allergic response from these cells (see Table 1).

It has been known for more than a decade, that degranulation of mast cells and basophils can be induced by a variety mechanisms that do not involve IgE, e.g. mast cell secretion has been observed in response to bee and vespid venoms, or to drugs like codeine, morphine, or anaphylotoxins, and physical stimuli like pressure, heat, cold and sunlight [20,21]. Cross-linking by lectins that interact with carbohydrate residues on IgE or the receptors, oxygen radicals and proteases will induce the release of mediators [20]. Pollen and grain dust contain lectins in addition to proteases, which are also found in fungal spores, secreted by parasites, and several well characterised proteolytic enzymes have been isolated and cloned from house dust mites [22].

6

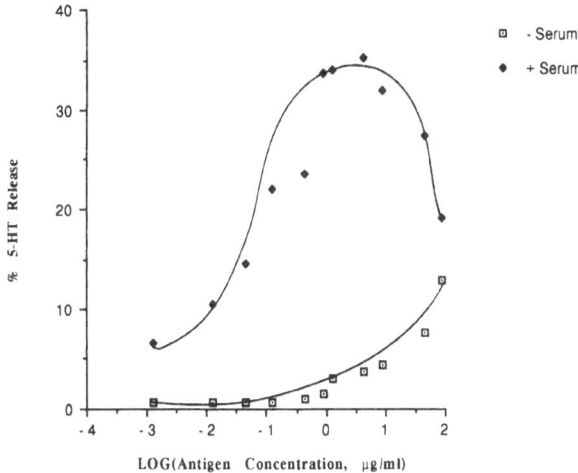

Release of Tritiated 5-Hydroxytryptamine (5-HT) Following Incubation With Bee Venom Phospholipase A2

LOG(Antigen Concentration, μg/ml)

□ - Serum
♦ + Serum

% 5-HT Release

Figure 3. Transfected RBL-2H3 clones were incubated for 24h with or without serum (1:10 dilution) from a bee venom sensitive individual (EC) in the presence of 10^{-6}M dexamethazone and [^{3}H] 5-hydroxytryptamine. After washing, cells were challenged with antigen in the presence of 100mM 5' (N-ethylcarboxyamido)-adenosine (NECA).

[^{3}H] 5-Hydroxytryptamine was measured after a 15 min incubation period, release was corrected for background and expressed relative to the total [^{3}H] 5-hydroxytryptamine incorporated. For experimental details see ref.[18].

The molecular mechanism by which these agents trigger cell secretion is unknown. The pharmacologically active mediators that are released from the activated cell-line in response to these agents include histamine, serotonin, proteases, leukotrienes and cytokines [17]. Of particular interest is the cytokine interleukin (IL)4, which is known to be essential for the induction of IgE by B-lymphocytes, and IL-6, which further enhances the IgE response induced by IL-4. Although IL-4 initiates the transcription of ε-chain germ line transcripts, a second B-cell activating signal, which may be involved in promoting recombinational accessibility is necessary in order to induce immunoglobulin gene class-switching [15]. *In vivo,* this signal is usually delivered to B-cells following interaction of the T-cell receptor/CD3 complex with MHC class II antigens. In the absence of such a stimulus, it can be replaced by a hormone/receptor interaction or a mitogenic stimulus [15].

Activated mast cells and basophils also secrete at least three proteases, the physiological function of which is unknown. These enzymes are serine endoproteases with a trypsin-like specificity. We have recently shown that after secretion following an immunological, non-immunological or oxide mediated stimulus, proteases released from the activated RBL cell-line can induce mediator secretion from cells of their kind (Helm *et al.,* paper in preparation). This secondary burst of the release of inflammatory mediators can be attenuated by the inclusion of protease inhibitor or substrates for serine

7

Table 1. Antigen-induced mast cell mediator release from RBL-2H3 cells in the absence of sensitization with antigen-specific IgE.

Mediators Measured

Antigen	5-HT	Protease	β-Hexosaminidase
Venoms, bee/wasp (1% suspensions)	++++	+++++	++++
Phospholipase A2 (bee venom), 1μg/ml	+	+	+
Phospholipase C (*B. cereus*) 1μg/ml	++++	++++	ND
House Dust Mite extracts, (freshly prepared) 1% suspensions	+++	+++	ND
House Dust Mite extracts, commercial sources, 1% suspensions	-	+/-	ND
Condom extract (1:400) (*Hevea brasiliensis*)	++	++	ND
Aspirin (25mg/ml)	+++	+++	ND
Influenza virus (5% suspension)	+	+	ND
Herpes simplex virus (5% suspension)	+	+	ND

Parent RBL-2H3 cells or clones transfected with the α-chain of the human high-affinity receptor complex were plated out in 24-well plates at 2×10^5 cells/well as described previously[18]. For the determination of antigen-induced mediator release, plates were incubated at 37°C for 15 min, cooled on ice, the supernatant was removed, spun at 200 x g (1 min) before liquid scintillation counting to measure [^{3}H]-5-hydroxytryptamine (5-HT) release. The ImmunoTech histamine enzyme immunoassay was used to quantify histamine release, and hydrolysis of toluene sulphonyl methyl ester was employed to monitor protease release [for exp. details see ref.[18]].

+ ~ 3-8% mediator release
++ ~ 8-15% mediator release
+++ ~ 15-25% mediator release
++++ ~ 25-45% mediator release

proteases like the synthetic substrate p-toluenesulphonyl-L-arginine methyl ester (TAME) or the human IgE-derived pentapeptide (HEPP) [23] in the medium bathing the cells. This observation provides a self-evident explanation for the observed therapeutic effect of administration of HEPP to patients suffering from e.g. allergic rhinitis, initially attributed to be due to competitive inhibition of the IgE/FcεR1 interaction, but later held untenable [24].

It is well established that serine proteases mimic a number of insulin- and growth factor mediated processes including the initiation of cell division in quiescent cells [25,26]. Collectively this activity and the above mentioned observations suggest that IgE-independent stimuli can activate cells of mast cell/basophil lineage to deliver the IL-4 signal together with a mitogenic signal, thus providing a combination of stimuli which should be sufficient for the induction of IgE synthesis in B-cells. Under such conditions, any antigen present might be expected to give rise to the synthesis of an antibody of the IgE isotype.

ACKNOWLEDGEMENTS

I wish to thank my collaborators, D. Moreira, N. Rhodes, A.P.M. Wilson, Y. Ling and K. Ahmad for making results available prior to publication. This work was supported in parts by joint awards from the Medical Research Council, the Science and Engineering Research Council/Biotechnology Directorate, and the Department of Trade and Industry in collaboration with Euro/DPC, Glyn Rhonwy, Llanberis, Caernarfon, Gwynedd, G.B.

REFERENCES

1. K. Ishizaka, T. Ishizaka, and Hornbrook, M.M., Physicochemical properties of human reaginic antibody. J. Immunol. 97:840 (1966).

2. K. Ishizaka, IgE-binding factors from rat T-lymphocytes. Ann. Rev. Immunol. 2:269, (1984).

3. A. Capron, J.P. Dessaint, M. Capron, M. Joseph, J.C. Ameisen and A.B. Tonnel, From parasites to allergy: a second receptor for immunoglobulin E. Immunology Today 7:15 (1986).

4. B.A. Helm, Y. Ling, N. Rhodes, E.A. Padlan, Impact of Biotechnology on Allergy Research: Can structural and functional studies on components of the IgE receptor/effector system assist the design and M. Okuda eds. Hogrefe & Huber Publ. Kyoto (1992).

5. B.A. Helm, Y. Ling, K. Ahmad, E.A. Padlan, A modelling study of the interaction of human IgE with the α- chain of the high-affinity receptor. Seventh International Symposium on the Immunobiology of Proteins and Peptides, Abstract 7, (1992).

6. A. Nissim, H. Jouvin, and Z. Esshar, Mapping of the high-affinity receptor Fcϵ binding site to the third constant region domain of IgE. EMBO 10:101 (1991).

7. A. Henry, C. Chan, H.J. Gould and M.H. Jouvin, N-terminal border of the FcϵRI site on human IgE. Abstract W-5, 32. 8th International Congress of Immunology, Budapest, Hungary (1992).

8. D. Vercelli, B. Helm, P. Marsh, E. Padlan, R. Geha and H. Gould, The B-cell binding site on human immunoglobulin E. Nature 338:649 (1989).

9. J. Slattery, D. Holowka, and B. Baird, Segmental flexibility of receptor-bound immunoglobulin E. Biochemistry 24:7810 (1985).

10. E. Nagy, I. Berczi and A. Sehon, Growth inhibition of murine mammary carcinoma by monoclonal IgE antibodies specific for mammary tumor virus. Cancer Immunol. Immunother 34:63, (1991).

11. W.O.C.M. Cookson, The atopic response and chromosome 11, in "Progress in Allergy and Clinical Immunology" 2. T. Miyamato and M. Okuda eds. Hogrefe & Huber Publ. Kyoto (1992).

12. Y. Ohno, K. Tanno, K. Yamauchi and T. Takishima, Gene expression and production of tumor necrosis factor by a rat basophilic leukaemia cell-line with IgE receptor triggering. Immunology 70:88 (1990).

13. W.R. Hendersen, E. Chi and S.J. Klebanoff, Eosinophil peroxidase induced mast cell secretion, J. Exp. Med. 152:265 (1980).

14. F.C. Mills, G. Thyrophronitis, F. Finkelman, and E.E. Max, Ig μ-ϵ isotype switch in IL-4 treated human lymphoblastoid cells. J. Immunol. 149:1075.

15. D. Vercelli and R. Geha, Regulation of IgE synthesis in human: A tale of two signals. Allergy and Clinical Immunology 88:285 (1989).

16. O. Abeshsira-Amar, M. Gilbert, M. Joliy, J. Theze and D. Jankovic, Il-4 plays a dominant role in the differential development of TH_0 into TH_1 and TH_2 cells. J. Immunol. 148:3820 (1992).

17. M. Plaut, J.H. Pierce, J.C. Watson, J. Hanley-Hyde, R.P. Nordan and W. E. Paul, Mast cell lines produce lymphokines in response to cross-linkage of FcϵR1 or to calcium ionophores. Nature 339:64 1989).

18. A.P.M. Wilson, C.E. Pullar, A.M. Camp and B.A. Helm, Human IgE mediates stimulus secretion coupling in rat basophilic leukaemia cells transfected with the α-chain of the human high-affinity receptor, Eur. J. Immunol. 23:240 (1993).

19. U. Blank, C. Ra, L. Miller, K. White, H. Metzger and J.P. Kinet, Complete structure and expression in transfected cells of the high-affinity IgE receptor. Nature 337:187 (1989).

20. D. Lagunoff and T.W. Martin, Agents that release histamine from mast cells. Ann. Rev. Pharmacol. Toxicol. 23:331 (1983).

21. M. Ichihashi and Y. Funasaka, Solar Allergy: A role of Inhibition Spectrum in Solar Urticaria. in Progress in Allergy and Clinical Immunology ". T. Miyamato and M. Okuda eds. Hogrefe & Huber publ. Kyoto (1992).

22. W. Greene, J.G. Cyster, K.Y. Chua, R.M. O'Brien and W. Thomas, IgE and IgG binding of peptides expressed from fragments of cDNA encoding the major house dust mite allergen Der p 1. J. Immunol. 147:3768 (1991).

23. R.N. Hamburger, Peptide inhibition of the Prausnitz-Küstner reaction. Science 189: 389 (1975).

24. G.S. Hahn, Immunoglobulin-derived drugs. Nature 324:283 (1986)

25. D.H. Carney and D.D. Cunningham, Role of specific cell-surface receptors in thrombin stimulated cell division. Cell 15:1341 (1978).

26. S. Tamura, Y. Fujita-Yamaguchi and J. Larner, Insulin-like effect of trypsin on the phosphorylation of adipocyte insulin receptor. J. Biol. Chem 258:14749 (1983).

IMMUNOGENETIC ASPECTS OF IgE-MEDIATED RESPONSES

Shau-Ku Huang, Ming Yi, Megumi Kumai, and David G. Marsh

Johns Hopkins University School of Medicine
Johns Hopkins Asthma and Allergy Center
5501 Hopkins Bayview Circle, Baltimore, MD 21224

INTRODUCTION

Atopic allergy is a multifactorial disease, and is dependent on several genetic determinants as well as a complex array of environmental factors. A clear *causal* relationship has been established between the clinical expression of specific allergies and the induction of specific IgE antibody (Ab) responses.[1] Several studies have provided evidence for at least two types of genetic control of IgE responses in humans, namely, non-*MHC*-linked control of the overall production of IgE and *MHC*-linked control of specific immune responses.[2] Since T cells and T-cell cytokines are important in the regulation of IgE synthesis, a detailed investigation of the role of T cells in controlling the synthesis of IgE, and the genetic regulation of various IgE-modulating cytokine genes should provide some useful approaches toward solving the complex problem of the genetic control of IgE production. We have focused on the immunogenetic controls of specific Ab responsiveness to allergens, where the relevant genes clearly include MHC class II (HLA-D), TcR specificity-determining genes. The allergens we studied included a group of "minor" inhalant allergens (relative molecular mass, $M_r = ca$ 4400-5000).

SPECIFIC IMMUNE RESPONSIVENESS TO THE *AMB* V HOMOLOGUES FROM *AMBROSIA* (RAGWEED) POLLENS

The *Amb* V homologues ($M_r = ca$ 4400-5000) provide a particularly good model system for genetic studies. Because of the ultra-limiting Ag doses under natural conditions of allergen exposure,[3] and the relatively simple structures of these molecules, immune recognition may often be limited to single MHC-binding sites, a consideration which greatly facilitates genetic analysis.[4] These allergens have been sequenced at the amino acid and DNA levels, expressed as recombinant proteins and their three-dimensional structures have been analyzed by NMR and computer modelling: *Amb a* V[5-9] from *Ambrosia artemisiifolia* (short ragweed), *Amb t* V[10-13] from *A trifida* (giant ragweed) and *Amb p* V from *A psilostachya* (western ragweed) (ref. 14 and Marsh et al unpub). The genetic basis of immune responsiveness to these allergens has been extensively studied. In populations naturally exposed to short ragweed pollen, the presence of specific IgE and IgG Abs to *Amb a* V was strongly associated with HLA-DR2Dw2 (Table 1).[15-17] Following ragweed immunotherapy with the different ragweed species, IgG Ab responses to all three *Amb* V molecules was also strongly associated with DR2Dw2.[18]

Table 1. Significant associations of HLA with specific antibody responsiveness toward highly purified pollen allergens in atopic Caucasoid subjects.

Systematic Name	M_r	Major HLA Association	Subjects +ve	−ve	Overall p values*
Amb a V	5,000	DR2/Dw2	36/38 (95%)	30/139 (22%)	$<10^{-9}$
Amb t V	4,400	DR2/Dw2	6/7 (86%)	1/20 (5%)	$<10^{-3}$

* p values are given only for the most significant associations for analyses of IgE Ab data, except that IgG Ab data from immunized patients were used for *Amb t* V.

In order to achieve further definition of the HLA-D gene polymorphism at the molecular level, we have studied the polymorphic second exons of HLA-DRB and DQB genes that encode Ag-binding portions on the respective Class II molecules.[19,20] Results showed that virtually all people producing IgE Ab to *Amb a* V possessed the DRB1 and DRB5 sequences associated with DR2.2. To determine which DR molecule is involved, $\alpha\beta1$ or $\alpha\beta5$, we studied the blocking effect of a monoclonal antibody, Hu30 (which is specific for all $\alpha\beta1$, but not $\alpha\beta5$, heterodimers having the DR2 haplotype), on *Amb a* V-specific T-cell proliferative responses.[21] We found that this MAb inhibited Ag presentation by DR2.2 and DR2.12 APCs, suggesting that the respective DR$\alpha\beta1$ heterodimers are the principal Class II molecules involved in *Amb a* V presentation. We compared the sequences of DR$\beta1$*1501

(DR2.2-associated) and DRβ1*1502 (DR2.12-associated) which can present $Amb\ a$ V, versus DRβ1*1601 (DR2.21) and DRβ1*1602 (DR2.22) which can not. These two groups of DR2-associated β1 sequences differ notably from each other at amino acid residues 67, 70 and 71, all of which are postulated to be involved in binding to the Ag and/or TcR.

The finding that immune responsiveness toward $Amb\ t$ V was also significantly associated with HLA-DR2.2 suggests that DR2.2 may control the recognition of all (or part) of the homologous structures on these molecules. Ag competition studies using $Amb\ a$ V-specific T-cell clones and a polyclonal T-cell line demonstrated that both $Amb\ t$ V and $Amb\ p$ V inhibit the presentation of $Amb\ a$ V by DR2.2$^+$ APCs (Fig. 1), suggesting all three members of the Amb V allergen homologues are DR2.2-restricted (Ref. 22 and Huang et al, unpublished). Taken together, these results suggest that the same Ia molecule involves in the immune recognition of a similar MHC binding site on all three Amb V molecules. Furthermore, using an $Amb\ a$ V-specific polyclonal T-cell line and overlapping synthetic peptides, a dominant Ia/T cell epitope was localized on the C-terminus of $Amb\ a$ V (Huang et al, unpublished). This peptide contains $Amb\ a$ V-specific residues and sequence shared by both $Amb\ a$ V and $Amb\ t$ V.

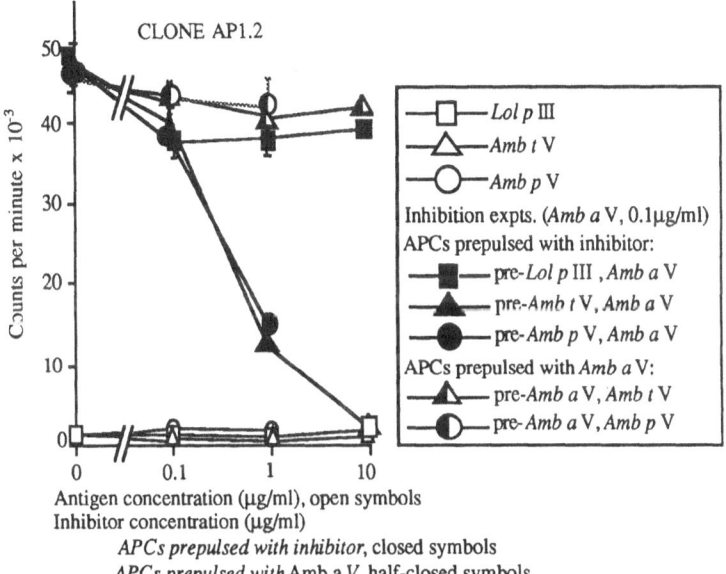

Figure 2. Inhibition of $Amb\ a$ V presentation by $Amb\ t$ V and $Amb\ p$ V. $Lol\ p$ III was used as the control Ag. For direct stimulation with Ags, T cells were stimulated in triplicate with or without Ags in the presence of irradiated autologous PBMCs for three days. Varying concentrations of highly purified Amb V allergens ($Amb\ a$ V, or $Amb\ t$ V, or $Amb\ p$ V; or a control Ag, $Lol\ p$ III) were added at the initiation of the culture. For blocking experiments, autologous APCs were preincubated with or without each of the $Amb\ a$ V homologues (or $Lol\ p$ III as control Ag) at 0.1-10 µg/ml in culture medium for 10 hrs and washed three times. The prepulsed APCs were then cultured with cloned T cells in the presence or absence of $Amb\ a$ V (0.1 µg/ml) for 60 hrs. Reciprocal experiments were run in parallel wherein autologous APCs were prepulsed with 0.1 µg/ml of $Amb\ a$ V followed by culturing with T cells and each of the homologues. The cell cultures were then pulsed with 1 µCi of [^{3}H]-thymidine, harvested 18 hrs later, and counted.

13

Using the polymerase chain reaction (PCR) and DNA sequencing techniques, we have analyzed the T-cell receptor (TcR) gene usage in an *Amb a* V-specific T-cell clone (AP1.2) and two polyclonal T-cell lines from two unrelated allergic patients. Results showed that first, multiple $V\beta$ and $V\alpha$ gene usages are observed in polyclonal T-cell lines; and second, the AP1.2 β gene sequence is dominant in a polyclonal *Amb a* V-specific T-cell line and an additional T-cell line from an unrelated allergic patient. Our subsequent FACS analysis using $V\beta$-specific antibodies confirmed our PCR results. These results suggest a heterogeneous population of T cells recognizing the same *Amb a* V peptide fragment; and among these T cells, the usage of clone AP1.2 β gene appears to be dominant.

SUMMARY

The reported significant HLA association is consistent with the codominant expression of MHC-linked *Ir* genes. Similar studies are needed in the case of the other HLA-immune response associations. It seems unlikely that any unique HLA-D genetic sequence will be found only among subjects responding to a particular Ag. It is likely, however, that we will find that a particular sequence is a necessary, but not sufficient, requirement for responsiveness to a particular antigenic epitope. Previous family studies,[23] in which failed to observe parent-to-child transmission of specific immune responsiveness to a particular allergen, suggest that further genetic and/or environmental factors are required for the expression of specific immune responsiveness. These factors include variations in the degree of antigenic exposure. Of particular importance is the need for Ag-specific TcR genes to be expressed in the mature T-cell repertoire. In study of the specific immune response, there have been several studies of the protein and DNA sequences of allergenic proteins; therefore, in regard to understanding the genetics of specific immune responsiveness to allergens and its relationship to atopic diseases, rapid advances can be anticipated over the next several years.

REFERENCE

1. K. Ishizaka and T. Ishizaka. "Human reaginic antibodies and immunoglobulin E," J. Allergy 42:330 (1968).
2. D.G. Marsh, "Historical introduction, *In:* "Genetic and Environmental Factors in Clinical Allergy," D.G. Marsh and M.N. Blumenthal, Eds, University of Minnesota Press, Minneapolis (1990).
3. D.G. Marsh, "Allergens and the genetics of allergy," *In:* "The Antigens," M. Sela, ed, Academic Press, New York (1975).
4. D.G. Marsh, "Allergy: A model for studying the genetics of human immune response," *In:* "Molecular and Biological Aspects of the Acute Allergic Reaction," S.G.O. Johansson, K. Strandberg, B. and Uvnas, eds, Plenum Publishing Co., New York (1976).

5. C.B. Lapkoff and L. Goodfriend, "Isolation of a low molecular weight ragweed pollen allergen; Ra5," *Int. Arch. Allergy Appl. Immunol.* 46:215 (1974).

6. L.E. Mole, L. Goodfriend, C.B. Lapkoff, J.M. Kehoe, and J.D. Capra, "The amino acid sequence of allergen Ra5," *Biochem.* 14:1216 (1975).

7. B. Ghosh, M.P. Perry, T. Rafnar, and D.G. Marsh, "Molecular cloning and sequence analysis of the gene encoding the *Amb a* V allergen from short ragweed (*Ambrosia artemisiifolia*) pollen," Submitted.

8. T. Rafnar, B. Ghosh, S.K. Huang, W.J. Metzler, L. Mueller, and D.G. Marsh, "Expression of cystine-rich ragweed allergens in *E. coli*: confirmation of the structural and immunological identity of recombinant and native *Amb t* V for epitope studies," *J. Biol. Chem.* (1992). In press.

9. W.J. Metzler, K. Valentine, M. Roebber, D.G. Marsh, and L. Mueller, "Proton resonance assignments and three-dimensional solution structure of the ragweed allergen *Amb a* V by nuclear magnetic resonance spectroscopy" *Biochemistry.* (1992) In press.

10. L. Goodfriend, A.M. Choudhury, D.G. Klapper, K.M. Coulter, G. Dorval, J. DelCarpio, and C.K. Osterland, "Ra5G, a homologue of Ra5 in giant ragweed pollen: isolation, HLA–DR–associated activity and amino acid sequence," *Mol. Immunol.* 22:899 (1985).

11. M. Roebber, D.G. Klapper, L. Goodfriend, W.B. Bias, S.H. Hsu, and D.G. Marsh, "Immunochemical and genetic studies of *Amb.t.*V (Ra5G), an Ra5 homologue from giant ragweed pollen," *J. Immunol.* 134:3062 (1985).

12. B. Ghosh, M.P. Perry, and D.G. Marsh, "Cloning the cDNA encoding the *Amb t* V allergen from giant ragweed (*Ambrosia trifida*) pollen," *Gene* 101: 231 (1991).

13. W.J. Metzler, K. Valentine, M. Roebber, M. Friedrichs, D.G. Marsh, and L. Mueller, "Solution structures of ragweed allergen *Amb t* V" *Biochemistry* 31: 5117 (1992).

14. D.G. Marsh, P. Zwollo, L. Freidhoff, D.B.K. Golden, A.A. Ansari, E.E. Kautzky, D.A. Meyers, and C.L. Holland, "Studies of human immune response to the *Amb* V (Ra5) homologues," *J. Allergy Clin. Immunol.* 85: 201 (1990).

15. D.G. Marsh, S.H. Hsu, M. Roebber, E.E. Kautzky, L.R. Freidhoff, D.A. Meyers, M.K. Pollard, and W.B. Bias, "HLA–Dw2: a genetic marker for human immune response to short ragweed pollen allergen Ra5. I. Response resulting primarily from natural antigenic exposure," *J. Exp. Med.* 155:1439 (1982).

16. K.M. Coulter, G.D. Dorval, and L. Goodfriend, "Genetic control of IgE antibody responses in humans: the *Amb a* V (Ra5) model," *In:* D.G. Marsh and M.N. Blumenthal, eds, "Genetic and Environmental Factors in Clinical Allergy," Univ. of Minnesota Press, Minneapolis (1990).

17. M.N. Blumenthal, D. Marcus-Bagley, Z. Awdeh, B. Johnson, E.J. Yunis, and C.A. Alper, "HLA-DR2, [HLA-B7, SC31, DR2], and HLA-B8, SC01, DR3] haplotypes

distinguish subjects with asthma from those with rhinitis only in ragweed pollen allergy," *J. Immunol.* 148:411 (1992).

18. D.G. Marsh, D.A. Meyers, L.R. Freidhoff, E.E. Kautzky, M. Roebber, P.S. Norman, S.H. Hsu, and W.B. Bias, "HLA-Dw2: a genetic marker for human immune response to short ragweed pollen allergen Ra5. II. Response after ragweed immunotherapy," *J. Exp. Med.* 155:1452 (1982).

19. D.G. Marsh, P. Zwollo, S.K. Huang, B. Ghosh, and A.A. Ansari, "Molecular studies of human response to allergens" *Cold Spring Harbor Symp. Quant. Biol.* 54: 459 (1990).

20. P. Zwollo, E.E. Kautzky, A.A. Ansari, S.J. Scharf, H.A. Erlich, and D.G. Marsh, "Molecular studies of human immune response genes for the short ragweed allergen, *Amb a* V. Sequencing of HLA-D second exons in responders and non-responders," *Immunogenet* 46:1050 (1991).

21. S.K. Huang, P. Zwollo, and D.G. Marsh, "Class II MHC restriction specificity of human T-cell responses to a short ragweed pollen allergen, *Amb a* V," *Eur. J. Immunol.* 21:1469 (1991).

22. S.K. Huang and D.G. Marsh, "Human T-cell responses to ragweed allergens, *Amb* V homologues" *Immunology* 73:363 (1991).

23. L.R. Freidhoff, E.E. Kautzky, D.A. Meyers, S.H. Hsu, W.B. Bias, and D.G. Marsh, "Association of HLA–DR3/Dw3 and total serum immunoglobulin E level with human immune response to *Lol p* I and *Lol p* II allergens in allergic subjects," *Tissue Antigens* 31:211 (1988).

STRUCTURE AND FUNCTION OF THE LOW AFFINITY IgE RECEPTOR

Daniel H. Conrad, Kim A. Campbell, William C. Bartlett,
Coles M. Squire and Steven E. Dierks

Department of Microbiology and Immunology
Virginia Commonwealth University
Box 678, MCV Station
Richmond, VA 23298

INTRODUCTION

Signal transduction through cell surface receptors such as those specific for antigen, lymphokines or immunoglobulin (Ig) is an important event for many facets of cell growth and differentiation. The group of receptors which interact with the Fc region of Ig are involved in a number of immune functions, including phagocytosis, antibody dependent cellular cytotoxicity and release of inflammatory mediators (see[1] for review). The Fc receptors that interact with immunoglobulin E (IgE) have been divided into two predominant receptors. $Fc_\epsilon RI$, or the high affinity receptor for IgE, is found predominantly on mast cells and basophils and is involved in cytokine and allergic mediator release from these cells[1]. Discovered somewhat later, the $Fc_\epsilon RII$, or the low affinity receptor for IgE, is expressed on a wide variety of hematopoietic cell types. The finding that the $Fc_\epsilon RII$ was synonymous with the B cell differentiation antigen CD23[2,3] led to a merging of two areas of investigation for this protein; the study of the role of the $Fc_\epsilon RII$/CD23 (the names are used interchangeably) in B cell activation/differentiation and the study of the role of the $Fc_\epsilon RII$ in IgE production. This review chapter will concentrate on summarizing recent developments regarding the $Fc_\epsilon RII$ with regard to both the structure and function of this molecule. Several other recent reviews on this subject are also available[4-6].

STRUCTURE OF THE $Fc_\epsilon RII$

Initial studies in both the human and mouse systems demonstrated that the $Fc_\epsilon RII$ was a single glycosylated polypeptide of 45 (human)[7] or 49 (mouse)[8] kDa. Development of monoclonal antibodies to the $Fc_\epsilon RII$ allowed the demonstration that a soluble form (termed soluble $Fc_\epsilon RII$ or $sFc_\epsilon RII$) also existed; the $sFc_\epsilon RII$ was smaller that the intact molecule, and has been clearly demonstrated to result from proteolytic degradation of the intact $Fc_\epsilon RII^{9,10}$. The degradation is evidently progressive as several different sizes of $sFc_\epsilon RII$ have been shown and evidence indicates that at least the first cleavage event may be autoproteolytic[11]. The most common $sFc_\epsilon RII$ sizes reported are 37, 33, 29, and 25 kDa and 38, 28 and 25 kDa for human and mouse $sFc_\epsilon RII$ respectively[10,12].

Regulation of $Fc_\epsilon RII$ Expression

Early studies demonstrated that increased levels of B cell $Fc_\epsilon RII$ expression were seen when IgE was cultured with lymphocytes[13] or when *in vivo* levels of IgE were elevated[14]. Mechanistic studies demonstrated that this was the result of protection from degradation into $sFc_\epsilon RII^{15}$. In addition, the $Fc_\epsilon RII$ is one of a number of proteins in which increased expression is seen when the $Fc_\epsilon RII$ expressing cell is exposed to the appropriate cytokine. Interleukin-4 (IL-4) increases $Fc_\epsilon RII$ expression on B cells[16,17], monocytes[18] and certain eosinophil cell lines[19], while interferon-γ (IFN-γ) increased $Fc_\epsilon RII$ expression on $CD5^+$ B cells[20] and monocytes[18]. In contrast IFN-γ antagonizes expression of the $Fc_\epsilon RII$ on conventional ($CD5^-$) B cells in both man[21] and mouse[22]. The mechanism of IL-4 upregulation appears to be an increased transcription rate ([23] and E. Studer and D. Conrad unpublished observations). The highest level of $Fc_\epsilon RII$ expression (termed $Fc_\epsilon RII$ superinduction) is seen when B cells are polyclonally activated in the presence of IL-4 either by T cells[24] or by lipopolysaccharide (LPS)[25]; $Fc_\epsilon RII$ levels 20-50 fold above constitutive expression are then usually seen.

$Fc_\epsilon RII$ Cloning

The cDNA for the human and mouse $Fc_\epsilon RII$ was cloned independently by three[16,26,27] and two[28,29] groups respectively. The predicted protein sequence indicated that the $Fc_\epsilon RII$ was a type II membrane protein with the amino terminus cytoplasmic. In contrast to all other presently cloned Fc receptors, the $Fc_\epsilon RII$ is not a member of

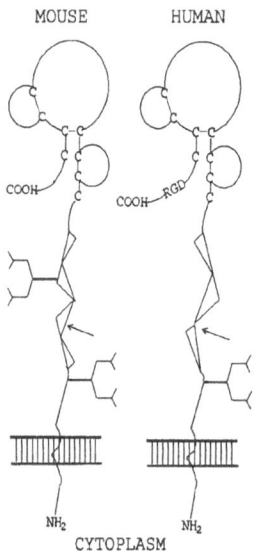

FIGURE 1. Model of the human and mouse $Fc_\epsilon RII$. The model is modified from an earlier version[4]. Differences between the human and mouse $Fc_\epsilon RII$ are illustrated. The arrow indicates the site of the initial cleavage to produce $sFc_\epsilon RII$. See text for discussion.

18

the Ig gene superfamily. Rather the carboxyl terminal sequence exhibited a domain that demonstrated homology with the family of C-(for calcium dependent) type animal lectins. In spite of this homology, there is conflicting evidence whether lectin activity plays any role in the interaction with IgE. IgE binding is calcium dependent[30], however deglycosylated IgE was shown to interact equally well (as native IgE) with the $Fc_\epsilon RII$[31]. In contrast, others have described some inhibition of IgE binding by certain sugars[32]. The C-type lectin family is extensive and includes other membrane proteins such as the asialoglycoprotein receptor and the B cell activation antigen Lyb2 (CD72)[33] both of which exhibit a similar overall architecture as the $Fc_\epsilon RII$. In addition, the C-type lectin cassette is also found in a group of cell adhesion proteins that have been termed selectins [34] and is found in soluble proteins such as mannose binding proteins (MBP)[35]. In human (but not mouse (see Fig. 1)) the carboxyl terminus of the lectin domain contains an inverted "RGD" (arginine-glycine-glutamic acid) sequence; such a sequence even in inverted form can be involved in interaction with the integrin family of cell adhesion receptors[36]. However, as yet this RGD sequence has not been shown to play a role with the $Fc_\epsilon RII$.

The lectin domain is separated from the predicted transmembrane sequence by a relatively hydrophobic sequence (termed the stalk region) containing internal homology in the form of 21 amino acid homologous repeats (3 in human, 4 in mouse); these repeats are indicated by triangles in Fig. 1 and are each encoded by a separate exon in the genomic organization of the $Fc_\epsilon RII$[37,38]. Using a pattern search strategy, Beavil et al.[39] demonstrated that the stalk region has a heptad repeat pattern that is characteristic of protein sequences that exhibit an α-helical coiled-coil structure. The cleavage sites for $sFc_\epsilon RII$ production are located in this heptad repeat region ([40] and W.C Bartlett and D. Conrad, unpublished observations). As is discussed below we have developed additional evidence that supports the existence of such a structure at least for the intact $Fc_\epsilon RII$.

$Fc_\epsilon RII$ Isoforms and $Fc_\epsilon RII$ Expression

In humans, RNase protection studies have demonstrated the existence of two isoforms, termed $Fc_\epsilon RIIa$ and $Fc_\epsilon RIIb$[41]. These isoforms result from the use of alternative transcription initiation sites which results in differences in the 5' untranslated region and the predicted first 6 amino acids; thus, the protein differences are entirely cytoplasmic. The $Fc_\epsilon RIIa$ isoforms is relatively B cell specific; it is constitutively produced and is upregulated by IL-4. The $Fc_\epsilon RIIb$ isoform is dependent upon IL-4 for its expression and is found on a variety of hematopoietic cells types included monocytes[41], T cells[42] and $CD5^+$ B cells[43]. Recently Yokota et al.[44] demonstrated that these two isoforms had differential capacities with respect to endocytosis and phagocytosis. The $Fc_\epsilon RIIa$ form was more efficient in endocytosis while the $Fc_\epsilon RIIb$ form would allow more efficient phagocytosis of IgE coated particles.

In mouse, only the $Fc_\epsilon RIIa$ isoform has been found[45], perhaps explaining the more restricted pattern of expression seen for the $Fc_\epsilon RII$ in the mouse system. To date $Fc_\epsilon RII$ expression has been confirmed only on B cells and follicular dendritic cells (FDC)[46], although low level $Fc_\epsilon RII$ expression is evidently seen on Th2 lines subsequent to their activation[47]. In addition, a structurally similar molecule that reacts with polyclonal but not monoclonal anti-$Fc_\epsilon RII$ is seen on $CD8^+$ T cells from IgE plasmocytoma bearing mice[48]. The B cell expression in mouse is primarily limited to conventional ($CD5^-$) B cells. The $CD5^+$ as well as the $Mac1^+$ ($CD5^+$ "sister") B cells do not express the $Fc_\epsilon RII$ constitutively and furthermore, IL-4 does not induce its expression[49]. $Fc_\epsilon RII$ expression can be induced, however, under superinduction conditions (LPS plus IL-4). Autoimmune mouse lines such as NZB and MRL *lpr/lpr* show greatly increased levels of these $Fc_\epsilon RII^-$ B cells and this lack of $Fc_\epsilon RII$ expression, especially after exposure to IL-4, has proven to be a superior method for isolation of these B cells[50]. Waldschmidt and coworkers have further analyzed the $Fc_\epsilon RII^-$ and $Fc_\epsilon RII^+$ B cells for isotype switching and proliferative capacities. While some quantitative differences were seen, the most striking difference was seen when peritoneal $Fc_\epsilon RII^-$ B cells were examined. These B cells do not exhibit the capacity to make significant (possibly zero) levels of IgE upon stimulation with LPS and IL-4[50]. These studies are ongoing, but it is evident that at least in the mouse the capacity to express $Fc_\epsilon RII$ is a useful method to study B cell subsets.

$Fc_\epsilon RII$ Self-association

Early studies in mouse demonstrated, using crosslinking reagents, that the $Fc_\epsilon RII$ exhibited a capacity to self associate; the molecular weight was indicative of a trimer[51]. The finding (discussed above) that the stalk region of the $Fc_\epsilon RII$ would be predicted to be in an α-helical coiled-coil provided a model for this self association[39,52]. Our recent studies have indicated that it is the self associated $Fc_\epsilon RII$ that interacts with IgE with significant (biologically relevant) affinity[53]. These studies were performed as in the example shown in Fig. 2 where it is seen that labeled (^{125}I) $Fc_\epsilon RII$ that had lost its ability to cross-link also had lost its ability to bind to an IgE adsorbent.

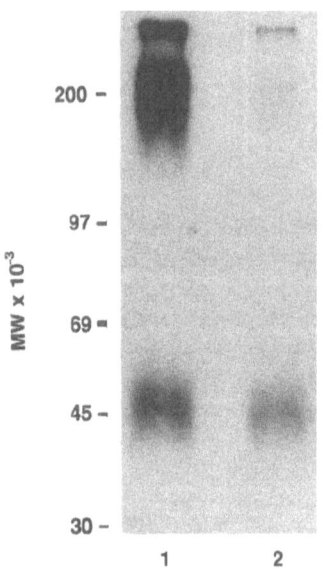

FIGURE 2. Crosslinking of purified mouse $Fc_\epsilon RII$ before and after adsorption with IgE-affigel. ^{125}I-$Fc_\epsilon RII$ was isolated from surface labeled $Fc_\epsilon RII$ transfected CHO cells by affinity chromatography on IgE-adsorbants. After elution, one-third of the ^{125}I-$Fc_\epsilon RII$ was cross-linked with *bis*-sulfosuccimidyl-superimidate (BS^3) and the remaining material absorbed with IgE affigel. Subsequently the non-bound ^{125}I-$Fc_\epsilon RII$ was crosslinked in the same manner. SDS-PAGE analysis of the initial crosslinked eluate (lane 1) and the adsorbed crosslinked effluent (lane 2) is shown. In the audioradio-graph, migration positions of mw standards are shown at left. Figure is based on reference[53].

This suggests that the $Fc_\epsilon RII$ interacts with IgE in a multivalent fashion; the high affinity would be the result of two of the lectin heads interacting with (presumably) symmetrical sites on the IgE molecule. A cartoon to illustrate this method of interaction is shown in Fig. 3.

Note that with the $Fc_\epsilon RII$ coiled-coil trimer two of the lectin heads interact with a single IgE molecule; the third is available for interaction with a separate IgE. This availability explains earlier data in which $Fc_\epsilon RII$ in non-ionic detergent solution was

20

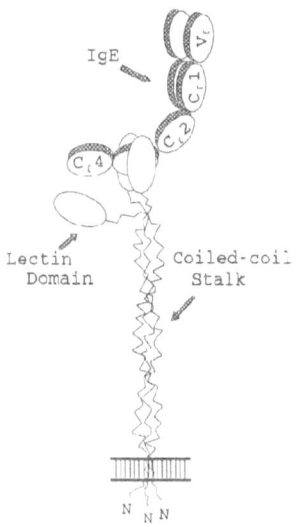

IgE

$C_\epsilon 4$ · $C_\epsilon 2$ · $C_\epsilon 1$ · V_ϵ

Lectin Domain

Coiled-coil Stalk

N N N

FIGURE 3. Model of trimeric Fc$_\epsilon$RII interacting with IgE. Dual point binding of two lectin domains to the Cϵ3 domain is shown. Present data are consistent with interaction of the Fc$_\epsilon$RII with this domain of IgE[31,57]. The lectin domain and stalk, potentially in an α-helical coiled-coil configuration are labeled. Model is based on reference[52].

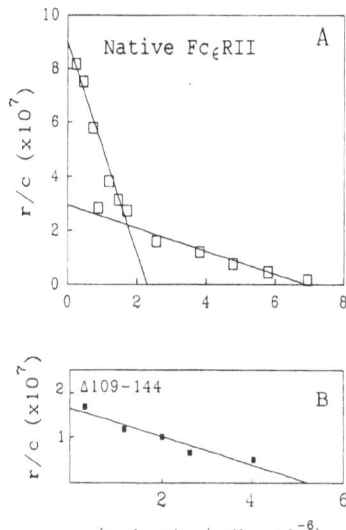

r/c $(\times 10^7)$

Native Fc$_\epsilon$RII   A

$\Delta 109-144$ B

r (molecules/cell x 10^{-6})

FIGURE 4. Scatchard analysis of ^{125}I-IgE binding to native vs stalk mutant Fc$_\epsilon$RII. CHO cells transfected with either the native Fc$_\epsilon$RII (A) or Δ109-144 stalk mutant Fc$_\epsilon$RII (refers to amino acids deleted) (B) were subjected to saturation analysis using increasing amounts of ^{125}I-IgE. Non-specific binding was corrected by subtracting the counts bound to untransfected CHO cells and the amount of IgE was corrected for the amount of bindable IgE, determined with excess Fc$_\epsilon$RII$^+$ cells. Value for the high affinity binding by the native Fc$_\epsilon$RII was 2.37 x 10^7 M^{-1} and 2.59 x 10^6 M^{-1} for the low affinity. The mutant Fc$_\epsilon$RII exhibited a single affinity of 1.9 x 10^6 M^{-1}. Figure is based on reference[53].

shown to interact with more than one IgE molecule[54]. The data is also in line with studies that have demonstrated, using appropriate Fc$_\epsilon$RII mutants, that the site of interaction of the Fc$_\epsilon$RII with IgE is contained within the lectin domain[55,56]. Additional studies involving Fc$_\epsilon$RII stalk mutants in which one or more of the homologous repeat regions were deleted have further strengthened this model. Fig. 4 shows saturation analysis of IgE binding to either recombinant murine Fc$_\epsilon$RII or a Fc$_\epsilon$RII stalk mutant expressed in chinese hamster ovary (CHO) cells. Note that the Fc$_\epsilon$RII exhibits a dual affinity for mouse IgE, an expected result with the multimeric $Fc_\epsilon RII$, since both dual point and single point interaction with IgE, depending on the IgE concentration, would be anticipated. Similar dual affinity results have been previously suggested for the human $Fc_\epsilon RII$[58]. In contrast, the stalk deletion mutant shown only a single affinity of interaction and in other studies (data not shown), it was it was demonstrated by cross-linking studies that the deletion mutant no longer self-associated. Thus, the higher affinity seen is the result of avidity with two lectin domains interacting with a single IgE molecule.

The development of the deletion mutants have allowed the re-examination of the murine $sFc_\epsilon RII$ for interaction with IgE. Previous studies did not find evidence for significant $sFc_\epsilon RII/IgE$ interaction[15], but, given that the lectin domain was entirely contained within the $sFc_\epsilon RII$ and that the this domain has been strongly implicated in interacting with IgE[55,56] this result was quite difficult to understand. This was re-examined using purified $sFc_\epsilon RII$ and CHO cells expressing the stalk deletion mutant described in Fig. 4. Clear dose-dependent inhibition of ^{125}I-IgE binding to the stalk mutant was observed (data not shown); in contrast the $sFc_\epsilon RII$ had only a minimal ($<15\%$) effect on IgE binding to the non-modified $Fc_\epsilon RII$. Thus, both the human[9] and the mouse $sFc_\epsilon RII$ interact with IgE; however, the $1-5 \times 10^6 M^{-1}$ affinity indicates that this interaction, in contrast to IgE interacting with the intact $Fc_\epsilon RII$, would have little biologic significance. This would suggest that the $sFc_\epsilon RII$ production is a method of regulating the IgE (and other ligand(s) -- see below) -- $Fc_\epsilon RII$ interaction (due to loss of significant binding) and would indicate that the biologically important molecule is the intact, multimeric $Fc_\epsilon RII$. In spite of this, several activities of the human but not mouse $sFc_\epsilon RII$ have been reported.

FUNCTIONS OF THE $Fc_\epsilon RII$

The $Fc_\epsilon RII$ has been implicated to date in a variety of functional activities. Indeed the variety has lead to a search for alternate ligands for the $Fc_\epsilon RII$, since some of these functions clearly do not involve IgE. The proposed functions, as well as our current understanding of the ligands interacting with the $Fc_\epsilon RII$ are discussed below.

Involvement of the $Fc_\epsilon RII$ in IgE Production

There are a number of studies suggesting, especially in the human system, that the $Fc_\epsilon RII$ is involved in the regulation of IgE synthesis. The "switch factor" for IgE is clearly IL-4 (see[59] for review), indicating that the mechanism for $Fc_\epsilon RII$ involvement would be with respect to differentiation of IgE switched B cells to IgE production. Sherr et al.[60] demonstrated that IgE complexes suppressed IgE synthesis from both B cells and an IgE producing plasmocytoma cell line and this group subsequently demonstrated that this inhibition resulted in inhibition of mRNA for the secreted but not membrane form of IgE[61]. Other groups have noted that some mab anti-$Fc_\epsilon RII$ will inhibit IgE synthesis[62,63]. The $sFc_\epsilon RII$ may have some activity in this regard; depending upon the form of $Fc_\epsilon RII$ used a modest enhancement[64,65] or inhibition[65] respectively was seen with IgE production from IL-4 stimulated human PBL. Interestingly, the inhibitory activity was seen with a $sFc_\epsilon RII$ not previously described, where a carboxyl terminal portion was lost as well as the usual amino terminal degradation. In contrast, $sFc_\epsilon RII$ effects on murine IgE production have not been found[24,66].

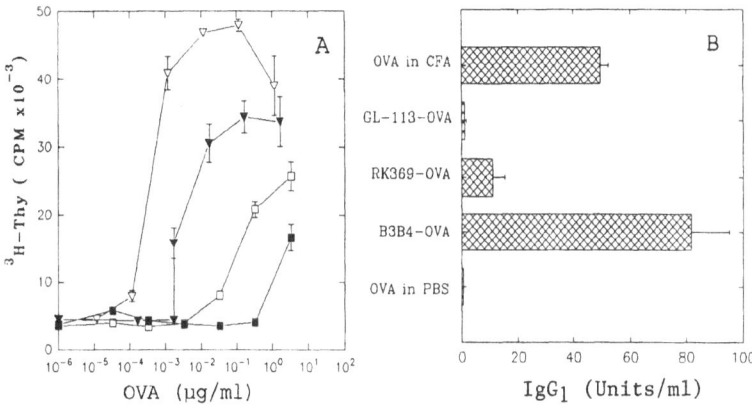

FIGURE 5. Activity of anti-Fc$_\epsilon$RII-antigen (ovalbumin) conjugate both *in vitro* (A) and *in vivo* (B). Anti-Fc$_\epsilon$RII-ovalbumin (OVA) conjugates (1:1 ratio) were prepared as described[69]. In (A) the *in vitro* activity of this conjugate (Δ) as well as control conjugates, Pgp-1 (RK3G9[70])-OVA ($\triangledown$), a nonspecific rat IgG$_{2a}$(GL-113)-OVA ($\square$) or unconjugated OVA (■) alone, were determined. The OVA specific Th2 cell line BB6.5[71] was irradiated and incubated with purified C57Bl/6 B cells and the indicated increasing amount of conjugate--OVA levels are shown. B cell proliferation was measured to determine activity by pulsing for 12 hrs with ^{3}H-thymidine after 60 hrs of culture. In (B) 100 μg of the indicated form of conjugate was injected intradermally into BALB/c mice on day 0 and again on day 35. Serum was collected on day 44 and assayed via antigen specific ELISA [71,72] as described. Complete Freund's/OVA injected animals received a single injection and the serum collected at day 21 and assayed in the same manner.

Involvement of the Fc$_\epsilon$RII in Enhancement of Antigen Processing

A new function of the Fc$_\epsilon$RII was described in 1989 by Kehry *et al.*[67] for the mouse system and subsequently, by Pirron *et al.*[68] for human Fc$_\epsilon$RII. A strong enhancement of antigen presentation was observed when IgE-antigen complexes were allowed to interact with Fc$_\epsilon$RII$^+$ B cells. Indeed the efficiency of antigen presentation, measured by the induction of proliferation of antigen specific T cells, approached that seen when antigen was targeted to surface immunoglobulin on B cells. As an alternative to IgE, we have used anti-Fc$_\epsilon$RII-antigen conjugates [71] and have again observed excellent enhancement of antigen presentation (Fig. 5). Interestingly, the anti-Fc$_\epsilon$RII-antigen conjugates also resulted in significantly increased immunogenicity (as determined by measuring antigen specific IgG$_1$ serum levels) when *in vivo* injections were performed, indicating that this enhancement is operative both *in vitro* and *in vivo*.

Involvement of the Fc$_\epsilon$RII in B Cell Activation

The identification of the Fc$_\epsilon$RII as the B cell activation antigen CD23^2 was the first indication of an involvement of the Fc$_\epsilon$RII in B cell activation. John Gordon's laboratory had described evidence that certain anti-Fc$_\epsilon$RII (CD23) could modulate B cell activation[73] and this group subsequently found that the intact $Fc_\epsilon RII$ and to a variable extent the $sFc_\epsilon RII$ would enhance mitogen induced human B cell activation[74]. No co-stimulatory activity for the mouse $sFc_\epsilon RII$ was seen with B cells stimulated with either anti-IgD or LPS[66]. Recently, an involvement in germinal centers has also been suggested. As stated earlier, in germinal centers both the B cell and FDC[46] are $Fc_\epsilon RII^+$. Liu et al.[75] described evidence that $sFc_\epsilon RII$, in conjugation with IL-1, inhibits the apoptosis that is seen with germinal center B cells. The $sFc_\epsilon RII$ plus IL-1 treated B cells have the morphology of plasmablasts.

A second role for the $Fc_\epsilon RII$ in B cell activation has been implicated by studies examining the co-crosslinking of B cell surface immunoglobulin (sIg) and the $Fc_\epsilon RII$. This has been examined in two ways with differing results, both using the murine system. Waldschmidt et al.[76] prepared an IgE isotype switch variant of the anti-IgD mab 10.4.22 and noted that the IgE anti-IgD had a substantially increased capacity to induce B cell proliferation. Our studies were performed using anti-IgD coupled to DNP (dinitrophenol)-dextran[77]. Anti-IgD dextran has been previously shown to be a powerful B cell stimulant[78] and the anti-IgD-DNP-dextran was equally potent. However, when IgE-anti-DNP was added to the system, inhibition of B cell proliferation occurred[77], especially at the higher inputs of anti-IgD-DNP-dextran (Fig. 6). This

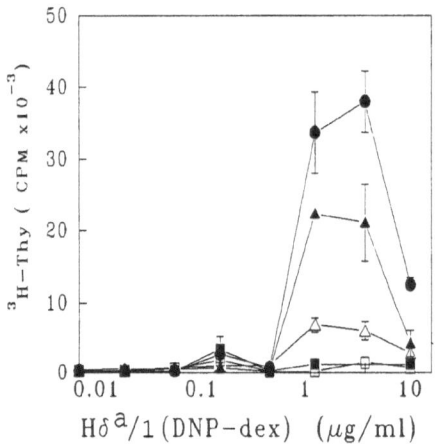

FIGURE 6. Co-crosslinking of sIgD and the $Fc_\epsilon RII$ can inhibit B cell activation. Purified BALB/c B cells were activated with the anti-IgD mab $H\delta^a/1$[79] coupled to DNP-dextran (●) as described[77]; proliferation was determined by pulsing with ^{3}H-thymidine. Proliferation was increasingly inhibited by adding the indicated amounts of IgE anti-DNP (0.01 (△); 0.1 (▲); 1.0 (■) or 10 (□) μg/ml respectively. Data is based on reference[77].

suggests that the $Fc_\epsilon RII$ can act in a manner analogous to the B cell $Fc_\gamma RII$, where a similar co-crosslinking (of the sIg and $Fc_\gamma RII$) resulted in inhibition of B cell activation[80]. However, the different results with the IgE anti-IgD discussed above remain unexplained. The most logical explanation for the discrepancy would be the level of co-crosslinking of the $Fc_\epsilon RII$ and sIg -- the IgE anti-IgD would be anticipated to give a significantly lower level of co-crosslinking than the anti-IgD-DNP-dextran plus IgE anti-DNP system. Thus, at low levels of co-crosslinking B cell activation may be actually enhanced, while at higher levels activation is suppressed.

Cytokine Activities of the $sFc_\epsilon RII$ Plus IL-1

An exciting and unexpected activity for the human $sFc_\epsilon RII$ has been described in recent years. Supernatants of B lymphoblastoid cells were shown to have cytokine like activities with regard to maturation of $CD7^+$ T cell precursors[81]. Subsequently, this activity was shown to be recapitulated when recombinant $sFc_\epsilon RII$ was used in conjunction with IL-1, suggesting new functions for the $sFc_\epsilon RII$[82]. In addition, $sFc_\epsilon RII$ plus IL-1 also enhanced the proliferation of $CD4^+$ T cells[83] and myeloid cell precursors in bone marrow[84], suggesting that the $sFc_\epsilon RII/IL-1$ combination has an IL-3 like function on human hematopoietic cells[85]. A search for similar activities in the mouse system was unsuccessful[66], suggesting that this function of the $Fc_\epsilon RII$ may be surprisingly species restricted. In addition, the identity of the $sFc_\epsilon RII$ ligand remains unclear at present, although an intriguing possibility is CD21 (see below).

$Fc_\epsilon RII$ Ligands

The "classic" ligand for the $Fc_\epsilon RII$ is, of course, IgE and it was by virtue of this interaction that the $Fc_\epsilon RII$ was first discovered[86]. This interaction led to the proposal that the $Fc_\epsilon RII$ was involved in IgE regulation and evidence for this was discussed above. However, the new activities described for the $Fc_\epsilon RII$, especially the cytokine like properties of the $Fc_\epsilon RII$ when used in conjunction with IL-1, led to the search for an alternate ligand. Recently, the Glaxo group described intriguing evidence that CD21 is a ligand for CD23[87]. CD21 also functions as the complement receptor, type 2 (CR2) and, in humans, is the recognition unit for Epstein Barr virus (EBV) (see[88] for review). To perform these studies, this group first developed a liposome binding assay in which intact $Fc_\epsilon RII$ was first allowed to incorporate into liposomes[89]. This reagent was shown to specifically interact with $CD21^+$ cells, potentially in a lectin dependent manner, since the interaction was blocked with fucose-1-phosphate. Subsequent studies demonstrated that the binding of the liposome/$Fc_\epsilon RII$ reagent was also blocked by certain anti-CD21 monoclonal antibodies[87]. Furthermore these mab anti-CD21 also influenced IgE production with IL-4 stimulated human PBL, suggesting that the $Fc_\epsilon RII/CD21$ interaction was relevant to IgE production; it is not presently clear if this reflects B-B or B-T interaction. Importantly, the binding activity was observed for the intact $Fc_\epsilon RII$ and not the $sFc_\epsilon RII$, indicating that the multimeric $Fc_\epsilon RII$ interacts more effectively with CD21, just as it does with IgE[87]. The other CD21 ligands, C3dg and EBV, interact at different but closely associated sites[90]. The site of $Fc_\epsilon RII$ interaction with CD21 is not presently known. It is not presently known if the $Fc_\epsilon RII/CD21$ interaction is also in involved in the cytokine activities discussed above or if still other $Fc_\epsilon RII$ ligands exist. Certainly the potential for $Fc_\epsilon RII$ to act as a lectin would suggest other ligands.

REFERENCES

1. J.V. Ravetch and J-P. Kinet, Fc receptors. *Ann.Rev.Immunol.* 9:457-492, 1991.
2. K. Yukawa, H. Kikutani, H. Owaki, K. Yamasaki, A. Yokota, H. Nakamura, E.L. Barsumian, R.R. Hardy, M. Suemura and T. Kishimoto, A B cell-specific differentiation antigen CD23, is a receptor for IgE ($Fc_\epsilon R$) on lymphocytes. *J.Immunol.* 138:2576-80, 1987.
3. J-Y. Bonnefoy, J-P. Aubry, C. Peronne, J. Wijdenes and J. Banchereau, Production and characterization of a monoclonal antibody specific for the human lymphocyte low affinity receptor for IgE: CD23 is a low affinity receptor for IgE. *J.Immunol.* 138:2970-78, 1987.
4. D.H. Conrad, $Fc_\epsilon RII/CD23$: The low affinity receptor for IgE. *Ann.Rev.Immunol.* 8::623-45, 1990.
5. G. Delespesse, U. Suter, D. Mossalayi, B. Bettler, M. Sarfati, H. Hofstetter, E. Kilcherr, P.

Debre and A. Dalloul, Expression, structure, and function of the CD23 antigen. *Adv.Immunol.* 49:149-191, 1991.

6. M.L. Richards and D.H. Katz, Biology and chemistry of the low affinity IgE receptor (FcϵRII/CD23). *CRC Crit.Rev.Immunol.* 11:65-86, 1991.

7. G.C. Meinke, A.M. Magro, D.A. Lawrence and H.L. Spiegelberg, Characterization of an IgE receptor isolated from cultured B type lymphoblastoid cells. *J.Immunol.* 121:1321-1326, 1978.

8. D.H. Conrad and L.H. Peterson, The murine lymphocyte receptor for IgE. I. Isolation and characterization of the murine B cell Fc epsilon receptor and comparison with Fc epsilon receptors from rat and human. *J.Immunol.* 132:796-803, 1984.

9. T. Nakajima and G. Delespesse, Relationship between human IgE-binding factors (IgE-BF) and lymphocyte receptors for IgE. *J.Immunol.* 139:848-54, 1987.

10. M. Rao, W.T. Lee and D.H. Conrad, Characterization of a monoclonal antibody directed against the murine B lymphocyte receptor for IgE. *J.Immunol.* 138:1845-51, 1987.

11. M. Letellier, T. Nakajima, G. Pulido-Cejudo, H. Hofstetter and G. Delespesse, Mechanism of formation of human IgE-binding factors (Soluble CD23): III. Evidence for a receptor (FcϵRII)-associated proteolytic activity. *J.Exp.Med.* 172:693-700, 1990.

12. M. Sarfati, T. Nakajima, H. Frost, E. Kilchherr and G. Delespesse, Purification and partial biochemical characterization of IgE-binding factors secreted by a human B lymphoblastoid cell line. *Immunology* 60:539-545, 1987.

13. S-S.A. Chen, J.A. Bohn, F-T. Liu and D.H. Katz, Murine lymphocytes expressing Fc receptors for IgE (FcϵR). I. Conditions for inducing FceR+ lymphocytes and inhibition of the inductive events by suppressive factor of allergy (SFA). *J.Immunol.* 127:166-173, 1981.

14. R. Vander-Mallie, T. Ishizaka and K. Ishizaka, Lymphocytes bearing receptors for IgE. VIII. Affinity of mouse IgE for Fcϵ receptor on mouse B lymphocytes. *J.Immunol.* 128:2306-2312, 1982.

15. W.T. Lee, M. Rao and D.H. Conrad, The murine lymphocyte receptor for IgE IV. The mechanism of ligand-specific receptor upregulation on B Cells. *J.Immunol.* 139:1191-1198, 1987.

16. H. Kikutani, S. Inui, R. Sato, E.L. Barsumian, H. Owaki, K. Yamasaki, T. Kaisho, N. Uchibayashi, R.R. Hardy, T. Hirano, S. Tsumasawa, F. Sakiyama, M. Suemura and T. Kishimoto, Molecular Structure of Human Lymphocyte Receptor for immunoglobulin E. *Cell* 47:657-665, 1986.

17. S.A. Hudak, S.O. Gollnick, D.H. Conrad and M.R. Kehry, Murine B cell stimulatory factor 1 (interleukin 4) increases expression of the Fc receptor for IgE on mouse B cells. *Proc.Natl.Acad.Sci.USA* 84:4606-10, 1987.

18. D. Vercelli, H.H. Jabara, B.W. Lee, N. Woodland and R.S. Geha, Human recombinant interleukin-4 induces FcϵR2/CD23 on normal human monocytes. *J.Exp.Med.* 167:1406-1416, 1988.

19. M. Hosoda, S. Makino, T. Kawabe, Y. Maeda, S. Satoh, M. Takami, M. Mayumi, K-I. Arai, H. Saitoh and J. Yodoi, Differential regulation of the low affinity Fc receptor for IgE (FcϵR2/CD23) and the IL-2 receptor (Tac/p55) on eosinophilic leukemia cell line (Eol-1 and Eol-3). *J.Immunol.* 143:147-152, 1989.

20. S. Fournier, G. Delespesse, M. Rubio, G. Biron and M. Sarfati, CD23 antigen regulation and signaling in chronic lymphocytic leukemia. *J.Clin.Invest.* 89:1312-1321, 1992.

21. J-P. Galizzi, H. Cabrillat, F. Rousset, C. Menetrier, J. DeVries and J. Banchereau, IFN-γ and prostaglandin E2 inhibit IL-4 induced expression of FceR2/CD23 on B lymphocytes through different mechanisms without altering binding of IL-4 to its receptor. *J.Immunol.* 141:1982-1988, 1988.

22. D.H. Conrad, T.J. Waldschmidt, W.T. Lee, M. Rao, A.D. Keegan, R.J. Noelle, R.G. Lynch and M.R. Kehry, Effect of B cell stimulatory factor-1 (Interleukin 4) on Fcϵ and Fcγ receptor expression on murine B lymphocytes and B cell lines. *J.Immunol.* 139:2290-2296, 1987.

23. U. Suter, G. Texido and H. Hofstetter, Expression of Human Lymphocyte IgE Receptor (FcϵRII/CD23) Identification of FcϵRIIa Promoter and its Functional Analysis in B lymphocytes. *J.Immunol.* 143:3087-3092, 1989.

24. A.D. Keegan, C.M. Snapper, R. VanDusen, W.E. Paul and D.H. Conrad, Superinduction of the murine B cell FcεRII by T helper cell clones. Role of Interleukin-4.*J.Immunol.*142:3868-3874, 1989.

25. D.H. Conrad, A.D. Keegan, K.R. Kalli, R. Van-Dusen, M. Rao and A.D. Levine, Superinduction of low affinity IgE receptors on murine B lymphocytes by lipolysaccharide and IL-4. *J.Immunol.* 141:1091-1097, 1988.

26. K. Ikuta, M. Takami, L.W. Kim, T. Honjo, T. Miyoshi, Y. Tagaya, T. Kawabe and J. Yodoi, Human lymphocyte Fc receptor for IgE: Sequence homology of its cloned cDNA with animal lectins. *Proc.Natl.Acad.Sci.USA* 84:819-823, 1987.

27. C. Ludin, H. Hofstetter, M. Sarfati, C.A. Levy, U. Suter, D. Alaimo, E. Kelchherr, H. Frost and G. Delespesse, Cloning and expression of the cDNA coding for a human lymphocyte IgE receptor. *EMBO J.* 6:109-114, 1987.

28. B. Bettler, H. Hofstetter, M. Rao, W.M. Yokoyama, F. Kilchherr and D.H. Conrad, Molecular structure and expression of the murine lymphocyte low affinity receptor for IgE (FcεRII). *Proc.Natl.Acad.Sci.USA* 86:7566-7570, 1989.

29. S.O. Gollnick, M.L. Trounstine, L.C. Yamashita, M.R. Kehry and K.W. Moore, Isolation,characterization and expression of cDNA clones encoding the mouse Fc receptor for IgE (FcεRII). *J.Immunol.* 144:1974-1882, 1990.

30. M.L. Richards, F-T. Liu and D.H. Katz, The induction of murine B cell Ia by IgE-antigen complexes is dependent on protein synthesis and preceded by class II mRNAaccumulation. *Cell.Immunol.* 128:198-208, 1990.

31. D. Vercelli, B. Helm, P. Marsh, E. Padlan, R.S. Geha and H. Gould, The B-cell binding site on human immunoglobulin E. *Nature* 338:649-651, 1989.

32. G. Delespesse, M. Sarfati, C.Y. Wu, S. Fournier and M. Letellier, The low-affinity receptor for IgE. *Immunol.Rev.* 125:77-97, 1992.

33. E. Nakayama, I. vonHoegen and J. Parnes, Sequence of the Lyb-2 B-cell differentiation antigen defines a gene superfamily of receptors with inverted membrane orientation. *Proc.Natl.Acad.Sci.USA* 86:1352-1356, 1989.

34. M. Bevilacqua, E. Butcher, B. Furie, M. Gallatin, M. Gimbrone, J. Harlan, K. Kishimoto, L. Lasky, R. McEver, J. Paulson, S. Rosen, B. Seed, M. Siegelman, T. Springer, L. Stoolman, T. Tedder, A. Varki, D. Wagner, I. Weissman and G. Zimmerman, Selectins: A family of adhesion receptors. *Cell* 67:233, 1991.

35. K. Drickamer, M.S. Dordal and L. Reynolds, Mannose-binding proteins isolated from rat liver contain carbohydrate-recognition domains linked to collagenous tails. *J.Biol.Chem.* 261:6878-6886, 1986.

36. S.K. Akiyama and K.M. Yamada, Synthetic peptides competitively inhibit both direct binding to fibroblast and functional biological assays for the purified cell binding domain of fibronectin. *J.Biol.Chem.* 260:10402-10405, 1985.

37. U. Suter, R. Bastos and H. Hofstetter, Molecular structure of the gene and the 5'-flanking region of the human lymphocyte immunoglobulin E receptor. *Nuc.Acids Res.* 15:7295-7308, 1987.

38. M.L. Richards, D.H. Katz and F.-T. Liu, Complete genomic sequence of the murine low affinity Fc receptor for IgE: Demonstration of alternative transcripts and conserved sequence elements. *J.Immunol.* 147:1067-1074, 1991.

39. A.J. Beavil, R.L. Edmeades, H.J. Gould and B.J. Sutton, α-Helical coiled-coil stalks in the low-affinity receptor for IgE (FcεRII/CD23) and related C-type lectins. *Proc.Natl.Acad.Sci.USA* 89:753-757, 1992.

40. M. Letellier, M. Sarfati and G. Delespesse, Mechanisms of formation of IgE-binding factors (soluble CD23)--I. FcεR II bearing B cells generate IgE-binding factors of different molecular weights. *Mol.Immunol.* 26:1105-1112, 1989.

41. A. Yokota, H. Kikutani, T. Tanaka, R. Sato, E.L. Barsumian, M. Suemura and T. Kishimoto, Two species of human Fcε receptor II (FcεRII/CD23): Tissue-specific and IL-4-specific regulation of gene expression. *Cell* 55:611-618, 1988.

42. M. Nonaka, D.K. Hsu, C.M. Hanson, F. Aosai and D.H. Katz, Cloning of cDNA coding for low-affinity Fc receptors for IgE on human T lymphocytes. *Int.Immunol.* 1:254-259, 1989.

43. S. Fournier, I.D. Tran, U. Suter, G. Biron, G. Delespesse and M. Sarfati, The *in vivo*

expression of type B CD23 mRNA in B-chronic lymphocytic leukemic cells is associated with an abnormally low CD23 upregulation by IL-4: Comparison with their normal cellular counterparts. *Leuk.Res.* 15:609-618, 1991.

44. A. Yokota, K. Yukawa, A. Yamamoto, K. Sugiyama, M. Suemura, Y. Tashiro, T. Kishimoto and H. Kikutani, Two forms of the low-affinity Fc receptor for IgE differentially mediate endocytosis and phagocytosis: Identification of the critical cytoplasmic domains. *Proc.Natl.Acad.Sci.USA* 89:5030-5034, 1992.

45. D.H. Conrad, C.A. Kozak, J. Vernachio, C.M. Squire, M. Rao and E.M. Eicher, Chromosomal location and isoform analysis of mouse $Fc\epsilon RII/CD23$. *Mol.Immunol.* 1992 (In Press).

46. K. Maeda, G.F. Burton, D.A. Padgett, D.H. Conrad, T.F. Huff, A. Masuda, A.K. Szakal and J.G. Tew, Murine Follicular dendritic cells (FDC) and low affinity Fc-receptors for IgE ($Fc\epsilon RII$). *J.Immunol.* 148:2340-2347, 1992.

47. M. Sandor, T. Gajewski, J. Thorson, J.D. Kemp, F.W. Fitch and R.G. Lynch, $CD4^+$ murine T cell clones that express high levels of immunoglobulin binding belong to the interleukin 4-producing T helper cell type 2 subset. *J.Exp.Med.* 171:2171-2176, 1990.

48. A. Mathur, D.H. Conrad and R.G. Lynch, Characterization of the murine T cell receptor for IgE ($Fc\epsilon RII$). Demonstration of shared and unshared epitopes with the B cell $Fc\epsilon RII$. *J.Immunol.* 141:2661-2667, 1988.

49. T.J. Waldschmidt, F.G.M. Kroese, L.T. Tygrett, D.H. Conrad and R.G. Lynch, The expression of B cell surface receptors III. The murine low affinity IgE receptor is not expressed on Ly 1 or "Ly 1 like" B cells. *Int.Immunol.* 3:305-315, 1991.

50. T. Waldschmidt, K. Snapp, T. Foy, L. Tygrett and C. Carpenter, B-cell subsets defined by the $Fc\epsilon R$. *Ann.NY Acad.Sci.* 651:84-98, 1992.

51. W.T. Lee and D.H. Conrad, The murine lymphocyte receptor for IgE.III.Use of chemical cross-linking reagents to further characterize the B lymphocyte Fc epsilon receptor. *J.Immunol.* 134:518-525, 1985.

52. H. Gould, B. Sutton, R. Edmeades and A. Beavil, $CD23/Fc_\epsilon RII$: C-type lectin membrane protein with a split personality. *Monogr.Allergy* 29:28-49, 1991.

53. S.E. Dierks, W.C. Bartlett, R.L. Edmeades, H.J. Gould, M. Rao and D.H. Conrad, The oligomeric nature of the murine $Fc\epsilon RII/CD23$: Implications for function. *Submitted for publication* 1992.

54. W.T. Lee and D.H. Conrad, The murine lymphocyte receptor for IgE.II.Characterization of the multivalent nature of the B lymphocyte receptor for IgE. *J.Exp.Med.* 159:1790-1795, 1984.

55. B. Bettler, R. Maier, D. Ruegg and H. Hofstetter, Binding site for IgE of the human low affinity $Fc\epsilon$ receptor $Fc\epsilon RII/CD23$) is confined to the domain homologous with animal lectins. *Proc.Natl.Acad.Sci.USA* 86:7118-7122, 1989.

56. B. Bettler, G. Texido, S. Raggini, D. Rüegg and H. Hofstetter, Immunoglobulin E-binding site in Fc_ϵ receptor ($Fc_\epsilon RII/CD23$) identified by homolog-scanning mutagenesis. *J.Biol.Chem.* 267:185-191, 1992.

57. S. Schwarzbaum, A. Nissim, I. Alkalay, M.C. Ghozi, D.G. Schindler, Y. Bergman and Z. Eshhar, Mapping of murine IgE epitopes involved in IgE-Fc_ϵ receptor interactions. *Eur.J.Immunol.* 19:1015-1023, 1989.

58. F.M. Melewicz, J.M. Plummer and H.L. Spiegelberg, Comparison of the Fc receptors for IgE on human lymphocytes and monocytes. *J.Immunol.* 129:563-569, 1982.

59. W.E. Paul, Interleukin-4: A prototypic immunoregulatory lymphokine. *Blood* 77:1859-1870, 1991.

60. E. Sherr, E. Macy, H. Kimata, M. Gilly and A. Saxon, Binding the low affinity $Fc\epsilon R$ on B cells suppresses ongoing human IgE synthesis. *J.Immunol.* 142:481-489, 1989.

61. A. Saxon, M. Kurbe-Leamer, K. Behle, E.E. Max and K. Zhang, Inhibition of human IgE production via $Fc\epsilon R$-II stimulation results from a decrease in the mRNA for secreted but not membrane ϵ H chains. *J.Immunol.* 147:4000-4006, 1991.

62. J. Pene, F. Rousset, F. Briere, I. Chretein, J-Y. Bonnefoy, H. Spits, T. Yokota, N. Arai, K-I. Arai and J. Banchereau, IgE production by human lymphocytes is induced by interleukin 4 and suppressed by interferons γ and α and prostaglandin E2. *Proc.Natl.Acad.Sci.USA* 85:6880-6884, 1988.

63. M. Sarfati and G. Delespesse, Possible role of human lymphocyte receptor for IgE (CD23) or its soluble fragments in the in vitro synthesis of human IgE. *J.Immunol.* 141:2195-2199, 1988.

64. J. Pene, I. Chretein, F. Rousset, F. Briere, J-Y. Bonnefoy and J. DeVries, Modulation of IL-4-induced human IgE production in vitro by IFN-γ and IL-5: the role of soluble CD23 (sCD23). *J.Cell.Biochem.* 39:253-264, 1989.

65. M. Sarfati, B. Bettler, M. Letellier, S. Fournier, M. Rubio-Trujillo, H. Hofstetter and G. Delespesse, Native and recombinant soluble CD23 fragments with IgE suppressive activity. *Immunology* 76:662-667, 1992.

66. W.C. Bartlett and D.H. Conrad, Murine soluble FcϵRII: A molecule in search of a function. *Res.Immunol.* 143:431-436, 1992.

67. M.R. Kehry and L.C. Yamashita, Fcϵ receptor II (CD23) function on mouse B cells: Role in IgE dependent antigen focusing. *Proc.Natl.Acad.Sci.USA* 86:7556-7560, 1989.

68. U. Pirron, T. Schlunck, J.C. Prinz and E.P. Rieber, IgE-dependent antigen focusing by human B lymphocytes is mediated by the low-affinity receptor for IgE. *Eur.J.Immunol.* 20:1547-1551, 1990.

69. A. Lees, S.C. Morris, G. Thyphronitis, J.M. Holmes, J.K. Inman and F.D. Finkelman, Rapid stimulation of large specific antibody responses with conjugates of antigen and anti-IgD antibody. *J.Immunol.* 145:3594-3600, 1990.

70. M. Rao, R. Knox and D.H. Conrad, Characterization of Pgp-1 antigen on murine B lymphocytes using a new anti-Pgp-1 monoclonal antibody. *Hybridoma* 10:281-284, 1990.

71. C.M. Squire, E. Studer, A. Lees, F.D. Finkelman and D.H. Conrad, Antigen presentation is enhanced by targeting antigen to the FcϵRII by antigen-anti-FcϵRII conjugates. *Submitted for publication* 1992.

72. C.M. Snapper, F.D. Finkelman and W.E. Paul, Differential regulation of IgG1 and IgE synthesis by interleukin 4. *J.Exp.Med.* 167:183-196, 1988.

73. J. Gordon, A.J. Webb, G.R. Guy and L. Walker, Triggering of B lymphocytes through CD23: epitope mapping and studies using antibody derivatives indicate an allosteric mechanism of signaling. *Immunology* 60:517-521, 1987.

74. J. Cairns and J. Gordon, Intact, 45-kDa (membrane) form of CD23 is consistently mitogenic for normal and transformed B lymphoblasts. *Eur.J.Immunol.* 20:539-543, 1990.

75. Y.-J. Liu, J.A. Cairns, M.J. Holder, S.D. Abbot, K.U. Jansen, J-Y. Bonnefoy, J. Gordon and I.C.M. MacLennan, Recombinant 25-kDa CD23 and interleukin 1α promote the survival of germinal center B cells: Evidence for bifurcation in the development of centrocytes rescued from apoptosis. *Eur.J.Immunol.* 21:1107-1114, 1991.

76. T.J. Waldschmidt and L.T. Tygrett, The low affinity IgE Fc receptor (CD23) participates in B cell activation. In: , Plenum Press, New York and London: 1992, p. In press.

77. K.A. Campbell, A. Lees, F.D. Finkelman and D.H. Conrad, Co-crosslinking FcϵRII/CD23 with B cell surface immunoglobulin modulates B cell activation. *Eur.J.Immunol.* 22:2107-2112, 1992.

78. M. Brunswick, F.D. Finkelman, P.F. Highet, J.K. Inman, H.M. Dintzis and J.J. Mond, Picogram quantities of anti-Ig antibodies coupled to dextran induce B cell proliferation. *J.Immunol.* 140:3364-3372, 1988.

79. I.M. Zitron and B.L. Clevinger, Regulation of murine B cells through surface immunoglobulin. I. Monoclonal anti-delta antibody that induces allotype-specific proliferation. *J.Exp.Med.* 152:1135-1146, 1980.

80. N.E. Phillips and D.C. Parker, Cross-linking of B lymphocyte Fc gamma receptors and membrane immunoglobulin inhibits anti-immunoglobulin-induced blastogenesis. *J.Immunol.* 132:627-632, 1984.

81. M.D. Mossalayi, J.C. Lecron, P. Goube de Leforest, G. Janossy, P. Debré and J. Tanzer, Characterization of prothymocytes with cloning capacities in human bone marrow. *Blood* 71:1281-1287, 1988.

82. M.D. Mossalayi, J.-C. Lecron, A.H. Dalloul, M. Sarfati, J-M. Bertho, H. Hofstetter, G. Delespesse and P. Debré, Soluble CD23 (FcϵRII) and interleukin 1 synergistically induce early human thymocyte maturation. *J.Exp.Med.* 171:959-964, 1990.

83. J.-M. Bertho, C. Fourcade, A.H. Dalloul, P. Debré and M.D. Mossalayi, Synergistic effect of

interleukin 1 and soluble CD23 on the growth of human $CD4^+$ bone marrow-derived T cells. *Eur.J.Immunol.* 21:1073-1076, 1991.

84. M.D. Mossalayi, M. Arock, J-M. Bertho, C. Blanc, A.H. Dalloul, H. Hofstetter, M. Sarfati, G. Delespesse and P. Debré, Proliferation of early human myeloid precursors induced by interleukin-1 and recombinant soluble CD23. *Blood* 75:1924-1927, 1990.

85. M.D. Mossalayi, A.H. Dalloul, C. Fourcade, M. Arock and P. Debré, Soluble CD23 is a potent cytokine for early human haematopoietic precursors. *Bull.Inst.Pasteur* 89:139-146, 1991.

86. D.A. Lawrence, W.O. Weigle and H.L. Spiegelberg, Immunoglobulins cytophilic for human lymphocytes, monocytes, and neutrophils. *J.Clin.Invest.* 55:268-275, 1975.

87. J.-P. Aubry, S. Pochon, P. Graber, K.U. Jansen and J.-Y. Bonnefoy, CD21 is a ligand for CD23 and regulates IgE production. *Nature* 358:505-507, 1992.

88. D.T. Fearon and J.M. Ahearn, Complement receptor type 1 (C3b/C4b receptor; CD35) and complement receptor type 2 (C3d/Epstein-Barr virus receptor; CD21). *Curr.Top.Microbiol.Immunol.* 153:83-98, 1990.

89. S. Pochon, P. Graber, M. Yeager, K. Jansen, A.R. Bernard, J.-P. Aubry and J.-Y. Bonnefoy, Demonstration of a second ligand for the low affinity receptor for immunoglobulin E (CD23) using recombinant CD23 reconstituted into fluorescent liposomes. *J.Exp.Med.* 176:389-397, 1992.

90. D.R. Martin, A. Yuryev, K.R. Kalli, D.T. Fearon and J.M. Ahearn, Determination of the structural basis for selective binding of Epstein-Barr virus to human complement receptor type 2. *J.Exp.Med.* 174:1299-1311, 1991.

CHARACTERIZATION OF THE HUMAN IgE Fc - FcεRIα INTERACTION

Jarema P. Kochan, Michael Mallamaci, Alasdair Gilfillan,
Vincent Madison*, and Mitali Basu

Department of Bronchopulmonary Research and the
*Department of Molecular Sciences, Hoffmann-La Roche Inc.
Nutley, N.J. 07110, USA

INTRODUCTION

Type I immediate hypersensitivity is mediated by the aggregation of IgE occupied high affinity IgE receptors (FcεRI) by antigen. This aggregation event triggers the release of a variety of inflammatory mediators, including histamine, leukotrienes, and cytokines that contribute to the allergic condition. The FcεRI is composed of at least three different subunits α, β and γ, in the rodent and the human (reviewed in Ravetch and Kinet, 1991) (figure 1). The FcεRIα subunit is involved in binding IgE, even in the absence of the β and γ subunits (Hakimi et al, 1990, Blank et al, 1991). The functions of the β and γ subunits are under investigation, and there is evidence demonstrating their role in signal transduction.

The fundamental step in the allergic reaction involves the binding of IgE to the FcεRI. In the 1960s, evidence was presented that the only immunoglobulin which was responsible for triggering the allergic response (as measured by the Prausnitz-Küstner, P-K, reaction) was IgE (reviewed in Ishizaka, 1970). This reaction could be inhibited by myeloma IgE, which was not specific for a particular antigen. The identity of IgE as the critical component for sensitization was also demonstrated by the inability of IgE depleted serum to induce an allergic response. Therefore, the binding of IgE to mast cells/basophils is the crucial step in eliciting an allergic response. Subsequent studies confirmed and expanded these initial observations, and research was initiated on characterizing the immunological and biochemical phenomenon associated with the allergic response.

What has been established by the earlier studies as well as more recent studies, is that the initiating event in the allergic response is the binding of IgE to the mast cell and basophil

FcϵRIα. In the absence of IgE bound to the mast cell surface, there is no antigen specific allergic response. The ability to understand the binding interaction between FcϵRIα and IgE Fc at the molecular level may provide a potential means of blocking allergic responses. In this article, we will review some of our more recent studies that have been directed towards characterizing the interaction between the human IgE Fc and the human FcϵRIα.

THE IgE Fc BINDING SITE

The ability of the Fc portion of the IgE molecule to bind to the FcϵRIα was first demonstrated almost 25 years ago (reviewed in Dorrington and Bennich, 1978). Since then, a considerable effort has been directed towards the identification of the site on the IgE Fc molecule which interacts with the FcϵRIα. The use of molecular biology has facilitated the overproduction and characterization of IgE Fc fragments which bind to the FcϵRIα. These fragments have been characterized by several groups (reviewed in Burton and Woof 1992) and it has been reported that the smallest IgE Fc fragment which is still capable of interacting with the FcϵRIα encompasses amino acid residues 301 - 363 (Helm et al, 1988, Helm et al, 1989). The ability to determine the crystal structure of this fragment would represent a significant advance towards the detailed characterization of the IgE Fc - FcϵRIα interaction. Our own studies in purifying a recombinant E. coli-derived IgE Fc fragment spanning residues 301 - 376 were not successful (The reasons for the lack of success are not clear, but the refolding of proteins is subject to many variations). In order to define the smallest IgE Fc fragment which was capable of binding FcϵRIα, and one that could also be used for detailed structural analysis, a new approach was developed. While several studies had utilized eukaryotic expression to characterize the domains involved in FcϵRIα binding, the intact IgE proteins were always produced (Nissim et al, 1991, Nissim and Eshhar, 1992). These proteins have certain advantages, however they are probably too large for detailed structural studies.

To circumvent this problem, we have developed a eukaryotic expression system, whereby different IgE Fc fragments can be expressed and analyzed. In this approach, regions of the IgE Fc are fused to a signal peptide and transfected into COS cells, resulting in the expression of a protein which is usually secreted. These recombinant IgE Fc fragments can then be analyzed for their binding to the FcϵRIα, utilizing a solid phase IgE binding assay (SPEBA). In this manner, it is possible to identify the regions of the IgE Fc which are essential for the binding to the FcϵRIα. The smallest fragment which binds to the FcϵRIα is encoded by amino acid residues 330-547 (Basu et al, 1993). Attempts to make this fragment smaller by deletion of residues from either the N- or C- termini results in the production of protein which does not bind FcϵRIα as determined by the SPEBA. A comparison of the different regions of the IgE Fc which have been demonstrated, by various studies, to be essential for FcϵRIα binding are presented in figure 1. The studies on the chimeric IgE molecules demonstrating that the Cϵ3 domain is essential for FcϵRIα binding, could not rule out the contribution of the Cϵ4 domain in the binding interaction. Our inability to demonstrate FcϵRIα binding of IgE Fc fragments which lack the Cϵ4 domain provides further support for the role of this domain in either facilitating the conformation of the

Cϵ3 domain, or contributing along with Cϵ3 in the binding interaction. Further experimentation by site-directed mutagenesis is required to resolve this issue.

Two different IgE Fc fragments, IgE Fc315-547 and IgE Fc330-547 have now been overexpressed in CHO cells and purified to homogeneity (figure 2). The fragments differ by fifteen amino acids, including Cys328, which forms the covalent interchain disulfide bond. Although only the IgE Fc315-547 contains Cys328, both fragments behave as functional dimers. These physical properties have been determined by gel filtration and sedimentation equilibrium studies. These results demonstrate that the Cys 328 residue is not essential for dimer formation,

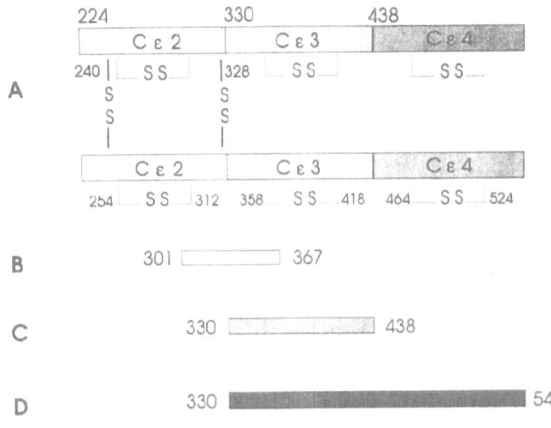

Figure 1 - Schematic representation of the IgE Fc regions which are involved in binding FcϵRIα. A. Intact IgE Fc dimer showing the relevant amino acid positions. B. Smallest *E. coli*-derived fragment which has been demonstrated to bind receptor (Helm et al, 1989). C. The region of the IgE Fc which has been demonstrated to be important in binding receptor from studies on intact chimeric IgE Fc molecules (Nissim et al, 1992). D. The smallest IgE Fc fragment which has been demonstrated to bind receptor from expression in eukaryotic cells (Basu et al., 1993).

but further studies have demonstrated that it forms the intermolecular disulfide bond. Although IgE Fc330-547 does not contain an interchain disulfide, we could not find any evidence for a monomeric species of this protein. Both IgE Fc315-547 and IgE Fc330-547 were able to bind FcϵRIα with high affinity, and their binding characteristics were very similar to native IgE. Based on these studies, we conclude that the binding of IgE Fc to the FcϵRIα requires a molecule which contains both the Cϵ3 and Cϵ4 domains. The ability to produce a homogeneous preparation of IgE Fc in large quantities should facilitate the resolution of the three-dimensional structure of the IgE Fc molecule.

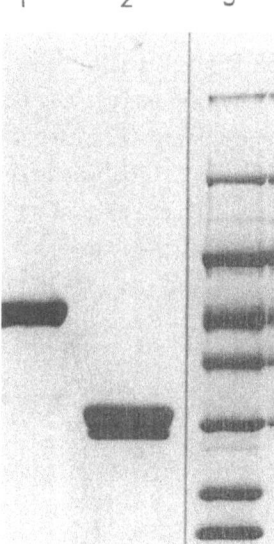

Figure 2 - Purified recombinant IgE Fc fragments. Purified recombinant IgE Fc from CHO cell supernatants were analyzed by SDS-PAGE under non-reducing conditions. Lane 1, IgE Fc315-547, lane 2, IgE Fc330-547, lane 3, molecular weight standards.

THE FcϵRIα BINDING SITE

Compared to the IgE Fc molecule, very few studies have been reported on the characterization of the FcϵRIα binding site. The demonstration that the FcϵRIα subunit was capable of binding IgE with high affinity in the absence of the other FcϵRI subunits was demonstrated by two different approaches. Hakimi et al, 1990, constructed a chimeric FcϵRIα subunit which consisted of the FcϵRIα extracellular domain, fused to the p55 IL-2 receptor transmembrane and cytoplasmic domains. The chimeric polypeptide was efficiently expressed on the cell surface in the absence of the other two subunits, and bound IgE with high affinity. The other approach by Blank et al, 1991, involved the construction of a soluble form of human FcϵRIα, which was efficiently secreted by eukaryotic cells. This form of the receptor was also capable of binding IgE with high affinity. Curiously, the rat and mouse FcϵRIα's were not secreted under conditions where the human was secreted. The reason for these differences has not yet been determined. Both of these studies conclusively demonstrated that the extracellular domain of the FcϵRIα subunit was sufficient for high affinity IgE binding.

Only two studies have been reported on the characterization of the FcϵRIα binding domains to date. One of these involved the generation of monoclonal antibodies against the human FcϵRIα (Riske et al, 1991). Two different classes of monoclonal antibodies were isolated, inhibitory and non-inhibitory. The non-inhibitory mAbs were able to bind FcϵRIα-IgE complexes, did not inhibit the binding of IgE to the FcεRIα, and were able to stimulate the release of histamine from freshly isolated human basophils. The inhibitory mAbs were able to inhibit IgE binding to FcεRIα, did not bind FcεRIα-IgE complexes, and were able to stimulate the release of histamine from freshly isolated human basophils. The ability of the inhibitory mAbs to stimulate histamine release from the human basophils was reduced when compared to the non-inhibitory class, probably due to the occupation of FcεRIα by IgE. MAb mapping of the reactivity with FcεRIα peptides indicated that one of the inhibitory mAbs reacted with a peptide spanning amino acid residues 125-140. These results suggested that the inhibitory mAbs recognized domain II of the native FcεRIα. The results could not determine whether the inhibition of IgE binding was due to direct overlap, steric inhibition or induced conformational changes.

The second study by Hogarth et al, 1992, used a form of homologue scanning mutagenesis, where regions of the human FcεRIα were inserted into the human FcγRII, and the ability of the chimeric receptors to bind IgE was determined. The chimeras were capable of binding multimeric IgE, but the binding of monomeric IgE (high affinity IgE binding) could not be demonstrated. The ability to bind the multimeric IgE was localized to three different contiguous regions which make up the entire domain II. The contribution of these regions in domain II could not be discerned.

In order to obtain information about the high affinity binding of IgE to the FcεRIα subunit, we have constructed and characterized a series of chimeric receptors. Our initial strategy involved domain substitutions, which were subsequently followed by more defined homologue scanning mutagenesis to identify smaller stretches of amino acids which were required for IgE binding. The ability to bind both human and rat IgE was examined, as well as the ability to be recognized by anti-FcεRIα mAbs. The results of our domain substitution studies are summarized in figure 3 (data is compiled from Mallamaci et al, 1993). Several conclusions can be made from examining the data. The human FcεRIα domain II is essential for high affinity binding since substitution by either the rat FcεRIα or the human FcγRIIIA domain II results in the loss of IgE binding. However, domain I can be substituted by either the rat FcεRIα or the human FcγRIIIA domain I without a significant effect on the apparent affinity for IgE binding. All of the human anti-FcεRIα mAbs which inhibit the binding of IgE map to domain II, while all of the non-inhibitory mAbs map to domain I. In contrast to the human FcεRIα, two of the inhibitory mAbs to the rat FcεRIα map to domain I, while one recognizes an epitope made up of both domains I and II. The reason for the differences of the epitopes which are recognized by the anti-human or anti-rat FcεRIα mAbs is not entirely clear. The data indicates that in the human FcεRIα, domain II plays an important role in IgE binding, but does not exclude the role of domain I in IgE binding. For the rat FcεRIα, the data suggests the importance of domain I and domain II in binding IgE.

To provide a more detailed understanding of the interaction of the human FcεRIα with IgE, homologue scanning mutagenesis was performed. In this procedure, stretches of ten to fifteen amino acid residues were replaced from the homologous regions of the human FcγRIIIA, and the binding of IgE and the different anti-FcεRIα mAbs monitored. Two different substitutions in domain I resulted in the loss of IgE binding. These regions spanned amino acid residues 35 - 46 and 80 - 92. Although the chimeric receptors were expressed on the cell surface as determined by immunofluorescence analysis with anti-FcεRIα peptide antisera, neither the inhibitory or non-inhibitory mAbs recognized these two constructs. These observations suggest that it is possible that these regions are important for IgE binding, but they may also be very important for the conformation of the receptor, thus the lack of mAb binding.

In domain II, three different regions were identified as being important for IgE binding, 118 - 129, 136 - 150, and 148 - 162. In contrast to the domain I chimeras, all of the three domain II chimeras were recognized by all of the non-inhibitory mAbs. The chimera encompassing residues 148 - 162 was also recognized by the inhibitory mAbs, but the other two chimeras were not recognized by any of the inhibitory mAbs. These results allow us to conclude that residues 148 - 162 are important for IgE binding, as substitution of these residues by those from FcγRIIIA result in the loss of IgE binding. However, this substitution does not affect the interaction with the inhibitory mAbs demonstrating that the inability to bind IgE is not associated with a drastic change in the conformation of these chimeras. The inability of the chimeras encompassing residues 118 - 129 and 136 - 150 to bind IgE can be interpreted in at least two

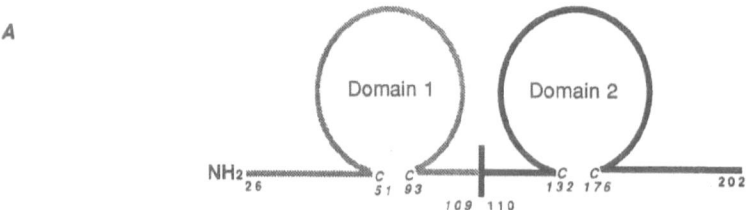

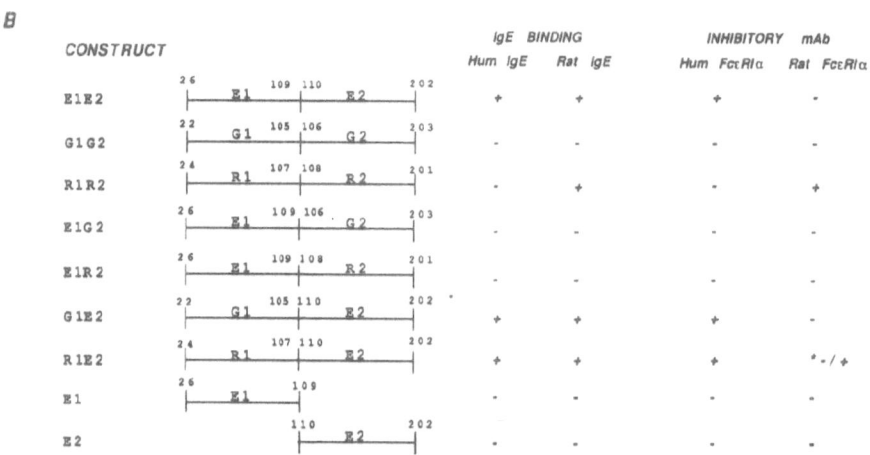

Figure 3 - Characterization of the FcεRIα domains in binding IgE. A. Schematic representation of the FcεRIα subunit indicating the positions of the amino acid residues. B. Summary of the binding characteristics of the different chimeric receptors. "E", "G" and "R" refer to the domains of the human FcεRIα subunit, the human FcγRIIIA, and the rat FcεRIα subunit respectively. 1 and 2 indicate the domain of the subunit, and the positions of the ends of each domain are marked. IgE binding and monoclonal antibody binding were determined as described in Mallamaci et al., 1993. "+" and "-" refer to binding or no binding. "*" indicates that two of the inhibitory anti-rat FcεRIα mAbs recognized the chimera, while one mAb did not.

different ways. These regions of the FcεRIα may be important for binding IgE, and also span the inhibitory mAb epitopes, or the lack of reactivity with the inhibitory mAbs might indicate a significant change in conformation leading to the loss of IgE binding. It is also possible that both of the explanations combined lead to the loss of IgE binding.

A three-dimensional model for the FcεRIα was constructed based upon the known structure of the CD4 molecule. The location of the regions in the FcεRIα which have been identified to affect the binding of IgE have been highlighted on the model presented in figure

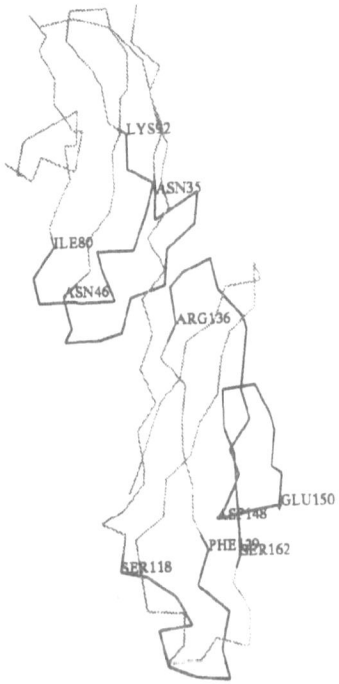

Figure 4 - Model of the FcεRIα subunit. The regions of the FcεRIα subunit which were identified by homologue scanning mutagenesis are highlighted on the Cα backbone of the FcεRIα model.

4. It must be emphasized that the contributions of the different regions to the binding of IgE has been inferred from the analysis of the properties of the different chimera. It is entirely possible that some of the regions exert their effect by affecting the configuration of the FcεRIα subunit. While the nature of the IgE binding residues awaits to be confirmed by further experimentation, the location of the regions which affect IgE binding, suggests that the interdomain region of the FcεRIα plays an important role in binding IgE.

While this kind of analysis will certainly be helpful in the study of the FcεRIα subunit and the nature of its interaction with the IgE Fc molecule, the ultimate information will be derived from the X-ray crystallographic structure of the FcεRIα. Work in our laboratories is in progress towards this end.

SUMMARY

A significant amount of progress has been achieved on characterizing the interaction of the IgE Fc molecule with the FcεRIα. However, there is yet no definitive structural information which precisely defines the nature of this interaction. It is clear that this information will only be provided by the resolution of the X-ray crystallographic structures of the IgE Fc molecule, the FcεRIα subunit extracellular domain, and the IgE Fc - FcεRIα complex. It is anticipated that these structures will be determined in the near future, and that they may provide some insight into the development of potential therapeutics effective in the management of IgE-mediated allergic diseases.

REFERENCES

Basu, M., Hakimi, J. Dharm, E., Kondas, J., Tsien, W.-H., Pilson, R.S., Lin, P., Gilfillan, A., Haring, P., Braswell, E.H., Nettleton, M.Y., and Kochan, J.P. 1993. *J. Biol. Chem.* 268:13118-13127.

Blank, U., Ra, C., and Kinet, J.P. 1991. *J. Biol. Chem.* 266:2639-2646.

Burton, D.R., and Woof, J.M. 1992. *Advances in Immunology* 51:1-84.

Dorrington K.J. and Bennich H.H. 1978. *Immun. Rev.* 41: 3-25.

Hakimi J., Seals C., Kondas J.A., Pettine L., Danho W. and Kochan J. 1990. *J. Biol. Chem.* 265: 22079-22081.

Helm, B., Marsh, P., Vercelli, D., Padlan, E., Gould, H., and Geha, R. 1988. *Nature* 331:180-183.

Helm., Kebo, D., Vercelli, D., Glovsky, M., Gould, H., Ishizaka, K., Geha, R., and Ishizaka, T. 1989. *Proc. Natl. Acad. Sci. USA* 86:9465-9469.

Hogarth, P.M., Hulett, M.D., Ilerino, F.L., Tate, B., Powell, M.S., and Brinkworth, R.I. 1992. *Immunological Reviews* 125:21-35.

Ishizaka, K. 1970. *Ann. Rev. Medicine* 15:187-200.

Mallamaci, M., Chizzonite, R., Griffin, M., Nettleton, M., Hakimi, J., Tsien, W.H., Kochan, J.P. 1993. *J. Biol. Chem.* 268:22076-22083.

Nissim, A., Jouvin, M.-H. and Eshhar, Z. 1991. *EMBO Journal* 10, 101-107.

Nissim, A., and Eshhar, Z. 1992. *Mol. Immunol.* 29:1065-1072.

Ravetch, J.V., and Kinet, J.P. 1991. *Ann. Rev. Immunol.* 9:457-492.

Riske F., Hakimi J, Mallamaci M., Griffin M., Pilson B., Tobkes N., Lin P., Danho W., Kochan J. and Chizzonite R. 1991. *J. Biol. Chem.* 266, 11245-11251.

THE ANALYSIS OF MAST CELL FUNCTION *IN VIVO* USING MAST CELL-DEFICIENT MICE

Barry K. Wershil[1, 2] and Stephen J. Galli[1]

[1]Departments of Pathology, Beth Israel Hospital and Harvard Medical School, and the Division of Experimental Pathology, Beth Israel Hospital
[2]The Combined Program in Pediatric GI and Nutrition, The Children's Hospital and Massachusetts General Hospital, Boston, MA 02215

INTRODUCTION

Mast cells represent rich sources of histamine, heparin, proteases, and many other biologically active mediators (reviewed in Metcalfe, Kaliner, and Donlon, 1981; Schwartz and Austen, 1984; Galli and Lichtenstein, 1988). Mast cells also express surface receptors for immunoglobulins of the IgE class which, when occupied by intact IgE, prime the cells to release stored and newly generated mediators in response to encounters with specific antigens (reviewed in Ishizaka, 1988; Metzger, Kinet, Blank, et al., 1989). The remarkable variety of mediators potentially derived from mast cells, and the recognition that mast cells can release these mediators in response to stimulation by many different signals in addition to IgE and specific antigen (e.g., anaphylatoxins, certain neuropeptides, basic compounds derived from leukocytes, and components of insect venoms), has led to speculation that mast cells may play important roles in diverse immunologic, pathologic and physiologic processes (reviewed in Metcalfe, Kaliner, and Donlon, 1981; Schwartz and Austen, 1984; Galli, Dvorak, and Dvorak, 1984).

Unfortunately, proof of the importance of mast cells in specific immunological or pathological responses has been difficult to obtain (reviewed in Galli, 1987; Wershil and Galli, 1991). Many "mast cell-associated" mediators are also produced by other cell types. As a result, even convincing evidence that one of these mediators has a critical role in a particular biological response is not, by itself, sufficient evidence that the mast cell is important in that response. Pharmacological experiments employing antagonists of individual mast cell-associated mediators, or inhibitors of mast cell activation, can provide important information about the role of these mediators, and/or the effects of these drugs, in the process under study. However, such experiments may not necessarily elucidate the role of the mast cell itself. Indeed, the extent to which efforts at pharmacological inhibition of mast cell activation can clarify the role of the mast cell depends upon the strength of the evidence documenting that the effects of the drugs used are selective for the mast cell.

Immunobiology of Proteins and Peptides VII
Edited by M.Z. Atassi, Plenum Press, New York, 1994

Analysis of mast cell function *in vivo* is further complicated by the fact that mast cell activation usually results in the elaboration of multiple mediators, which may have diverse, and sometimes even opposing, biological effects. The net biological consequences of the combined activities of all of these mediators may be difficult to predict. Finally, it is now widely accepted that individual mast cell populations can express significant variation in multiple aspects of phenotype, including morphology, mediator content and response to drugs and secretogogues; such evidence of "mast cell heterogeneity" suggests that phenotypically distinct mast cells may have different roles in health and disease (reviewed in Enërback, 1986; Galli, 1990).

W OR *SL* MUTANT MAST CELL-DEFICIENT MICE

The dominant white-spotting (*W*) locus on mouse chromosome 5, and the steel (*Sl*) locus on chromosome 10, had been studied by developmental biologists, hematologists and geneticists for decades before it was clear that mutations at these loci influenced mast cell development. Indeed, mutations at *W* or *Sl* affect several important developmental programs, including gametogenesis, pigmentation and hematopoiesis (Russell, 1979; Silvers, 1979b). Thus, most *W* or *Sl* mutations render the homozygous mutant animals sterile, devoid of coat pigmentation, and anemic (Russell, 1979; Silvers, 1979b). In 1978 and 1979, Kitamura et al. reported the important observations that mutations at *W* (Kitamura, Go, and Hatanaka, 1978) or *Sl* (Kitamura and Go, 1979) also profoundly affect the development of mast cells. Subsequently, analysis of mice with mutations at *W* or *Sl* have provided many important insights into mast cell development, heterogeneity and function (Galli and Kitamura, 1987; Galli et al, 1992).

The *W* or *Sl* mutant mice most commonly used in studies of mast cell development and function are WBB6F$_1$-*W/W^v* (*W/W^v*) or WCB6F$_1$-*Sl/Sl^d* (*Sl/Sl^d*) mice (reviewed in Galli and Kitamura, 1987; Kitamura, Nakayama, and Fujita, 1989). The genetic background and certain other important characteristics of *W/W^v* and *Sl/Sl^d* mice are depicted in Table 1. For the purposes of the present discussion, it is important to emphasize that *W/W^v* and *Sl/Sl^d* mice are virtually devoid of mature, morphologically identifiable mast cells in all organs and anatomical sites examined (reviewed in Galli and Kitamura, 1987; Kitamura, 1989, Galli, 1990). Thus, *W/W^v* and *Sl/Sl^d* mice virtually lack both the "connective tissue-type mast cells" (CTMC) characteristic of the normal skin, peritoneal cavity, and gut muscularis propria, and the "mucosal mast cells" (MMC) characteristic of the mucosa of the gastrointestinal tissues.

Much of the early understanding of the nature of the defects caused by mutations at *W* or *Sl* was based on studies of transplantation of bone marrow among mutant and normal mice, and on attempts to reconstruct hematopoietic and stromal cells interactions *in vitro* (reviewed in Russell, 1979; Dexter and Moore, 1977). Such analyses indicated that the anemia and other hematologic defect(s) in *W/W^v* mice reflected a problem intrinsic to the affected lineages, whereas the defect(s) in the *Sl/Sl^d* mice influenced the microenvironments necessary for the migration, proliferation, and/or survival of the affected lineages (Russell, 1979). For example, bone marrow transplantation of cells from the congenic normal +/+ mice cured the macrocytic anemia of *W/W^v* animals, but was without effect in *Sl/Sl^d* mice.

Kitamura, Go, and Hatanaka (1978) showed that transplantation of congenic +/+ bone marrow cells also repaired the mast cell deficiency of *W/W^v* mice. By contrast, the mast cell

Table 1. Characteristics of genetically mast cell-deficient WBB6F$_1$$W/W^v$ or WCB6F$_1$-Sl/Sl^d mice.*

Characteristics	WBB6F$_1$-W/W^v	WCB6F$_1$-Sl/Sl^d
Genetic Background^{++}	WB/Re-W/+ x C57BL/6-W^v/+	WC/Re-Sl/+ x C57BL/6-Sl^d/+
Mutations	c-kit (Chromosome 5)	a ligand for c-kit# (Chromosome 10)
Site of defect resulting in mast cell deficiency	Stem cells/mast cell lineage itself	Tissue microenvironments required for mast cell development
Repair of mast cell deficiency with congenic +/+ bone marrow cells or mast cells	Yes	No
Repair of mast cell-defiency by injection of the c-kit ligand, SCF	No	Yes
Mast cells (% of +/+ level in adult, disease-free mice)	<1%†	<1%†
Mast cells develop in skin at sites of chronic idiopathic or PMA-induced dermatitis	Yes‡	No‡

* Both WBB6F$_1$-W/W^v and WCB6F$_1$-Sl/Sl^d mice also have a macrocytic anemia, lack hair pigmentation and are sterile (Russell, 1979; Silvers, 1979b).
$^{++}$ The four possible genotypes of the F$_1$ mice (W/W^v, W/+, W^v/+, +/+, or Sl/Sl^d, Sl/+, Sl^d/+, +/+) ca be distinguished according to coat color. The W/W^v and Sl/Sl^d mice are white with black eyes, th congenic +/+ mice are black, the various heterozygotes can be distinguished by the pattern of spotting
This c-kit ligand has been named: SCF (stem cell factor), KL (kit ligand), MGF (mast cell grow factor), and Steel factor (see text).
† The skin of adult W/W^v orSl/Sl^d mice contains rare mast cells (~0.3% the number present in congen +/+ mice). No mast cells at all have been observed in the following organs/sites: bone marrov spleen, thymus, brain, heart, lung, kidney, urinary bladder, liver, stomach, ileum, cecum, mesenter peritoneal cavity, hindlimb skeletal muscle, uterus.
‡ See Galli, Arizono, Murakami, et al. (1987) and Gordon and Galli (1990a).
(Modified from Galli, 1990)

deficiency of Sl/Sl^d mice was refractory to bone marrow transplantation from the congenic +/+ mice. Moreover, transplantation of bone marrow from Sl/Sl^d mice repaired the mast cell deficiency of W/W^v animals (Kitamura and Go, 1979). These findings indicated that the the mast cell deficiency of W/W^v mice, like their anemia, reflected a defect intrinsic to the mast cell lineage, whereas the mast cell deficiency of the Sl/Sl^d mouse reflected a problem in the microenvironments necessary for normal mast cell development.

Recent work (reviewed in Galli, Geissler, Wershil, et al., 1992) demonstrated that the locus encodes the c-kit tyrosine kinase growth factor receptor and the Sl locus encodes a ligand for c-kit. Thus, the mast cell-deficiency in W/W^v mice reflects abnormalities in the production or function of the cells' c-kit receptors, whereas the mast cell deficiency of Sl/Sl^d mice reflects abnormalities in the production of the ligand for the c-kit receptor by cells in the microenvironment.

ANALYSIS OF MAST CELL FUNCTION USING W/W^v AND Sl/Sl^d MICE: AN APPROACH FOR IDENTIFYING AND QUANTIFYING THE ROLES OF MAST CELLS IN BIOLOGICAL RESPONSES

A very useful approach for identifying and quantifying the roles of mast cells in biological responses *in vivo* is to search for differences in the expression of biological responses in anatomical sites which differ only in that one site contains mast cells and the other does not (Galli and Kitamura, 1987; Wershil and Galli, 1991). Our general approach for evaluating the roles of mast cells in biological responses is shown in Table 2.

Table 2. General scheme for investigating mouse mast cell function *in vivo*.

1. Search for quantitative differences in the expression of biologic responses in genetically mast cell-deficient WBB6F$_1$-W/W^v and WCB6F$_1$-Sl/Sl^d mice and the congenic normal (+/+) mice.

2. Compare the responses in W/W^v mice and W/W^v mice which have received bone marrow transplantation from congenic +/+ mice.
   Note: This determines whether the response abnormally expressed in W/W^v mice has a bone marrow-dependent component.

3. Analyze the response in W/W^v mice selectively reconstituted with mast cells.
   Note: This determines whether the response abnormally expressed in W/W^v mice has a mast cell-dependent component.

4. Define the mechanism(s) by which mast cells contribute to the response.

(Modified from Galli, 1990)

Briefly, if W/W^v and Sl/Sl^d mice express a particular biological response differently than do the congenic normal (+/+) mice, this may reflect the virtual absence of mature mast cells in the mutants or some other effect of their W or Sl mutations. If the response of W/W^v mice is normalized after bone marrow transplantation from the congenic +/+ mice, this finding indicates that the mast cell and/or some other bone marrow-dependent lineage accounted for the abnormality in the response. To discriminate between contributions of the mast cell as opposed to other bone marrow-derived elements, one then should attempt to analyze W/W^v mice selectively repaired of their mast cell deficiency by the injection of cultured, growth factor-dependent mast cells derived from congenic +/+ mouse bone marrow cells (Nakano, Sonoda, Hayashi, et al., 1985; Galli and Kitamura, 1987; Galli, Wershil, Gordon, et al., 1989). Such adoptively transferred bone marrow-derived cultured mast cell (i.e., BMCMC) gradually acquire multiple phenotypic characteristics of the native mast cell populations present in the same anatomical sites in normal mice (reviewed in Galli, 1990).

The +/+ BMCMC populations used for repair of the mast cell deficiency of W/W^v mice ordinarily contain no detectable CFU-S, nor does the i.v. injection of W/W^v mice with up to 1.0×10^6 +/+ BMCMC correct the anemia of the recipients (Nakano, Sonoda, Hayashi, et al. 1985). However, the studies of Nakano, Waki, Asai, et al. (1989a, b) suggest that even a few stem cells with extensive proliferative potential present in a population of +/+ BMCMC adoptively transferred into W/W^v recipients might result in repair of the recipients' anemia and the production of myeloid and lymphoid cells of donor origin, i.e., under these circumstances, the repair of the recipients' mast cell deficiency would not be selective. In fact, the i.v. transfer of large numbers (e.g., $\geq 1.0 \times 10^7$) of +/+ BMCMC into W/W^v mice sometimes results in correction of the recipients' anemia, as well as repair of their mast cell deficiency (Wershil, Yano, and Galli, unpublished data). By contrast, when the +/+ BMCMC are injected intradermally, one routinely obtains selective local reconstitution of dermal mast cell populations, without such systemic effects as the correction of the W/W^v recipients' anemia or the appearance of mast cells in distant anatomical sites (Nakano, Sonoda, Hayashi, et al., 1985; Galli and Kitamura, 1987; Galli, Wershil, Gordon, et al., 1989). An additional advantage of W/W^v mice selectively repaired of their mast cell deficiency by local intradermal transplantation of +/+ mast cell populations is that the expression of biological responses can be evaluated in paired anatomical sites in the same mice, one site containing adoptively transferred mast cell populations and the other virtually devoid of mast cells. W/W^v mice locally and selectively repaired of their cutaneous mast cell deficiency therefore are especially valuable models for the investigation of mast cell function *in vivo*.

W/W^v mice injected with +/+ BMCMC intraperitoneally also may be used for studies of mast cell function (Qureshi and Jakschik, 1988; Ramos, Qureshi, Olsen, et al., 1990). However, i.p. injection of +/+ BMCMC generally results in the appearance of mast cells in other anatomical sites as well (Nakano, Sonoda, Hayashi, et al., 1985; Qureshi and Jakschik, 1988). We have recently reported that injection of +/+ BMCMC directly into the stomach wall at laparotomy can result in selective local repair of the gastric mast cell deficiency of W/W^v mice and can permit these animals to express an IgE-dependent inflammatory response in the stomach (Wershil, Wang, and Galli, 1991). In theory, approaches such as this can also be used to confirm evidence derived from studies performed with +/+ bone marrow-repaired W/W^v mice indicating that mast cells augment both the extent of ethanol-induced gastric damage *in vivo* (Galli, Wershil, Bose, et al., 1987) and the increase in short circuit current observed in isolated segments of small intestines after transmural electrical stimulation or in association with anaphylactic responses (Perdue, Masson, Wershil, et al., 1991).

Studies using the approach shown in Table 2 should include appropriate histological analysis to confirm the presence or absence of mast cells in the tissues of interest, to search for evidence of mast cell degranulation, and/or to determine whether the biological responses under investigation might be associated with the development of endogenous mast cell populations in mast cell-deficient mice (see below and Galli, Wershil, Yano, et al., 1989). In certain experiments, it may also be of interest to assess the phenotypic characteristics of mast cells potentially participating in the responses elicited in normal or mast cell-reconstituted mice. Several lines of evidence, much of it derived from studies in W/W^v and congenic normal mice, indicate that the phenotypic characteristics of mast cells can be regulated by a wide variety of microenvironmental factors (reviewed in Galli, 1990). Thus, some of the phenotype features of mast cells at the site of an immunological or pathological response may change during the expression of that response.

It is important to emphasize that while adult W/W^v or Sl/Sl^d mice are virtually devoid of mature mast cells, neither mutant totally lacks cells in the mast cell lineage. The skin of adult W/W^v or Sl/Sl^d mice contains rare morphologically identifiable mature mast cells (generally <0.5% of the number present in the skin of the adult congenic and +/+ mice) (reviewed in Galli and Kitamura, 1987). It is therefore possible that some responses dependent on the function of mature mast cells can be at least weakly expressed in the skin of W/W^v and Sl/Sl^d mice. By contrast, except as noted below, no mast cells whatsoever have been observed in multiple other tissues of adult W/W^v and Sl/Sl^d mice. On the other hand, the numbers of mast cell precursors present in the hematopoietic tissues and in the blood of W/W^v or Sl/Sl^d mice are similar to those of the congenic +/+ mice (reviewed in Kitamura, 1989; Kitamura, Nakayama, and Fujita, 1989). Even though several lines of evidence indicate that the numbers of mast cell precursors in the skin or peritoneal cavity of W/W^v or Sl/Sl^d mice ordinarily is very low (reviewed in Hayashi, Sonoda, Nakano, et al., 1985; Kanakura, Sonoda, Nakano, et al., 1987; Kitamura, 1989), certain inflammatory responses can result in the local development of large numbers of mature, connective tissue-type mast cells in the skin of W/W^v mice (Galli, Arizono, Murakami, et al., 1987; Gordon and Galli, 1990b). Moreover, Alizadeh and Murrel (1984) reported that infection with *Trichinella spiralis* was associated with the development of small numbers of mucosal mast cells in the intestines of W/W^v mice. Thus, mast cells can develop in the tissues of W/W^v mice under certain special circumstances *in vivo*. Even in biological responses which do not result in the appearance of recognizable mast cells in the tissues of W/W^v or Sl/Sl^d mice, the mast cell precursors or immature cells in the mast cell lineage which are present in these animals theoretically represent potential alternative effector cells in those responses which, in normal mice, might be significantly influenced by mature mast cells. At the moment, however, no actual example of such a possibility has yet been identified, nor is clear what type of functions might be expressed by representatives of very early stages in mast cell development.

THE USE OF MAST CELL-RECONSTITUTED W/W^v MICE TO DEMONSTRATE THAT MAST CELLS HAVE AN ESSENTIAL ROLE IN CERTAIN IGE-DEPENDENT BIOLOGICAL RESPONSES

Mast Cells are Essential for the Tissue Swelling, Fibrin Deposition, and Leukocyte Infiltration Associated with IgE-Dependent Cutaneous Responses

Studies employing mast cell-reconstituted W/W^v mice have identified three patterns of mast cell involvement in biological responses. In some reactions, mast cells appear to have an essential role, in that the responses are not detectably expressed in the absence of the mast cell. In other responses, mast cells appear to regulate the intensity and/or the kinetics of the response, but the reactions can be detectably expressed in the absence of mature mast cells. In yet other responses, no specific mast cell-dependent contribution has been identified.

IgE-dependent reactions were among the first to be investigated in mast cell-reconstituted W/W^v mice. Several lines of evidence indicate that mast cells represent a major source of the mediators responsible for both local and systemic expressions of IgE-dependent immediate hypersensitivity (reviewed in Metcalfe, Kaliner, and Donlon, 1981; Schwartz and Austen, 1984; Ishizaka, 1988; Galli and Lichtenstein, 1988). However, a variety of other cell types may also contribute to the pathogenesis of IgE-dependent reactions. These cells include basophils, which express $Fc_\varepsilon RI$ and can elaborate many mediators with effects similar to those of the mast cell (reviewed in Metcalfe, Kaliner, and Donlon, 1981; Schwartz and Austen, 1984; Ishizaka, 1988; Galli and Lichtenstein, 1988), and monocytes, macrophages, platelets and perhaps eosinophils, which can interact with IgE or complexes of IgE via $Fc_\varepsilon RII/CD23$ or by other mechanisms (reviewed in Kikutani, Yokota, Uchibayashi, et al., 1989; Capron, Capron, Grangette, et al.,

44

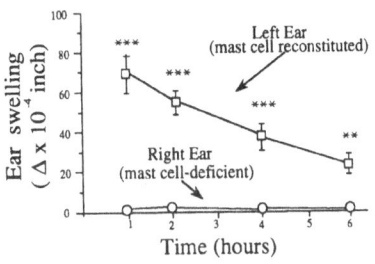

Time (hours)

Figure 1. The ear swelling associated with IgE-dependent cutaneous reactions in the mouse is mast cell-dependent. Ear swelling responses were determined in both mast cell-reconstituted (left, L) and mast cell-deficient (right, R) ears of WBB6F$_1$-W/W^v mice that had been injected > 10 wk before the experiment with WBB6F$_1$-+/+ BMCMC. Both ears were injected with IgE anti-DNP antibodies and, 1 day later, the mice were challenged with DNP$_{30-40}$-HSA i.v. Data represent mean ± s.e.m. (n = 4/point). Significant differences between values for L and R ears are shown as **P < 0.02, ***P<0.001. (From Wershil, Wang, Gordon, et al., 1991).

1988; Galli, Wershil, Gordon, et al., 1989). Genetically mast cell-deficient, congenic normal, and mast cell-reconstituted mast cell-deficient mice provide a means to discriminate between the roles of mature mast cells, as opposed to other potential effectors, in IgE-dependent responses.

At least three groups demonstrated that mast cell-deficient mice were unable to express IgE-dependent PCA responses (Thomas and Schrader, 1983; Askenase, Van Loveren, Kraueter-Kops, et al., 1983; Inagaki, Goto, Nagai, et al., 1986). To prove that this inability to express PCA reflected the mutants' mast cell deficiency, rather than another phenotypic abnormality of the mice, we investigated whether long-term selective local repair of the dermal mast cell deficiency of W/W^v mice rendered the animals competent to express IgE-dependent PCA (Wershil, Mekori, Murakami, et al., 1987).

Using the approach outlined in Table 2, we found that essentially all of the tissue swelling, augmented vascular permeability, and deposition of cross-linked ^{125}I-fibrin associated with PCA

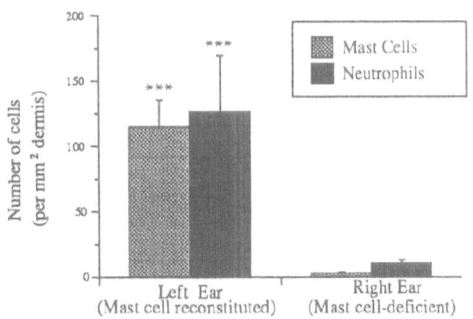

Figure 2. The neutrophil infiltration associated with IgE-dependent cutaneous reactions in the mouse is mast cell-dependent. The total numbers of mast cells and neutrophils were measured in the mast cell-reconstituted left (L) and mast cell-deficient right (R) ears of WBB6F$_1$-W/W^v mice that > 10 wk earlier underwent local reconstitution of the left ear with WBB6F$_1$-+/+ BMCMC. Data are from the same mice shown in Fig. 5, in which IgE had been injected into both ears 1 day before antigen challenge (mean ± s.e.m.). (From Wershil, Wang, Gordon, et al., 1991).

reactions was mast cell-dependent (Wershil, Mekori, Murakami, et al., 1987). More recently, we found that the early phase of tissue swelling which occurred at sites of PCA reactions in normal or mast cell-reconstituted mice was followed by an influx of granulocytes, composed predominantly of neutrophils, which reached maximal levels 6-12 hours after antigen challenge (Wershil, Wang, Gordon, et al., 1991). Moreover, even though tissue swelling at these sites diminished progressively from 2 to 24 hours after antigen challenge, swelling at the reaction sites remained significantly greater than that at control sites for at least 24 hours. Analyses of mast cell-reconstituted W/W^v mice indicated that both the persistent tissue swelling (Figure 1), and the neutrophil influx (Figure 2), observed 6 hours after antigen challenge were essentially entirely mast cell dependent.

Role of TNF-α in IgE-, and mast cell-dependent leukocyte infiltration

In many allergic patients, intradermal challenge with specific antigen or anti-IgE induces an immediate wheal and flare reaction which is followed, 4 to 8 hours later, by a period of persistent swelling and leukocyte infiltration termed the late phase cutaneous reaction (reviewed in Galli and Lichtenstein, 1988; Lemanske and Kaliner, 1988). Late phase reactions (LPRs) were initially studied in detail in the skin. However, it is now clear that late consequences of IgE-dependent reactions, notably including infiltration of the reaction sites with blood-borne leukocytes, also occur in the respiratory tract and other anatomical locations. Indeed, it has been argued compellingly that many of the clinically significant consequences of IgE-dependent reactions, in both the respiratory tract and the skin, reflect the actions of the leukocytes recruited to these sites during the LPRs, rather than the direct effects of the mediators released by mast cells and other cells at early intervals after antigen challenge (reviewed in Galli and Lichtenstein, 1988; Lemanske and Kaliner, 1988). Several lines of evidence derived from both clinical and animal studies suggested that the leukocyte infiltration associated with LPRs occurs as a result of mast cell degranulation. On the other hand, several other potential effector cells, including basophils, lymphocytes, monocytes, platelets, and perhaps eosinophils, also represent potential sources of pro-inflammatory mediators in IgE-dependent responses.

The study of Wershil, Wang, Gordon et al. (1991) indicates that, in the mouse, the tissue swelling and leukocyte infiltration associated with cutaneous LPRs are indeed mast cell-dependent. This model system also provided an opportunity to investigate potential mechanisms by which mast cells might recruit leukocytes to LPRs. As reviewed in detail in Wershil, Wang, Gordon et al. (1991), the influx of leukocytes into sites of mast cell degranulation might reflect the actions of several mast cell-derived mediators, including products of arachidonic acid oxidation, chemotactic peptides or proteins, and/or several different cytokines. Several groups demonstrated that immunologically stimulated mouse mast cells can produce a variety of cytokines (Plaut, Pierce, Watson, et al., 1989; Wodnar-Pilipowicz, Heusser, and Moroni, 1989; Burd, Rogers, Gordon, et al., 1989). Furthermore, the activation of IL-3-dependent or -independent cloned mouse mast cells via the FcεRI results in increased levels of mRNA for at least 6 cytokines with demonstrated or potential ability to elicit leukocyte infiltration: TNF-α, IL-1α, MIP-1α, MIP-1β, TCA3, and JE. IgE-dependent activation of cloned mouse mast cells also resulted in the release of bioactivity for TNF-α (Gordon and Galli, 1990a) and IL-1 (Burd, Rodgers, Gordon, et al., 1989).

46

However, studies of cytokine production by the populations of mature mouse mast cells which participate in expressions of biological responses *in vivo* have been rather limited. Indeed, the only detailed studies of cytokine production by mature mouse mast cells have focused on TNF-α. Young, Liu, Butler, et al. (1987) demonstrated that mouse peritoneal mast cells contained a cytolytic activity with many immunological and functional similarities to TNF-α. Gordon and Galli (1990b) demonstrated that the TNF-α-like cytolytic activity of mouse mast cells represented a product of the TNF-α gene, and that freshly isolated mouse peritoneal mast cells constitutively contained, on a per cell basis, approximately twice as much TNF-α bioactivity as do mouse peritoneal macrophages stimulated with LPS. Moreover, Gordon and Galli (1991) demonstrated that IgE-dependent activation of both *in vitro*-derived mast cells and freshly isolated peritoneal mast cells resulted in the rapid release of preformed stores of TNF-α followed by the synthesis and sustained release of large quantities of newly formed TNF-α.

TNF-α can be produced by many cell types besides the mast cell, including monocytes/macrophages, T cells, B cells, and neutrophils (reviewed in Beutler and Cerami, 1989; Gordon and Galli, 1990b; Gordon, Burd, and Galli, 1990). However, in mouse skin, challenge with IgE and specific antigen results in much higher levels of TNF-α mRNA when mast cells are present than when these cells are absent. Gordon and Galli (1991) injected an anti-DNP mouse monoclonal IgE into the skin of normal mice, W/W^v mice, and mast cell-reconstituted genetically mast cell-deficient W/W^v mice. The next day, DNP$_{30-40}$HSA was injected intravenously and, 45 minutes later, the reaction sites were harvested for Northern analysis. This study demonstrated that challenge with specific antigen induced substantially higher levels of TNF-α mRNA at skin sites containing mast cells than at identically challenged sites in mast cell-deficient W/W^v mice that were devoid of mast cells.

Taken together, the *in vitro* and *in vivo* studies of Gordon and Galli (1991) indicate that mast cells probably represent a biologically significant source of TNF-α during IgE-dependent responses and define a mechanism whereby stimulation of mast cells via the Fc$_\varepsilon$RI can account for both the rapid release of preformed stores of the cytokine and the sustained release of newly synthesized product. We have proposed that these experiments also identify TNF-α as the first representative of what may constitute a new class of mast cell-associated mediators (Gordon and Galli, 1991). Mast cell-associated mediators classically are divided into the preformed mediators, which are stored in the cells' cytoplasmic granules and are released upon exocytosis, and the newly synthesized mediators, which are not stored but are produced and secreted only after appropriate stimulation of the cells (Metcalfe, Kaliner, and Donlon, 1981; Schwartz and Austen, 1984; Galli and Lichtenstein, 1988). However, mast cells activated via the Fc$_\varepsilon$RI release TNF-α with a prolonged kinetics that reflects the contributions of both preformed and newly synthesized pools of product (Gordon and Galli, 1991).

Given the many potential mechanisms by which mast cells might influence leukocyte infiltration in LPRs, we were interested in determining the extent to which TNF-α might contribute to this aspect of the reaction. We therefore tested the ability of a polyclonal rabbit anti-TNF-α antiserum to interfere with the development of inflammation at sites of IgE-dependent cutaneous mast cell activation (Wershil, Wang, Gordon, et al., 1991). The rabbit antibody used in this study had been raised against recombinant mouse TNF-α and was able to inhibit completely the cytotoxicity of either recombinant mouse TNF-α or mast cell-derived

TNF-α against L929 cells *in vitro* (Gordon and Galli, 1990b). In addition, this antibody was able to inhibit by ~84% the neutrophil infiltration induced by the intradermal injection of recombinant mouse TNF-α into the skin of normal mice (Wershil, Wang, Gordon, et al., 1991). When compared to the effect of similar dilutions of normal rabbit serum, the rabbit anti-TNF-α was able to diminish by 47±7% the neutrophil infiltration measured 6 hours after elicitation of IgE-dependent cutaneous mast cell activation in the skin of normal mice (Figure 3).

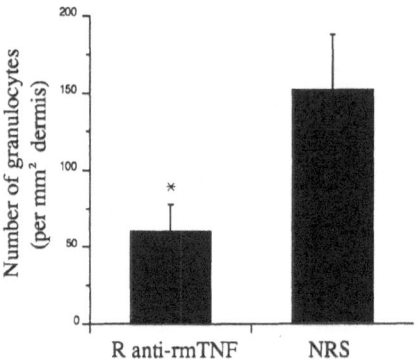

Figure 3. Inhibition of the neutrophil infiltration associated with IgE-dependent reactions by a rabbit anti-recombinant mouse TNF-a antiserum (R anti-rmTNF). Both ears of C57BL/6 mice were injected with IgE anti-DNP antibodies. The next day the left ears were injected with 20 ml of a 1:100 dilution of R anti-rm TNF antiserum and the right ears were injected with a 20 ml of a 1:100 dilution of normal rabbit serum (NRS) in the same vehicle. The mice were immediately challenged with DNP$_{30-40}$-HSA i.v. and sacrificed 6 hours later for measurement of neutrophil infiltration. Data represent mean ± s.e.m. (n=8). Significant differences between the left and right ears are shown as *P<0.05. (From Wershil, Wang, Gordon, et al., 1991).

Notably, our anti-TNF-α antiserum was more effective at diminishing the leukocyte infiltration elicited by recombinant mouse TNF-α than that observed after mast cell activation (Wershil, Wang, Gordon, et al., 1991). One possible explanation for this difference is that the intradermally injected recombinant mouse TNF-α was more accessible to the anti-serum than was endogenous TNF-α of mast cell origin. In mouse mast cells generated *in vitro*, a significant fraction of the TNF-α bioactivity is associated with the cells' cytoplasmic granules (Young, Liu, Butler, et al., 1987). If the same is true for dermal mast cells, the physical association of TNF-α with other components of the mast cell granule may have limited the anti-serum's access to the cytokine. Moreover, the recombinant mouse TNF-α was given as a single injection, whereas mast cells activated via the FcεRI *in vitro* release TNF-α over a prolonged period after stimulation (Gordon and Galli, 1991).

Our findings thus support the hypothesis that TNF-α contributes to the mast cell-dependent leukocyte infiltration associated with cutaneous LPRs. This result is not surprising, in light of many studies demonstrating that TNF-α can promote leukocyte infiltration by effects on both vascular endothelial cells and leukocytes (reviewed in Pober, 1987; Bevilacqua, Wheeler, Pober, et al., 1987; Beutler and Cerami, 1989). Moreover, using an organ culture system, Klein , Lavker, Matis, et al. (1989) demonstrated that IgE-dependent activation of human foreskin mast cells was followed by the expression of a TNF-α-inducible leukocyte adherence molecule (ELAM-1) on adjacent vascular endothelial cells. Klein, Lavker, Matis, et al. (1989) also demonstrated that antibodies to TNF-α inhibited the development of augmented expression of ELAM-1 in tissue fragments challenged with agents to induce mast cell activation. Although the cellular source of the TNF-α responsible for the augmented ELAM-1 expression observed in this model system was not initially identified, recent work by this group has demonstrated the presence of both immunoreactive TNF-α and TNF-α mRNA in human skin mast cells (Walsh, Trinchieri, Waldorf, et al., 1991). These findings support the possibility that mast cell-derived TNF-α may contribute to the leukocyte infiltration observed in IgE-dependent cutaneous LPRs in man.

On the other hand, it is very unlikely that TNF-α is the sole mediator by which mast cells influence leukocyte recruitment. Indeed, several mast cell products, acting alone or perhaps in concert with mediators from other cells that are recruited to these reactions, may contribute to the leukocyte infiltration which follows IgE-dependent mast cell activation. This possibility represents another explanation for the inability of our anti-TNF-α antibody to block completely the leukocyte infiltration at sites of IgE-dependent skin reactions (Wershil, Wang, Gordon, et al., 1991). This possibility is also supported by the work of Leung, Pober, and Cotran (1991), who analyzed the augmented expression of ELAM-1 by vascular endothelial cells at sites of intradermal antigen challenge of the skin *in vivo*, or in skin organ cultures challenged with allergen *in vitro*. The *in vivo* studies demonstrated that augmented ELAM-1 expression occurred concomitantly with the appearance of leukocyte infiltrates 3 to 4 hours after intradermal challenge in all atopic patients tested but in only 1 of 4 apparently normal controls. However, inhibition of the allergen-induced augmented ELAM-1 expression which was observed in the organ culture system required simultaneous treatment with anti-sera to both TNF-α and IL-1.

MAST CELL-LEUKOCYTE CYTOKINE CASCADES IN IgE-DEPENDENT AND OTHER INFLAMMATORY RESPONSES

While the evidence indicating that mast cells represent a significant source of many multifunctional cytokines is steadily increasing (reviewed in Gordon, Burd and Galli, 1990; Galli, Gordon and Wershil, 1991), it is important to emphasize that the mast cell represents just one among many potential sources of these biologically active compounds. In certain IgE-dependent responses, including some of those elicited in the skin, mast cells may represent the most significant initial source of multifunctional cytokines. But as these reactions evolve, additional sources of cytokines may be activated and/or recruited. For example, the neutrophils which are recruited to certain mast cell-dependent responses may represent a source of additional TNF-α (reviewed in Galli, Gordon, and Wershil, 1991). By contrast, other leukocytes recruited to sites of mast cell activation may represent sources of cytokines not produced in significant quantities by the mast cell itself. For example, certain populations of eosinophils have been shown to contain mRNA and immunoreactive TGF-α (Wong, Weller, Galli, et al., 1990), a cytokine which mast cells apparently do not produce.

It is likely that mast cell-derived cytokines, and cytokines derived from neutrophils, eosinophils, and other recruited leukocytes, participate importantly in a wide range of biological responses beyond those associated with IgE. Indeed, we feel that the pathogenesis of IgE-dependent reactions and many other mast cell-associated responses may be importantly influenced by a complex sequence of events that we have termed a mast cell-leukocyte cytokine cascade (Gordon, Burd, and Galli, 1990; Galli, Gordon, and Wershil, 1991). According to this hypothesis, certain important features of the reactions are initiated by mast cell-derived cytokines whereas later aspects of the reaction are critically regulated by cytokine activities derived from leukocytes recruited to the reaction site. The studies outlined briefly above indicate that mast cell-deficient and congenic normal mice, and W/W^v mice locally repaired of their dermal mast cell deficiency, represent excellent model systems in which to investigate the roles of mast cell-leukocyte cytokine cascades in the initiation, progression, and perhaps resolution of IgE-dependent and other biological reactions.

ACKNOWLEDGMENTS

We thank our many collaborators for their contributions to some of the studies reviewed herein. The work reviewed here has been supported in part by United States Public Health Service grants AI-22674, AI-23990, AI-33372, GM-45311, DK-33506 (subproject 6), DK-01543, AMGEN, Inc., and the Beth Israel Hospital Pathology Foundation, Inc.

REFERENCES

Alizadeh, H., and Murrell, K.D., 1984, The intestinal mast cell response to *Trichinella spiralis* infection in mast cell-deficient W/W^v mice, *J. Parasitol.* 70:767-773.

Askenase, P.W., Van Loveren, H., Kraueter-Kops, S., Yacov, Y., Meade, R., Theoharides, T.C., Nordlung, J.J., Scovern, H., Gershon, M.D., and Ptak, W., 1983, Defective elicitation of delayed-type hypersensitivity in W/W^v and Sl/Sl^d mast cell-deficient mice, *J. Immunol.* 131:2687-2894.

Beutler, B., and Cerami, A., 1989, The biology of cachectin/TNF-α primary mediator of the host response, *Annu. Rev. Immunol.* 7:625-655.

Bevilacqua, M.P., Wheeler, M.E., Pober, J.S., Fiers, W., Mendrick, D.L., Cotran, R.S., and Gimbrone, M.A., Jr., 1987, Endothelial-dependent mechanisms of leukocyte adhesion: regulation by interleukin 1 and tumor necrosis factor, *in*: "Leukocyte Emigration and Its Sequellae," H.Z. Movat, ed., S. Karger, New York.

Burd, P.R., Rogers, H.W., Gordon, J.R., Martin, C.A., Jayaraman, S., Wilson, S.D., Dvorak, A.M., Galli, S.J., and Dorf, M.E., 1989, Interleukin 3-dependent and -independent mast cells stimulated with IgE and antigen express multiple cytokines, *J. Exp. Med.* 170:245-257.

Capron, A., Capron, M., Grangette, C., and Dessaint, J.P., 1989, IgE and inflammatory cells, *in*: "IgE, Mast Cells and the Allergic Response. Ciba Foundation Symposium No. 147," D. Chadwick, D. Evered, and J. Whelan, eds., John Wiley and Sons, Ltd., Chichester, UK.

Dexter, T.M., and Moore, M.A.S., 1977, *In vitro* duplication of and 'cure' of haemopoietic defects in genetically anemic mice, *Nature.* 269:412-414.

Enerbäck, L., 1986, Mast cell heterogeneity: the evolution of the concept of a specific mucosal mast cell. *in*: "Mast Cell Differentiation and Heterogeneity," A.D. Befus, J. Bienenstock, and J.A. Denburg, eds., Raven Press, New York.

Galli, S.J., 1987, New approaches for the analysis of mast cell maturation, heterogeneity, and function, *Fed. Proc.* 46:1906-1914.

Galli, S.J., 1990, New insights into "the riddle of the mast cells": microenvironmental regulation of mast cell development and phenotypic heterogeneity, *Lab. Invest.* 62:5-33.

Galli, S.J., Arizono, N., Murakami, T., Dvorak, A.M., and Fox, J.G., 1987, Development of large numbers of mast cells at sites of idiopathic chronic dermatitis in genetically mast cell-deficient $WBB6F_1$-W/W^v mice, *Blood.* 69:1661-1666.

Galli, S.J., Dvorak, A.M., and Dvorak, H.F., 1984, Basophils and mast cells: morphologic insights into their biology, secretory patterns and function, *Prog. Allergy*.34:1-141.

Galli, S.J., Geissler, E.N., Wershil, B.K., Gordon, J.R., Tsai, M., and Hammel, I., 1992, Insights into mast cell development and function derived from analysis of mice carrying mutations at beige, *W/c-kit* or *Sl/SCF* (*c-kit* ligand) loci, *in*: "The Role of the Mast Cell in Health and Disease," M.A. Kaliner, and D.D. Metcalfe, eds., Marcel Dekker, New York.

Galli, S.J., Gordon, J.R., and Wershil, B.K., 1991, Cytokine production by mast cells and basophils, *Curr. Opinion Immunol.*, in press.

Galli, S.J., and Kitamura, Y., 1987, Animal model of human disease. Genetically mast cell-deficient W/W^v and Sl/Sl^d mice: their value for the analysis of the roles of mast cells in biological responses *in vivo*, *Am. J. Pathol.* 127:191-198.

Galli, S.J., and Lichtenstein, L.M., 1988, Biology of mast cells and basophils, *in*: "Allergy: Principles and Practice, 3rd ed.," E. Middleton, Jr., C.E. Reed, E.F. Ellis, N.F. Adkinson, Jr., and J.W. Yuninger, eds., Mosby, St. Louis, Missouri.

Galli, S.J., Wershil, B.K., Bose, R., Walker, P.A., and Szabo, S., 1987, Ethanol-induced acute gastric injury in mast cell-deficient and congenic normal mice. Evidence that mast cells can augment the area of damage, *Am. J. Pathol.* 128:131-140.

Galli, S.J., Wershil, B.K., Gordon, J.R., and Martin, T.R., 1989, Mast cells: immunologically specific effectors and potential sources of multiple cytokines during IgE-dependent responses, *in*: "IgE, Mast Cells and the Allergic Response. Ciba Foundation Symposium No. 147," D. Chadwick, D. Evered, and J. Whelan, eds., John Wiley and Sons, Ltd., Chichester, UK.

Galli, S.J., Wershil, B.K., Yano, H., Arizono, N., Gordon, J.R., and Murakami, T., 1989, Analysis of the roles of phenotypically distinct mast cell populations in non-immunological responses, *in*: "Mast Cell and Basophil Differentiation and Function in Health and Disease," S.J. Galli, and K.F. Austen, eds., Raven Press, New York.

Gordon, J.R., Burd, P.R., and Galli, S.J., 1990, Mast cells as a source of multifunctional cytokines, *Immunol. Today.* 11:458-464.

Gordon, J.R., and Galli, S.J., 1990a, Mast cells as a source of both preformed and immunologically inducible TNF-α/cachectin, *Nature.* 346:274-276.

Gordon, J.R., and Galli, S.J., 1990b, Phorbol 12-myristate 13-acetate-induced development of functionally active mast cells in W/W^v but not Sl/Sl^d genetically mast cell-deficient mice, *Blood.* 75:1637-1645.

Gordon, J.R., and Galli, S.J., 1991, Release of both preformed and newly synthesized tumor necrosis factor α (TNF-α)/cachectin by mouse mast cells stimulated by the Fc$_\varepsilon$RI. A mechanism for the sustained action of mast cell-derived TNF-α during IgE-dependent biological responses, *J. Exp. Med.* 174:103-107.

Hayashi, C., Sonoda, T., Nakano, T., Nakayama, H., and Kitamura, Y., 1985, Mast-cell precursors in the skin of mouse embryos and their deficiency in embryos of Sl/Sl^d genotype, *Devel. Biol.* 109:234-241.

Inagaki, N., Goto, S., Nagai, H., and Koda, A., 1986, Homologous passive cutaneous anaphylaxis in various strains of mice, *Int. Archs Allergy Appl. Immunol.* 81:58-62.

Ishizaka, T., 1988, Mechanisms of IgE-mediated hypersensitivity, *in*: "Allergy: Principles and Practice, 3rd ed.," E. Middleton, Jr., C.E. Reed, E.F. Ellis, N.F. Adkinson, Jr., and J.W. Yuninger, eds., Mosby, St. Louis, Missouri.

Kanakura, Y., Sonoda, S., Nakano, T., Fujita, J., Kuriu, A., Asai, H., and Kitamura, Y., 1987, Formation of mast-cell colonies in methylcellulose by mouse skin cells and development of mucosal-like mast cells from the cloned cells in the gastric mucosa of W/W^v mice, *Am. J. Pathol.* 129:168-176.

Kikutani, H., Yokota, A., Uchibayashi, N., Yukawa, K., Tanaka, T., Sugiyama, K., Barsumian, E.L., Suemura, M., and Kishimoto, T., 1989, Structure and function of Fc$_\varepsilon$ receptor II (Fc$_\varepsilon$RII/CD23): a point of contact between the effector phase of allergy and B cell differentiation, *in*: "IgE, Mast Cells and the Allergic Response. Ciba Foundation Symposium No. 147," D. Chadwick, D. Evered, and J. Whelan, eds., John Wiley and Sons, Ltd., Chichester, UK.

Kitamura, Y., 1989, Heterogeneity of mast cells and phenotypic changes between subpopulations, *Ann. Rev. Immunol.* 7:59-76.

Kitamura, Y., and Go, S., 1979, Decreased production of mast cells in Sl/Sl^d mice, *Blood.* 53:492-497.

Kitamura, Y., Go, S., and Hatanaka, S., 1978, Decrease of mast cells in W/W^v mice and their increase by bone marrow transplantation, *Blood.* 52:447-452.

Kitamura, Y., Nakayama, H., and Fujita, J.,1989, Mechanism of mast cell deficiency in mutant mice of W/W^v and Sl/Sl^d genotype. *in*: "Mast Cell and Basophil Differentiation and Function in Health and Disease," S.J. Galli, and K.F. Austen, eds., Raven, New York.

Klein, L.M., Lavker, R.M., Matis, W.L., and Murphy, G.F., 1989, Degranulation of human mast cells induces an endothelial antigen central to leukocyte adhesion, *Proc. Natl. Acad. Sci. USA* 86:8972-8976.

Lemanske, R.F., Jr., and Kaliner, M.A., 1988, Late phase allergic reactions, *in*: "Allergy: Principles and Practice, 3rd ed., E. Middleton, Jr., C.E. Reed, E.F. Ellis, N.F. Adkinson, Jr., and J.W. Yuninger, eds., Mosby, St. Louis, Missouri.

Leung, D.Y.M., Pober, J.S., and Cotran, R.S. ,1991, Expression of endothelial-leukocyte adhesion molecule-1 in elicited late phase allergic reactions, *J. Clin. Invest.* 87:1805-1809.

Metcalfe, D.D., Kaliner, M., and Donlon, M.A., 1981, The mast cell, *CRC Crit. Rev. Immunol.* 2:23-74.

Metzger, H., Kinet, J.-P., Blank, U., Miller, L., and Ra, C., 1989, The receptor with high affinity for IgE, *in*: "IgE, Mast Cells and the Allergic Response. Ciba Foundation Symposium No. 147," D. Chadwick, D. Evered, and J. Whelan, eds., John Wiley and Sons, Ltd., Chichester, UK.

Nakano, T., Sonoda, T., Hayashi, C., Yamatodani, A., Kanayama, Y., Yamamura, T., Asai, H., Yonezawa, Y., Kitamura, Y., and Galli, S.J., 1985, Fate of bone marrow-derived cultured mast cells after intracutaneous, intraperitoneal and intravenous transfer into genetically mast cell-deficient W/W^v mice. Evidence that cultured mast cells can give rise to both connective tissue-type and mucosal mast cells, *J. Exp. Med.* 162:1025-1043.

Nakano, T., Waki, N., Asai, H., and Kitamura, Y., 1989a, Effect of 5-fluorouracil on "primitive" hematopoietic stem cells that reconstitute whole erythropoiesis of genetically anemic W/W^v mice, *Blood.* 73:425-430.

Perdue, M.H., Masson, S., Wershil, B.K., and Galli, S.J., 1991, The role of mast cells in ion transport abnormalities associated with intestinal anaphylaxis. Correction of the diminished secretory response in genetically mast cell-deficient W/W^v mice by bone marrow transplantation, *J. Clin. Invest.* 87:687-693.

Pober, J.S., 1987, Effects of tumour necrosis factor and related cytokines on vascular endothelial cells, *CIBA Found. Symp.* 131:170-184.

Qureshi, R., and Jakschik, B.A., 1988, The role of mast cells in thioglycollate-induced inflammation, *J. Immunol.* 141:2090-2096.

Ramos, B.F., Qureshi, R., Olsen, K.M., and Jakschik, B.A., 1990, The importance of mast cells for the neutrophil influx in immune complex-induced peritonitis in mice, *J. Immunol.* 145:1868-1873.

Russell, E.S., 1979, Hereditary anemias of the mouse: a review for geneticists, *Adv. Genetics* 20:357-459.

Schwartz, L.B., and Austen, K.F., 1984, Structure and function of the chemical mediators of mast cells, *Prog. Allergy.* 34:271-321.

Silvers, W.K., 1979b, Dominant spotting, patch and rump-white; steel, flexed tail, splotch and varitint-waddler. *in*: "The Coat Colors of Mice. A Model for Gene Action and Interaction," Springer-Verlag, New York.

Thomas, W.R., and Schrader, J.W., 1983, Delayed hypersensitivity in mast cell-deficient mice, *J. Immunol.* 130:2565-2567.

Walsh, L.J., Trinchieri, G., Waldorf, H.A., Whitaker, D., and Murphy, G.F., 1991, Human dermal mast cells contain and release tumor necrosis factor a which induces endothelial leukocyte adhesion molecule-1, *Proc. Natl. Acad. Sci. USA.* 88:4222-4224.

Wershil, B.K., and Galli, S.J., 1991, Gastrointestinal mast cells. New approaches for analyzing their function in vivo, *Gastroenterol. Clin. N.A.* 20:613-627.

Wershil, B.K., Mekori, Y.A., Murakami, T., and Galli, S.J., 1987, ^{125}I-Fibrin deposition in IgE-dependent immediate hypersensitivity reactions in mouse skin. Demonstration of the role of mast cells using genetically mast cell-deficient mice locally reconstituted with cultured mast cells, *J. Immunol.* 139:2605-2614.

Wershil, B.K., Wang, Z., Gordon, J.R., and Galli, S.J., 1991, Recruitment of neutrophils during IgE-dependent cutaneous late phase responses in the mouse is mast cell dependent:

partial inhibition of the reaction with antiserum against tumor necrosis factor-alpha, *J. Clin. Invest.* 87:446-453.

Wershil, B.K., Wang, Z.-S., and Galli, S.J., 1991, Evidence of mast cell-dependent neutrophil infiltration during IgE-dependent gastric inflammation in the mouse: does this represent a gastric late phase reaction (LPR)?, *Gastroenterol.* 100:A625.

Wodnar-Filipowicz, A., Heusser, C.H., and Moroni, C., 1989, Production of the haemopoietic growth factors GM-CSF and interleukin-3 by mast cells in response to IgE receptor-mediated activation, *Nature.* 339:150-152.

Wong, D.T.W., Weller, P.F., Galli, S.J., Elovic, A., Rand, T.H., Gallagher, G.T., Chiang, T., Chou, M.Y., Matossian, K., McBride, J., and Todd, R., 1990, Human eosinophils express transforming growth factor-alpha, *J. Exp. Med.* 172:673-681.

Young, J.D.-E., Liu, C., Butler, G., Cohn, Z.A., and Galli, S.J., 1987, Identification, purification, and characterization of a mast cell-associated cytolytic factor related to tumor necrosis factor, *Proc. Natl. Acad. Sci. USA* 84:9175-9179.

THE IMMUNOGENETIC BASIS OF COLLAGEN INDUCED ARTHRITIS IN MICE: AN EXPERIMENTAL MODEL FOR THE RATIONAL DESIGN OF IMMUNOMODULATORY TREATMENTS OF RHEUMATOID ARTHRITIS

Gerald H. Nabozny and Chella S. David

Department of Immunology
Mayo Medical School
Rochester, MN 55905

INTRODUCTION

Elucidating the genetic influences and immunological processes involved in the induction and pathogenesis of rheumatoid arthritis (RA) has been an area of intense investigation over the past two decades. During this period, various experimental animal models of RA have been studied which have contributed greatly to our basic understanding of this disease. One model of particular interest is the mouse model of collagen induced arthritis (CIA). Following the initial description of the model in rats,[1] the successful induction of CIA in mice was achieved by Courtenay et al.[2] CIA is usually induced by the intradermal injection of heterologous or homologous type II collagen (CII) in complete Freund's adjuvant.[3,4] Both T and B cell reactivity to CII is detected and, approximately 4-6 weeks post-immunization, mice develop a polyarthritis initially characterized by erythema and edema followed by joint distortion and, in some cases, culminating in joint ankylosis.[5] Histologic examination of afflicted paws from mice with CIA reveals marked synovial proliferation, pannus formation and subsequent joint destruction with replacement by mononuclear cells.[6] These clinical and histopathologic features of CIA strongly resemble those observed in human RA. Of particular importance has been the fact that the availability of numerous inbred mouse strains has enabled scientists to undertake a systematic and detailed analysis of the genetic factors contributing to this disease. Thus, understanding the immunogenetic characteristics of murine CIA strengthens the potential of this model for developing specific immune intervention strategies in human RA.

MAJOR HISTOCOMPATIBILITY COMPLEX (MHC) GENE REGULATION OF CIA

Although numerous genetic components are involved in predisposing an individual to develop RA, it is clear that one critical element is associated with the HLA complex, in particular the HLA-DR4 serologic specificity.[7-10] In addition, among the genetic subtypes of HLA-DR4, only the Dw4 and Dw14 subtypes are prevalent in RA patients whereas other subtypes of DR4, like Dw10, do not associate with RA susceptibility.[11,12] Molecular analysis demonstrated that the major difference between the Dw10 and Dw4 vs. Dw14 subtypes is located in the third diversity region of the $DR\beta_1$ chain.[11] This finding has formulated the hypothesis that molecular conformations of this region dictates disease susecptibility to RA.[12]

Much like RA, susceptibility in murine CIA is associated with genes of the mouse MHC.[6] Induction of CIA can be consistently achieved in mice bearing only the $H-2^q$ and $H-2^r$ haplotypes.[6] Subsequent studies utilizing inbred H-2 recombinant strains demonstrated that this regulation maps to the Class II genes of the MHC.[6] Moreover, experiments involving the $H-2^q$ haplotype revealed that CIA susceptibility is specifically associated with the I-A subregion.[13]

Immunobiology of Proteins and Peptides VII
Edited by M.Z. Atassi, Plenum Press, New York, 1994

That MHC genes, in particular gene products of the class II region, are critical in dictating CIA susceptibility is shown through two lines of experimental evidence. In the first example, Wooley et al[14] illustrated the differing susceptibility of H-2^q and H-2^r strains to arthritis induction using CII from a variety of species. Whereas B10.Q (H-2^q) mice are susceptible to CIA induced with chick, bovine and deer CII, B10.RIII (H-2^r) mice are resistant to CIA induced with chick CII, but develop arthritis following immunization with bovine, porcine or deer CII. This finding indicated that multiple epitopes on CII exist which may display differing magnitudes of immunodominance in different CIA-susceptible haplotypes. A second example of MHC control of CIA susceptiblility has been shown by Gustafson et al.[15] Comparison of the H-2A$_\alpha$ and H-2A$_\beta$ chains of CIA-susceptible B10.Q and closely related but CIA-resistant B10.P strains revealed that both strains possessed identical H-2A$_\alpha$ chains but differed in their H-2A$_\beta$ chain by only 4 amino acids. These differences occur at residues 85, 86, 88 and 89 and it has been suggested that this region may be important in antigenic peptide-binding within the groove of the class II molecule.[15] Thus, the postulate has been put forth that the lack of the appropriate amino acid sequence in the B10.P H-2A$_\beta$ chain renders this strain unable to present the arthritogenic epitope(s), in turn conferring resistance to the induction of CIA. It is quite intriguing that his subtle change at the molecular level is analogous to the differences observed between Dw14 and Dw4 v.s. Dw10 subtypes in HLA-DR4 individuals. Moreover, these differences between the mouse H-2A$_\beta$^{p} and H-2A$_\beta$^{q} molecules lends credence to the pivotal role of molecular conformations of human MHC molecules in RA susceptibility.

CIA SUSCEPTIBILITY AND THE INFLUENCE OF T CELL RECEPTOR GENES

Clearly, possession of the appropriate MHC haplotype is an absolute requirement for CIA susceptibility. However, additional genes are also necessary. A group of genes which strongly influences arthritogenic responses to CII are the V$_\beta$ genes of the T cell receptor (TCR). That V$_\beta$ TCR genes are crucial in CIA was first suggested by Banerjee et al.[16] While evaluating various mouse strains for CIA susceptibility, it was discovered that SWR mice, despite bearing the H-2^q haplotype, remained resistant to CIA induction.[17] Behlke et al[18] subsequently demonstrated that SWR mice, along with SJL, C57/L and C57/Br animals, possessed a genomic deletion of up to 50% of the V$_\beta$ TCR genes (Figure 1). In light of this, Banerjee et al[16] hypothesized that the absence of T cells bearing a TCR specificity for arthritogenic determinants on CII was responsible for CIA resistance in SWR mice. Thus, SWR (TCR mutant, V$_\beta$^{a}) mice were mated with B10 animals, a strain which is CIA-resistant due to the H-2^b haplotype but bears a wild type (V$_\beta$^{b}) TCR repertoire. Analysis of F$_1$, F$_2$ and backcross progeny demonstrated a strong correlation between CIA-susceptibility and at least one copy of wild-type TCR genes was demonstrated.[16] Moreover, crosses between SWR (V$_\beta$^{a}) and C57/L (H-2^b, V$_\beta$^{a}) mice remained resistant to CIA.

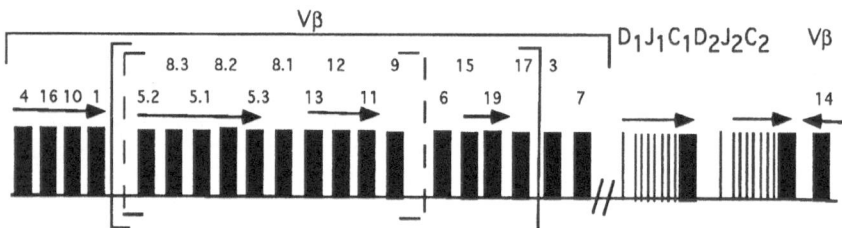

Figure 1. Illustration of the germline organization of the mouse V$_\beta$ TCR genes. The V$_\beta$ genes missing in the SWR strain is enclosed by the broken brackets. The deletion in the RIIIS/J strain is enclosed by the solid brackets.

An alternative mechanism for CIA resistance in the SWR strain may be due to the fact that SWR mice, in addition to their TCR deletion, possess a genetic deficiency in the C5 component of complement.[19] Watson et al[20] have suggested that this C5 deficiency is responsible for CIA resistance in these animals. To address the role of TCR and C5 genes in CIA, SWR mice were crossed with C3H.A animals. C3H.A, like SWR, are C5 deficient but bear the wild type V$_\beta$^{b} haplotype. F$_1$ and backcross analysis showed no correlation between CIA susceptibility and C5 sufficiency whereas the possession of the V_β^b haplotype appeared to be an absolute requirement.[21] This finding strengthened a role for TCR genes in regulating CIA. Interestingly, a similar study by Spinella et al[22] using DBA/1 (V_β^b, C5 sufficient) X SWR crosses revealed no correlation between CIA resistance and TCR deletions. A likely explanation for this observation could be the use of the DBA/1 strain in this study. Although DBA/1 animals would provide wild-type TCR and C5 genes in these experiments, the DBA/1 strain possesses exceptional susceptibility to CIA induction. For instance, spleen cells from naive DBA/1 mice proliferate to CII in vitro,[23] DBA/1 animals display strong responses to bacterial heat shock proteins (Dr. C.J. Krco, personal communication) and recently Nordling et al[24] reported the occurence of spontaneous arthritis in 6 month old male DBA/1 mice. Thus it is quite possible that additional genes provided by the DBA/1 genome contributed to these findings.

A role for TCR genes in CIA is further illustrated by the finding that CIA-resistant RIIIS/J mice, which bear a CIA-susceptible H-2^r haplotype, possess the largest known deletion of TCR V_β genes to date (Figure 1).[25] Similar to the findings using SWR mice, F_1, F_2 and backcross studies of B10 X RIIIS/J matings revealed a 100% correlation between CIA, H-2^r and the V_β^b haplotype.[26] The demonstration that CIA susceptibility in both the H-2^q and H-2^r haplotypes are heavily influenced by deletions within the TCR V_β gene locus implicates a role for particular V_β TCR genes in the induction and pathogenesis of CIA.

Although the gene complementation studies described above have been quite informative, the paradoxical findings between Banerjee et al[16] and Spinella et al[22] underscores the influence of non-MHC and non-TCR genes in dictating susceptibility to CIA. In an attempt to specifically examine the influence of TCR V_β gene deletions in CIA, we have bred B10.Q (H-2^q) and B10.RIII (H-2^r) mice congenic for the SWR (V_β^a) or RIIIS/J (V_β^c) V_β TCR deletion. Table 1 shows that B10.RIII-V_β^a mice display a delayed onset of arthritis compared to B10.RIII animals following immunization with porcine CII. This finding suggests that B10.RIII mice may possess a population of arthritogenic T cells which utilize a V_β TCR gene absent in V_β^a mice. However, the absence of such T cells, as in B10.RIII-V_β^a animals, has only a modest effect on the overall development of CIA. It is possible that the absence of one particular family of V_β TCR leads to the inability to recognize a particular CII epitope while the recognition of other subdominant epitopes remains intact. Such a situation could explain the delayed development of CIA in B10.RIII-V_β^a animals.

Table 1. CIA in B10.RIII vs. B10.RIII-V_β^a mice following immunization with porcine CII^a

		CIA	
Strain	Positive/Total	Onset (mean day $\pm$ SE)^a	Severity (mean score $\pm$ SE)^b
B10.RIII	26/27	26 $\pm$ 2	7.4 $\pm$ 0.6
		(p = 0.02)	
B10.RIII-V_β^a	31/36	36 $\pm$ 2	6.2 $\pm$ 0.5

^{a}Mice were immunized in the base of the tail with 100 µg porcine CII in CFA on day 0 and monitored for arthritis development three times per week for 12 weeks.
^{b}Mean value determined from arthritic mice only based on the scale described by Wooley et al.[6]

The influence of V_β TCR gene deletions in B10.Q congenic mice has also been quite informative. Table 2 shows that B10.Q-V_β^a mice, which bear the same V_β TCR deletion as the SWR strain, shows no difference in CIA susceptibility compared to B10.Q control animals. However, B10.Q-V_β^c mice, which lack four additional V_β TCR genes, displayed a marked resistance to CIA induction. These findings suggest that in B10.Q mice the genes absent in the V_β^a deletion plays little role in CIA susceptibility whereas TCR genes absent in the V_β^c deletion appear to be critical for the efficient induction of CIA. Clearly, this observation illustrates the influence of V_β TCR deletions in the induction and development of CIA. It is hoped that these V_β TCR congenic strains will provide a valuble tool in deciphering specific TCR usage in the recognition of arthritogenic eptiopes.

The suggestion of a role for specific TCR V_β genes in CIA, through the use of TCR deletion mutants, PCR technology[27] and transgenic mice,[28] has sparked numerous investigations into characterizing TCR usage in T cells recovered from inflamed RA synovium.[29-32] Although the results are unclear at this time, further studies should yield insight into the role of TCR genes in RA. Clearly, knowledge of TCR gene usage in CIA and RA will profoundly enhance our understanding of these diseases.

NON MHC AND NON TCR GENES IN CIA SUSCEPTIBILITY

In addition to MHC and TCR genes, other gene products can contribute to the regulation of CIA. One group of genes which regulates CIA via its influence on the TCR repertoire are products of the Minor lymphocyte stimulating (Mls) loci.[33] The Mls antigens are products of retroviral genes such as the mouse mammary tumor virus-7 gene encoding the Mls-1^a allele.[34] These gene products have a promiscuous ability to bind MHC class II molecules as well as stimulating a high frequency of T cells

Table 2. Resistance to CIA induction in B10.Q-V_β^c but not B10.Q-V_β^a mice following immunization with bovine CII[a].

			CIA	
Experiment	Strain	Positive/Total	Onset (mean day $\pm$ SE)[a]	Severity (mean score $\pm$ SE)[b]
1	B10.Q	18/22	41 ± 2	6.6 ± 1.0
	B10.Q-V_β^a	16/23	40 ± 3	4.2 ± 1.0
2	B10.Q	7/10	34 ± 5 ($\underline{p} = 0.05$)	5.0 ± 0.8 ($\underline{p} = 0.006$)
	B10.Q-V_β^c	4/13	51 ± 8	1.3 ± 0.3

[a]Mice were immunized in the base of the tail with 100 µg bovine CII in CFA on day 0 and monitored for arthritis development three times per week for 12 weeks.
[b]Mean value determined from arthritic mice only based on the scale described by Wooley et al.[6]

expressing particular V_β TCR genes. For example, antigen presenting cells from mice bearing the Mls-1^a allele can stimulate T cells which express the $V_\beta6$, 7, 8.1 or 9 TCR, whereas Mls-2^a antigens stimulate $V_\beta3$ positive T cells. Therefore, in an attempt to avoid self reactivity in vivo, T cells reactive to the particular Mls-1 allele are clonally deleted during thymic maturation.

A role for Mls genes in CIA was shown by Anderson et al.[35] Using F_1 offspring of a BALB.D2.Mls1^a(CIA resistant) X B10.Q cross, it was shown that these mice have a significantly lower incidence of CIA v.s. control (BALB/c X B10.Q)$_{F1}$ animals. Backcross analysis revealed a strong correlation with CIA resistance and the presence of the Mls-1^a allele. As mentioned earlier, Mls-1^a clonally deletes T cells bearing $V_\beta6$, 7, 8.1 and 9 TCRs. Thus, it was suggested that the absence of these T cells, due to clonal deletion in the context of Mls-1^a, led to reduced CIA susceptibility. It is quite interesting that B10.Q-V_β^c mice, which lack $V_\beta6$-bearing T cells, are also resistant to CIA induction.

Given the polygenic nature of RA and our current knowledge of CIA, it is clear that a number of other non-MHC genes are required for arthritis susceptibility. Characterizing these genes is a formidable task but recent technical advances may aid this avenue of research. Recently, scientists have been able to effectively study non-MHC genes linked to the susceptibility of type I diabetes in the nonobese diabetic (NOD) mouse.[36-38] By utilizing a new type of DNA marker called microsatellites, three genetic loci, located on chromosomes 1, 3 and 11, have been indentified which influence the development of diabetes in NOD mice.[36-38] Of particular interest is the finding that these loci are closely

58

linked to genes involved in the immune response such as the interleukin-1 receptor and the lymphokines interleukin-2, 4 and 5. It is tempting to speculate that immune system genes predisposing mice to type-1 diabetes likewise control susceptibility to other autoimmune diseases like CIA. We are currently employing microsatellite techniques to further characterize those non-MHC genes involved in CIA. It is our hope that this approach will yield exciting new information into the multigenic control of autoimmune arthritis.

T CELL EPITOPES OF TYPE II COLLAGEN

Identifying the epitopes involved in the induction and pathogenesis of any autoimmune disease is critical for a clear understanding of the mechanisms involved in the breakdown of self-tolerance. The contribution of CII antibodies in CIA development notwithstanding, numerous lines of experimental evidence makes it clear that the T cell is the major culprit in the induction and progression of CIA.[39-41] Thus, substantial research has focused on identifying T cell epitopes present on CII. Unfortunately progress in this effort has been slowed due to the large and complex nature of CII. Previous studies showed that a 245 amino acid cyanogen bromide peptide of chick CII, called CB11, could induce CIA in H-2^q mice.[42] Subsequent work revealed that a synthetic peptide corresponding to residues 122-147 of the CB11 molecule (residues 245-270 on intact CII) induced resistance to CIA induction when administered to neonatal animals.[43] The arthritogenic potential of this peptide has yet to be demonstrated but it has recently been shown that T cells can indeed proliferate to this region[44] and residues important for peptide binding to MHC class II molecules have been putatively identified.[44,45] A caveat to these studies, however, is that multiple arthritogenic epitopes on CII do exist.[14] Thus, concentrating solely on this epitope, although informative, may hinder our overall understanding of T cell recognitory mechanisms in CIA. Clearly, much work lies ahead in characterizing those T cell epitopes necessary for the induction of CIA. Knowledge of these epitopes will lay a solid foundation for developing specific immune intervention strategies at the interface of the TCR, MHC and arthritogenic determinant.

THE TRIMOLECULAR COMPLEX AS A THERAPEUTIC TARGET IN ARTHRITIS

As mentioned above, establishing a working knowledge of the interactions between the autoantigenic peptide, MHC class II molecule and TCR, the trimolecular complex, is vital for the rational design of specific immunosuppressive treatments for autoimmune diseases. This approach is quite attractive due to its potential of suppressing only the immune response of interest. The first indication that interfering with the trimolecular complex could suppress CIA was shown by Wooley et al.[46] Using monoclonal antibodies specific for class II MHC molecules, it was revealed that administration of these antibodies at the time of immunization with CII suppressed the development of CIA in both B10.Q and B10.RIII mice. Recently, we have generated antisera against synthetic peptides corresponding to specific regions of H-2A and H-2E molecules. Like conventional anti-class II antibodies, we observed that administration of the anti-class II peptide antisera could also suppress CIA induction (Table 3). This finding hints to the potential of using synthetic peptides of MHC molecules to generate site-specific antibodies for therapeutic use.

Another therapeutic target in CIA is the TCR. Due to the important role of particular V_β TCRs in CIA, elimination of T cells bearing these receptors may effectively ameliorate disease development. Chiocchia et al[47] suppressed CIA induction in DBA/1 mice by eliminating $V_\beta 8.1,2^+$ or $V_\beta 5.1,2^+$ T cells. More recently, Osman et al[48] successfully inhibited CIA induction by depleting $V_\beta 8.2^+$ cells. Immunization with synthetic peptides of TCR variable regions, which elicit anti-TCR responses, has shown some therapeutic effects in the rat model of experimental allergic encephalomyelitis (EAE).[49] Whether TCR peptide immunization would be effective in CIA remains to be determined.

A final approach at interfering with the trimolecular complex is by blocking the interaction of autoantigenic peptide with the MHC through the use of peptide analogs. Inhibition of EAE has been achieved using analogs of encephalitogenic epitopes of myelin basic protein.[50] Moreover, recent studies suggest that some peptide inhibitors may actually act as classical receptor antagonists.[51] Thus, elucidating the T cell epitopes on CII may allow investigators the opportunity to develop specific peptide inhibitors of CII for use in CIA.

59

Table 3. Modulation of CIA induction in B10.RIII (H-2^r) mice administered rabbit anti-H-2E$_\alpha^k$ 90-110 peptide antisera.

Treatment[a]		CIA	
	Postive/Total	Onset (mean day ± SE)	Severity (mean score ± SE)[b]
Rabbit anti-H-2E$_\alpha^k$ 90-110	10/19	35 ± 3	5.1 ± 1.1
		(p = 0.01)	(p = 0.02)
Normal Rabbit Serum	16/20	26 ± 2	7.7 ± 1.0

[a]B10.RIII mice received 250 μl of a rabbit antiserum generated against a peptide corresponding to the highly conserved residues 90-110 of the H-2E$_\alpha^k$ molecule i.p. on days -1,0,1. Control mice received an equivalent amount of normal rabbit serum. Both groups were immunized on day 0 with 100 μg porcine CII in CFA and monitored for arthritis development three times per week for 12 weeks.

[b]Mean value determined from arthritic mice only based on the scale described by Wooley et al.[6]

CLOSING COMMENTS

There is no doubt that great progress has been made in identifying the immunogenetic influences in the mouse model of CIA. Such findings have catalyzed investigations aimed at characterizing similar features in RA. Despite great progress, many questions and challenges remain. Thus, the mouse model of CIA should continue to further our understanding of the immunogenetic basis of not only autoimmune arthritis, but autoimmune diseases in general. It is hoped that these new findings will lead to the design of more effective treatments of autoimmune disease.

ACKNOWLEDGEMENTS

The authors would like to thank Mary Brandt for preparation of the manuscript.

REFERENCES

1. D.E. Trentham, A.S. Townes, and A.H. Kang, Autoimmunity to type II collagen : An experimental model of arthritis. J. Exp. Med. 146:857 (1977)

2. J.S. Courtenay, M.J. Dallman, A.D. Dayan, A. Martin, and B. Mosedale, Immunization against heterologous type II collagen induces arthritis in mice. Nature 285:666 (1980)

3. S. Banerjee, B-Y. Wei, K. Hillman, H.S. Luthra, and C.S. David, Immunosuppression of collagen-induced arthritis in mice with an anti-IL-2 receptor antibody. J. Immunol. 141:1150 (1988)

4. R. Holmdahl, L. Jansson, M. Andersson, and E. Larsson, Immunogenetics of type II collagen autoimmunity and susceptibility to collagen arthritis. Immunology 65:305 (1988)

5. P.H. Wooley, Collagen-induced arthritis in the mouse. Meth. Enzymol. 162:361 (1988)

6. P.H. Wooley, H.S. Luthra, J.M. Stuart, and C.S. David, Type II collagen induce arthritis mice: 1. Major histocompatibility complex (I-region) linkage and antibody correlates. J. Exp. Med. 154:688 (1981)

7. P. Stasny, Association of the B-cell alloantigen DRw4 with rheumatoid arthritis. N. Engl. J. Med. 298:869 (1978)

8. J. Tiwari, and P. Terasaki,"HLA and disease associations." Springer-Verlag, New York, (1967)

9. R.W. Karr, G.E. Rodey, T. Lee, and R.D. Schwartz, Association of HLA-DRw4 with rheumatoid arthritis in black and white patients. Arthritis Rheum. 23:1241 (1980)

10. J.H. Dobloug, O. Forre, E. Kass, and E. Thorsby, HLA antigens and rheumatoid arthritis: association between HLA-DR4/Dw4 positivity and IgM rheumatoid factor production. Arthritis Rheum. 23:309 (1980)

11. P. Gregersen, M. Shen, Q. Song, P. Merryman, S. Degar, and T. Seki et al. Molecular diversity of HLA-DR4 haplotypes. Proc. Nat. Acad. Sci. 83:2642 (1986)

12. P.K. Gregersen, J. Silver, and R.J. Winchester, The shared epitope hypothesis. An approach to understanding the molecular genetics of susceptibility to rheumatoid arthritis. Arthritis Rheum. 30:1205 (1987)

13. A. Huse, P.H. Wooley, H.S. Luthra, J.M. Stuart and C.S. David, New recombinants confirm mapping of the susceptibility to type II collagen induced arthritis in mice. Fed. Proc. 43:1820 (1984)

14. P.H. Wooley, H.S. Luthra, M.M. Griffiths, J.M. Stuart, A. Huse, and C.S. David, Type II collagen induced arthritis in mice. IV. Variations in immunogenetic regulation provide evidence for multiple arthritogenic epitopes on the collagen molecule. J. Immunol. 135:2443 (1985)

15. K. Gustafsson, M. Karlsson, L. Andersson, and R. Holmdahl, Structures on the I-A molecule predisposing for susceptibility to type II collagen-induced autoimmune arthritis. Eur. J. Immunol. 20:2127 (1990)

16. S. Banerjee, T.M. Haqqi, H.S. Luthra, J.M. Stuart, and C.S. David, Possible role of V_β T cell receptor genes in susceptibility to collagen induced arthritis in mice. J. Exp. Med. 167:832 (1988)

17. H.S. Luthra, P.H. Wooley, A.M. Dillon, S.K. Singh, W.P. Lafuse, and C.J. Krco, et al. Immunogenetics of collagen-induced arthritis (CIA) in mice: a model of autoimmune disease. Ann. N.Y. Acad. Sci. 475:361 (1986)

18. M.A. Behlke, H.S. Chow, K. Huppi, and D.Y. Loh, Murine T cell receptor mutants with deletions of ß-chain variable region genes. Proc. Nat. Acad. Sci. 83:767 (1986)

19. R.P. Erickson, D.K. Tachibana, L.A. Herzenberg, and L.T. Rosenberg, A single gene-controlling hemolytic complement and a serum antigen in the mouse. J. Immunol. 92:611 (1960)

20. W.C. Watson, and A.S. Townes, Genetic susceptibility to murine collagen II autoimmune arthritis. Proposed relationship to the IgG2 autoantibody subclass response, complement C5, major histocompatibilty complex (MHC) and non-MHC loci. J. Exp. Med. 162:1878 (1985)

21. S. Banerjee, G.D. Anderson, H.S. Luthra, and C.S. David, Influence of complement C5 and V_β T cell receptor mutations on susceptibility to collagen-induced arthritis in mice. J. Immunol. 142:2237 (1989)

22. D.G. Spinella, J.R. Jeffers, R.A. Reife, and J.M. Stuart, The role of C5 and the T cell receptor V_β genes in susceptibility to collagen-induced arthritis. Immunogenetics 34:23 (1991)

23. R. Holmdahl, M. Andersson, T.J. Goldschmidt, K. Gustafsson, L. Jansson, and J.A. Mo, Type II collagen arthritis in animals and provocations leading to arthritis. Immunol. Rev. 118:193 (1990)

24. C. Nordling, A. Karlsson-Parra, L. Jansson, R. Holmdahl, and L. Klareskog, Characterization of a spontaneously occurring arthritis in male DBA/1 mice. Arthritis Rheum. 35:717 (1992)

25. T.M. Haqqi, S. Banerjee, G.A. Anderson, and C.S. David, RIIIS/J (H-2^r)-an inbred mouse strain with a massive deletion of T cell receptor V_β genes. J. Exp. Med. 169:1903 (1989)

26. T.M. Haqqi, and C.S David, T-cell receptor V_β genes repertoire in mice possible role in resistance and susceptibility to type II collagen-induced arthritis. J. Autoimmunity 3:113 (1990)

27. T.M. Haqqi, G.D. Anderson, S. Banerjee, and C.S. David, Restricted heterogeneity in T-cell antigen receptor V_β gene usage in the lymph nodes and arthritic joints of mice. Proc. Natl. Acad. Sci. 89:1253 (1992)

28. L. Mori, H. Loetscher, K. Kakimoto, H. Bluethmann, and M. Steinmetz, Expression of a transgenic T cell receptor B chain enhances collagen-induced arthritis. J. Exp. Med. 176:381 (1992)

29. M.D. Howell, J.P. Diveley, K.A. Lundeen, A. Esty, S.T. Winters, D.J. Carlo, and S.W. Brostoff, Limited T-cell receptor ß-chain heterogeneity among interleukin-2 receptor-positive synovial T cells suggests a role for superantigen in rheumatoid arthritis. Proc. Natl. Acad. Sci. 88:10921 (1991)

30. A. Sottini, L. Imberti, R. Corla, R. Cattaneo, and D. Primi, Restricted expression of T cell receptor V_β but not V_α genes in rheumatoid arthritis. Eur. J. Immunol. 21:461 (1991)

31. G. Pluschke, G. Ricken, H. Taube, S. Droninger, I. Melchers, and H.H. Peter, et al. Biased T cell receptor V_α region repertoire in the synovial fluid of rheumatoid arthritis patients. Eur. J. Immunol. 21:2749 (1991)

32. X. Paliard, S.G. West, J.A. Lafferty, J.R. Clements, J.W. Kappler, P. Marrack, and B.L. Kotzin, Evidence for the effects of a superantigen in rheumatoid arthritis. Science 253:325 (1991)

33. J.W. Kappler, and P. Marrack, Self-tolerance eliminates T cells specific for Mls-modified products of the Major Histocompatibility Complex. Nature 332:35 (1990)

34. W.N. Frankel, C. Rudy, J.M. Coffin, and B.T. Huber, Linkage of Mls genes to endogenous mammary tumour viruses of inbred mice. Nature 349:526 (1991)

35. G.D. Anderson, S. Banerjee, H.S. Luthra, and C.S. David, Role of the Mls-1 locus and clonal deletion of T cells in susceptibility to collagen induced arthritis in mice. J. Immunol. 147:1189 (1991)

36. J.A. Todd, T.J. Aitman, R.J. Cornall, S. Ghosh, J.R.S. Hall, and C.M. Hearne, et al. Genetic analysis of autoimmune type 1 diabetes mellitus in mice. Nature 351:542 (1991)

37. R.J. Cornall, J-B. Prins, J.A. Todd, A. Pressey, N.H. DeLarato, L.S. Wicker, and L.B. Peterson, Type 1 diabetes in mice is linked to the interleukin-1 receptor and Lsh/Ity/Bcg genes on chromosome 1. Nature 353:262 (1991)

38. H-J. Garchon, P. Bedossa, L. Eloy, and J-F. Bach, Identification and mapping to chromosome 1 of a susceptibility locus for periinsulitis in non-obese diabetic mice. Nature 353:260 (1991)

39. R. Holmdahl, L. Klareskog, K. Rubin, J. Bjork, G. Smedegard, and R. Jonsson, Role of T lymphocytes in murine collagen induced arthritis. Agents and Actions 19:295 (1986)

40. J.T. Hom, L.D. Butler, P.E. Biedl, and A.M. Bendele, The progression of the inflammation in established collagen-induced arthritis can be altered by treatments with immunological or pharmacological agents which inhibit T cell activities. Eur. J. Immunol. 18:881 (1988)

41. K.G. Moder, H.S. Luthra, R. Kubo, M. Griffiths, and C.S. David, Prevention of collagen induced arthritis in mice by treatment with an antibody directed against the T cell receptor $\alpha\beta$ framework. Autoimmunity 11:219 (1992)

42. K. Terato, K.A. Hasty, M.A. Cremer, J.M. Stuart, A.S. Townes, and A.H. Kang, Collagen-induced arthritis in mice. Localization of an arthritogenic determinant to a fragment of the type II collagen molecule. J. Exp. Med. 162:637 (1985)

43. L.K. Myers, J.M. Stuart, J. Seyer, and A.H. Kang, Identification of an immunosuppressive epitope on type II collagen that confers protection vs. collagen induced arthritis. J. Exp. Med. 170:1999 (1989)

44. E. Michaelsson, M. Andersson, A. Engstrom, and R. Holmdahl, Identification of an immunodominant type-II collagen peptide recognized by T cells in $H-2^q$ mice: self tolerance at the level of determinant selection. Eur. J. Immunol. 22:1819 (1992)

45. L.K. Myers, K. Terato, J.M. Seyer, J.M. Stuart, and A.H. Kang, Characterization of a tolerogenic T cell epitope of type II collagen and its relevance to collagen-induced arthritis. J. Immunol. 149:1439 (1992)

46. P.H. Wooley, H.S. Luthra, W.P. Lafuse, A. Huse, J.M. Stuart, and C.S. David, Type II collagen-induced arthritis in mice. III. Suppression of arthritis by using monoclonal and polyclonal anti-Ia antisera. J. Immunol. 134:2366 (1985)

47. G. Chiocchia, M.C. Boisser, and C. Fournier, Therapy against murine collagen-induced arthritis with T cell receptor V_β-specific antibodies. Eur. J. Immunol. 21:2899 (1991)

48. G.E. Osman, M. Toda, O. Kanagawa, and L.E. Hood, Characterization of the T cell receptor repertoire causing collagen arthritis in mice. J. Exp. Med. 177:387 (1993)

49. A.A. Vandenbark, G. Hashim, and H. Offner, Immunization with a synthetic T-cell receptor V-region peptide protects against experimental autoimmune encephalomyelitis. Nature 341:541 (1989)

50. D.C. Wraith, D.E. Smilek, D.J. Mitchell, I. Steinman, and H.O. McDevitt, Antigen recognition in autoimmune encephalomyelitis and the potential for peptide-mediated immunotherapy. Cell 59:247 (1989)

51. M.T. De Magistris, J. Alexander, M. Coggeshall, A. Altman, F.C.A. Gaeta, H.M. Grey, and A. Sette, Antigen analog-major histocompatibility complexes act as antagonists of the T cell receptor. Cell 68:625 (1992)

SUPPRESSION OF EXPERIMENTAL AUTOIMMUNE MYASTHENIA GRAVIS BY EPITOPE-SPECIFIC NEONATAL TOLERANCE

Premkumar Christadoss*, Mohan Shenoy*, Minako Oshima+, and M. Zouhair Atassi+

*The Department of Microbiology, University of Texas Medical Branch Galveston, Texas 77550 and +The Department of Biochemistry, Baylor College of Medicine Houston, TX 77030

INTRODUCTION

An autoimmune response to muscle nicotinic acetylcholine receptor (AChR) culminates in myasthenia gravis (MG) (1,2). MG has been associated with certain HLA antigens (3), and linked to a polymorphism in the HLA-DQß subunit (4). Immunization of C57BL6(B6) mice with Torpedo californica AChR (T-AChR) produces an autoimmune disease mimicking MG called experimental autoimmune myasthenia gravis (EAMG). EAMG susceptibility has been mapped to the I-A subregion of the mouse major histocompatibility complex (MHC) (5,6). Both the I-Aα and the I-Aß genes contribute to EAMG pathogenesis (5-8). Further, expression of the I-E molecule on the cell surface of C57BL10.IEα^k transgenic mice could partially prevent the development of EAMG (9). Thus, the development of EAMG is primarily influenced by the Class II genes. Class II restricted AChR epitope reactive T helper (Th) cells are activated in MG patients and appear to contribute to the postsynaptic pathology (10-11). Th cells are also involved in EAMG pathogenesis (5,6, 12-17), and EAMG can either be prevented, or clinical remission of established disease induced, by in vivo administration of antibodies (Abs) to CD4 molecule (14). The in vitro lymphocyte proliferation to AChR generally correlates with EAMG susceptibility in congenic strains of mice, and is controlled by a Mendelian dominant gene at the I-A locus (12,13). Further, the lymphocyte proliferation is dependent on CD4+ Th cells and I-A molecule (18,19).

A gene conversion event between I-Eßb and I-Aßb in the bml2 (20,21) strain, which altered three amino acids in the C-terminal half of the first domain of I-Aßb (Ile-67 —> Phe; Arg-70 —> Gln; Thr-71 —> Lys) (22), resulted in resistance to EAMG development (6,7). The bml2 strain, when compared to the parental B6 strain (H-2b) not only generates a lower cellular and humoral immune responses to T-AChR, but also demonstrates reduced incidence of clinical EAMG (6,7). To study the effect of gene conversion at I-Aß positions 67, 70, and 71 on the T-cell responses to epitopes of T-AChR-α subunit, B6 and bml2 mice were primed with T-AChR, and the profiles of T-lymphocyte proliferation were determined with 18 synthetic overlapping peptides encompassing the entire extra-cellular portion of the T-AChR-α subunit (23). The immunodominant T cell and B cell epitopes for EAMG susceptible B6 mice have been mapped (16,17). The T cell epitopes reside within T-AChR-α subunit region 111-126, 146-162, and 182-198 (16). A marked reduction in the proliferative response of T-AChR primed bm12 lymphocytes to peptides α146-162 was observed, when compared with the corresponding response of B6 T-AChR-primed lymphocytes. Because bm12 resistance to EAMG correlates with reduced proliferative response to the otherwise immunodominant epitopes within α146-162 on the T-AChR-α subunit, the data implicated the importance of this epitope in EAMG pathogenesis. The pathogenic role of the dominant AChR-α subunit T cell epitopes were assesed by either immunzing or neonatally tolerizing B6 mice with the AChR-α subunit dominant T cell epitope and evaluating for MG pathogenesis.

MATERIALS AND METHODS

Acetylcholine Receptors and Synthetic Peptides

The T-AChR was prepared from the electric organs of Torpedo californica as described previously (24,25). Extract of B6 muscle AChR to be used as antigen for detecting autoantibodies was prepared according to a previously published method (25). The primary structure, synthesis, purification, and characterization of the eighteen overlapping peptides encompassing the entire extra-cellular part (residue 1-210) of the α subunits of T-AChR have been described in an earlier publication (23). The amino acid sequence of the 18 synthetic overlapping peptides of the T-AChR α subunit is provided in Fig. 1. In only one experiment AChR α chain peptide 182-198 of the human sequence (R-G-W-K--H-S-V-T-Y-S-C-C-P--D-T-P-Y) was used.

Peptide	Structure
α1-16	S-E-H-E-T-R-L-V-A-N-L-L-E-N-Y-N
α12-27	L-E-N-Y-N-K-V-I-R-P-V-E-H-H-T-H
α23-38	E-H-H-T-H-F-V-D-I-T-V-G-L-Q-L-I
α34-49	G-L-Q-L-I-Q-L-I-S-V-D-E-V-N-Q-I
α45-60	E-V-N-Q-I-V-E-T-N-V-R-L-R-Q-Q-W
α56-71	L-R-Q-Q-W-I-D-V-R-L-R-W-N-P-A-D
α67-82	W-N-P-A-D-Y-G-G-I-K-K-I-R-L-P-S
α78-93	I-R-L-P-S-D-D-V-W-L-P-D-L-V-L-Y
α89-104	D-L-V-L-Y-N-N-A-D-G-D-F-A-I-V-H
α100-115	F-A-I-V-H-M-T-K-L-L-L-D-Y-T-G-K
α111-126	D-Y-T-G-K-I-M-W-T-P-P-A-I-F-K-S
α122-138	A-I-F-K-S-Y-C-E-I-I-V-T-H-F-P-F-D
α134-150	H-F-P-F-D-Q-Q-N-C-T-M-K-L-G-I-W-T
α146-162	L-G-I-W-T-Y-D-G-T-K-V-S-I-S-P-E-S
α158-174	I-S-P-E-S-D-R-P-D-L-S-T-F-M-E-S-G
α170-186	F-M-E-S-G-E-W-V-M-K-D-Y-R-G-W-K-H
α182-198	R-G-W-K-H-W-V-Y-Y-T-C-C-P-D-T-P-Y
α194-210	P-D-T-P-Y-L-D-I-T-Y-H-F-I-M-Q-R-I

Figure 1. Structure of the synthetic overlapping peptides representing the extra-cellular part (residues 1-210) of the α subunit of the Torpedo californica AChR: The regions of overlaps between consecutive peptides are underlined. The single letter notations of the amino acids are: A, alanine; C, cysteine; D, aspartic acid; E, glutamic acid; F, phenylalanine; G, glycine; H, histidine; I, isoleucine; K, lysine; L, leucine; M, methionine; N, asparagine; P, proline; Q, glutamine; R, arginine; S, serine; T, threonine; V, valine; W, tryptophan; Y, tyrosine.

Immunization of Mice and Lymphocyte Proliferation Assay

Eight-week-old female B6 and bm12/KhEg mice (Jackson Laboratories, ME) were immunized subcutaneously at the base of the tail with T-AChR (20 µg/mouse) in PBS emulsified with an equal volume of complete Freund's adjuvant (CFA). Lymph node cells were harvested from the T-AChR-primed mice 7 days after the immunization. Cells were co-cultured in triplicate in flat bottom microtiter plates at a concentration of 6×10^5-7×10^5 cells/ well with various concentrations of T-AChR (1.5-6.0 µg/ml) or its peptides (10-40 µg/ml) in a final volume of 200 µl of RPMI 1640 medium supplemented with 1% normal mouse serum. Control wells were stimulated with Con A (1 ug/ml), LPS (500 ug/ml), unrelated protein (ovalbumin, 100 ug/ml) or unrelated synthetic peptide (sequence, ESSGTGIESSGTGI 10-40 µg/ml). After incubation for 3 days at 37°C in 5% CO_2 humidified incubator, the cultures were pulsed with [3H] thymidine (1 µCi/well) and harvested 16 hours later for counting by liquid scintillation.

Evaluation for Clinical MG and Measuring Serum Antibodies to Muscle AChR

Muscle weakness was graded blindly by at least two individuals. Grade 0 indicated no weakness either at rest, or after exercise consisting of 20-30 consecutive paw grips on cage top steel grids. Grade 1 was defined as normal strength at rest, but weak with chin on the floor and forelimb paralysis and inability to raise the head after exercise. Grade 2 was defined as grade 1 weakness at rest. Grade 3 was defined as moribundity, dehydration, or quadriplegia. To confirm that the muscle weakness was due to MG, neostigmine bromide and atropine sulphate were administered IP to mice with weakness and improvement in muscle strength was evaluated as before. Further, repetitive nerve stimulation test (electromyography; EMG) was performed in most of the mice with weakness to demonstrate decremental response according to a previously published method (27). Greater than 10% decrement in the 5th amplitude when compared with the second amplitude in the repetitive nerve stimulation test was considered positive.

Serum autoantibodies to mouse muscle AChR (M-AChR) was determined using an established protocol (14,25).

RESULTS

Gene Conversion at I-Aß Subunit Altered the T Cell Recognition of AChR-α Subunit T Cell Epitopes

The proliferative responses of T-AChR-primed lymphocytes of B6 and bml2 to challenge in vitro with the optimum dose of each of the 18 synthetic peptides or with T-AChR are summarized in Fig 2. As expected, T-AChR-primed lymphocytes of B6 mice responded better than those of bml2 when challenged with T-AChR. B6 and bml2 T-AChR-primed lymphocytes failed to elicit a good proliferative response when challenged with the peptides spanning the region α1-104, the first half of the extracellular domain of T-AChR-α subunit. T-AChR-primed B6 lymphocytes responded to the greatest extent (stimulation index, SI = 21.6) to the major immunodominant T-cell epitope α146-162. In contrast, T-AChR-primed bml2 lymphocytes mounted a considerably lower response (SI=6) to this peptide. Similarly, the proliferative response of the lymphocytes to peptide α182-198 was markedly lower for the cells of bml2 (SI=5) than those of B6 (SI=9). Furthermore, B6 responded to peptides α122-138 and α134-150 slightly better than bml2. The mutation of bml2 did not affect the T-cell response to the other major immunodominant region in B6, namely α111-126. In three separate experiments, we could demonstrate consistently the lower T-cell proliferative response of bml2 to peptides α146-162 and α182-198, when compared with T cell response of B6 to these peptides, and the preservation in bml2 of the T cell response to peptide α111-126. Finally, both B6 and bml2 T-AChR-primed lymph node cells responded equally well to ConA and LPS, but failed to respond to ovalbumin and the unrelated peptide (data not shown). If the magnitude of in vitro T-cell

67

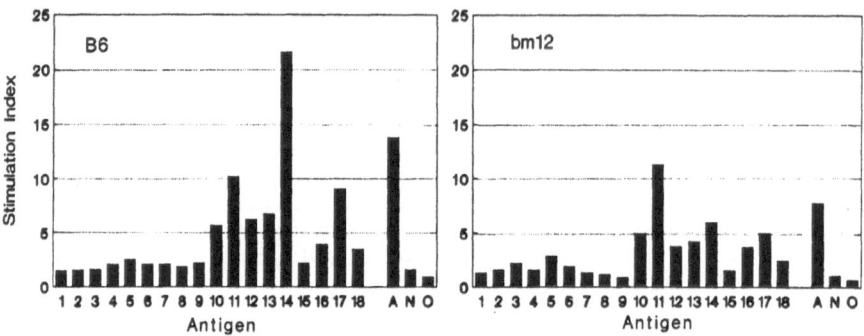

Figure 2. Proliferative responses to the synthetic AChR-α chain peptides of LNC from B6 mice and bm12 mice. Unstimulated cells gave the following cpm: B6, 6208; bm12, 6872. The diagram shows the stimulation index at the optimum challenge dose of each peptide: (1) α1-16; (2) α12-27; (3) α23-38; (4) α34-49;p (5) α45-60; (6) α56-71; (7) α67-82; (8) α78-93; (9) α89-104; (10) α100-115; (11) α111-126; (12) α122-138; (13) α134-150; (14) α146-162; (15) α158-174; (16) α170-186; (17) α182-198; (18) α194-210. Additional antigen letter symbols are: A, AChR; N, unrelated (to AChR) peptide; O, ovalbumin. Results were expressed as stimulation index (stimulation index = mean cpm incorporated by stimulated cells/mean cpm incorporated by unstimulated cells).

proliferative response predicts the in vivo outcome, these findings would indicate that epitopes within T-AChR regions α146-162 or α182-198 may be crucial in the development of EAMG.

Failure to Induce Clinical EAMG in B6 Mice with Immunodominant Peptides of the T-AChR-α Subunit

To demonstrate the involvement of the T-AChR-α subunit peptides in the development of EAMG, 16 adult B6 mice were immunized with 50 µg of peptide α146-162 in CFA into foot pads and shoulders (4 sites), and boosted three times at monthly intervals with 50 µg each, of respective peptide in CFA. These mice were bled from the tail at different periods, and evaluated for muscle weakness on a daily basis. Approximately three and a half months after the primary immunization all the mice were humanely terminated and their inguinal and axillary lymph node cells were pooled and lymphocyte proliferative response to respective peptides were assessed. Similarly, 10 B6 mice were immunized with peptide α182-198. None of the mice immunized with either peptide displayed clinical signs of EAMG after the boost. Therefore, two more immunizations with the respective peptides were given at monthly intervals in order to induce clinical MG. Still the animals failed to show obvious clinical signs of the disease. However, lymph node cells derived from these mice proliferated on challenge with the respective peptides in vitro, and furthermore, sera obtained from these mice 7 days after the second immunization with either peptide contained antibodies reactive to M-AChR (data not shown). Antibodies raised against these two peptides were not overtly pathogenic, since none of the mice immunized with either peptides developed obvious clinical EAMG. Therefore, the immunodominant AChR regions α146-162 and α182-198 have the capacity to stimulate T-cells to undergo proliferation and to induce B cells to produce antibodies, but these responses are non-pathogenic, and therefore, not effective in inducing obvious clinical MG.

Reducing the Incidence of Clinical MG by Induction of Neonatal Tolerance to T-AChR and T-AChR-α Subunit T Cell Epitope

As an alternative approach to investigate the involvement of T-AChR region α146-162 in the development of EAMG, less than 24 hr old neonatal B6 mice were injected subcutaneously in their back with 30-70 µg of soluble T-AChR, or 100 µg of T-AChR α146-162 or α182-198 (human sequence) using a 30 gauge needle. Neonatally unmanipulated B6 mice were kept as control. At 8 weeks of age, all the mice were immunized subcutaneously in the foot pads and the shoulders with 20 µg of T-AChR in CFA. These mice were boosted 30 days later with 20 µg of T-AChR in CFA. Peptide α182-198 of human sequence (26) was used as a control because it differs from the Torpedo sequence in three positions. The above groups of mice were observed daily for muscle weakness and bled for serum after the second immunization. 60% neonatally untreated mice developed muscle weakness and showed improvement in strength after anticholinesterase medication (Table I). One of the mice died due to severe muscle weakness, and therefore, EMG was not performed. Three of the remaining five mice with muscle weakness demonstrated EMG decrement.

Table 1. Effect of neonatal tolerance to AChR and AChR-α chain peptides in the development of clinical EAMG

Neonatal Injection	Muscle Weakness (Grade)						p=1
	0	1	2	3	Total(%)		
-	4	2	4	0	6/10	(60)	
T-AChR	8	0	1	0	1/9	(11)	0.042
α146-162	10	2	0	0	2/12	(16)	
α182-198	4	2	2	1	5/9	(55)	

(0.042 bracket linking the three upper rows)

1Fisher exact probability test was used to determine the statistical significance of the data obtained.

2The difference between the incidence of muscle weakness in non-tolerized and T-AChR or α146-162 tolerized mice is statistically significant at P = 0.04. The difference between the incidence of muscle weakness in non-tolerized and α182-198 tolerized mice did not attain statistical significance.

In contrast, only one of nine (11%) mice that were neonatally tolerized with T-AChR developed clinical MG and had a decremental response. The difference observed in clinical incidence of EAMG between neonatally T-AChR injected and control mice was statistically significant at p=0.04 (Fisher's exact probability test). Fifty-five percent of mice neonatally tolerized with peptide α182-198 (human AChR-α subunit sequence) developed clinical signs and demonstrated decremental response. In contrast, only 2/12 (16%) animals neonatally inoculated with peptide α146-162 (torpedo sequence) exhibited mild (1+) clinical signs, and both the animals with weakness failed to show decremental response. Also, these two animals developed muscle weakness very late after the second immunization with T-AChR. The difference in the incidence of muscle weakness between non-tolerized and peptide α146-162 tolerized mice was statistically significant at p=0.04. The clinical course of the disease is depicted in fig. 3. In two other experiments only 2/14 mice neonatally tolerized with peptide α146-162 developed muscle weakness, while, 24/47 (51%) neonatally unmanipulated B6 mice developed weakness. Altogether only 4 of 26 (15%) mice neonatally tolerized with α146-162 developed clinical MG. Thus, neonatal tolerance to the whole AChR molecule or to AChR-α subunit peptide α146-162 effectively reduced the incidence of clinical MG, when

69

tolerized mice were subsequently immunized with T-AChR in CFA. Our data do not rule out the role of Torpedo AChR α182-198 in EAMG pathogenesis, although the presence of three substitutions in the region α182-198 between T-AChR and mouse AChR (R-G-W-K-H-W-V-F-Y-S-C-C-P-D-T-P-Y) make this less likely. The human α-182-198 which also differs from Torpedo α182-198 in 3 amino acid served as an excellent control to demonstrate peptide specific tolerance. Also, this study does not rule out the role of the other immunodominant peptide α111-126 in the pathogenesis. This is the first evidence which demonstrates the involvement of the region within α146-162 in the development of EAMG in mice, and of the prevention of this autoimmune disease by epitope-specific neonatal tolerance with a synthetic peptide.

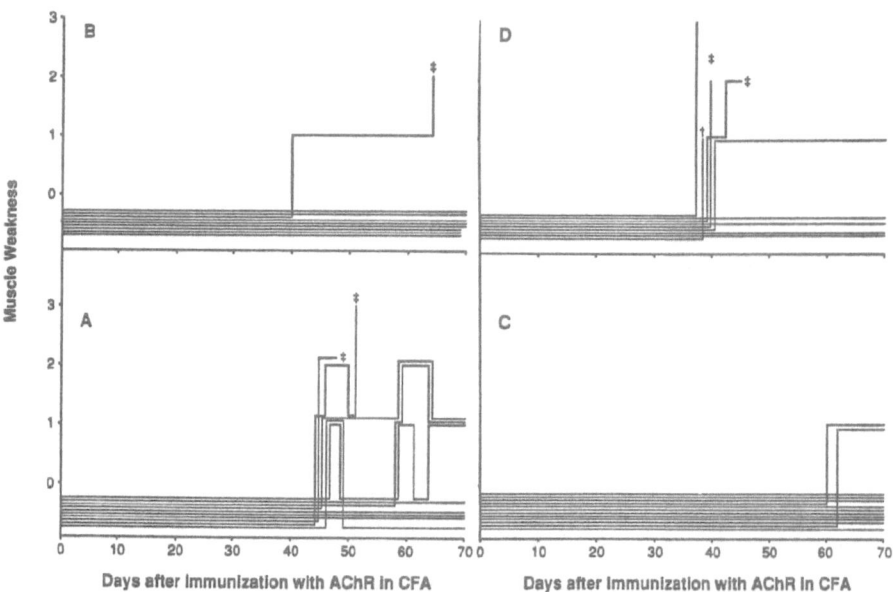

Figure 3. Clinical EAMG; time course and severity. A, B6 mice neonatally non-manipulated; B, injected with T-AChR; C, injected with peptide α146-162; and D, injected with peptide α182-198. When adult, mice were immunized with T-AChR in CFA on Day 0 and Day 30. + + mouse humanely killed because of severe weakness. + mouse died after developing weakness.

Effect of Neonatal Tolerance to T-AChR or T-AChR-α Subunit Peptides on Anti-AChR Antibody and In Vitro Lymphocyte Responses

The sera for anti-AChR antibody assay were taken a week after the second immunization with T-AChR in CFA; a time when mice start to show clinical signs of the disease. The anti-M-AChR antibody response was significantly (p=0.033; Student t test) reduced in mice neonatally tolerized with T-AChR when compared with the antibody response of non-tolerized mice (Table II). The difference in the antibody response between peptide α146-162 tolerized and non-tolerized mice was significant at p = 0.08. Although the mean serum anti-AChR antibody level was lower in mice tolerized with α182-198, the difference in antibody response between peptide α182-198 tolerized and non-tolerized mice did not attain statistical significance. Thus, neonatal tolerance to peptide α146-162 effectively suppressed the antibody response directed against a determinant within α146-162 (or against an entirely different region) in the mouse muscle AChR. The suppression of autoantibody response could be

70

TABLE 2. Effect of neonatal tolerance to T-AChR and T-AChR-α chain peptides on serum autoantibody to AChR and lymphocyte proliferative response

Neonatal Injection	Antibody to M-AChR[1] $(X10^{-10}M)$	3H Thymidine Uptake[2] D CPM + SEM	
		AChR	α146-162
——	6.89 $\pm$ 1.85	NT	NT
T-AChR	1.68 $\pm$ 0.89	NT	NT
α146-162 (Torpedo)	3.14 $\pm$ 1.31	NT	NT
α182-198 (human)	4.87 $\pm$ 1.73	NT	NT
——	NT	18.4 $\pm$ 4.7	2.8 $\pm$ 0.5
α146-162 (Torpedo)	NT	15.4 $\pm$ 10.3	1.3 $\pm$ 0.5

[1]Expressed as bungarotoxin binding sites precipitated per liter of serum.

[2]Values represent mean [3H] thymidine uptake of triplicate cultures (X10-3+SEM). [3H] Thymidine incorporation is expressed as counts per minute, with background (cells without antigen) values subtracted (D cpm).

NT - not tested

due to the induction of tolerance at the T helper cell level, because α146-162 is predominantly a T cell epitope in B6 mice (16).

To study the T cell tolerance to the region α146-162, in another experiment neonatally peptide α146-162 tolerized, and non-tolerized mice were immunized twice with T-AChR when adult. Lymph node cells were obtained at termination and cultured in the presence of T-AChR, or peptide α146-162, and the lymphocyte proliferative response was determined. Only a 17% reduction was observed in the lymphocyte proliferative response to the whole AChR in neonatally peptide α146-162 tolerized mice as compared to the proliferative response of non-tolerized mice (table II). However, there was a 54% reduction in the α146-162 specific lymphocyte proliferative response in α146-162 peptide tolerized mice, when compared to non-tolerized mice lymphocyte response to α146-162. Thus, T cell tolerance to peptide α146-162 could be partially achieved by neonatal injection of peptide α146-162. The lymphocyte response to T-AChR or peptide α146-162 was relatively low, because the lymphocyte response was measured more than two months after the primary immunization. In previous experiments reported in this paper maximal proliferative response to peptide α146-162 was seen when the assay was performed a week after immunization with T-AChR in CFA.

DISCUSSION

The exact nature of the control of T helper cell response exerted by the Ia molecule is still a matter of controversy. In determinant selection, the Ia molecule is believed to specifically bind some, but not all peptide antigens during presentation, thereby selecting the determinants that are to be presented by the antigen presenting cells (28,29). Therefore, the reduced lymphocyte proliferative response seen in bm12 to peptides α146-162 and α182-198 may be due to inefficient Aβbm12 binding and/or presentation of these epitopes to bm12 T cells. Alternatively, the low response seen in bm12 to these peptides may be due to the reduced frequency, or absence, of T cells responding to these peptides, due to clonal elimination or anergy. The bm12 mutation also

71

leads to lower expression of $A\beta bm12:A\alpha b$, by directly decreasing the efficiency of $\alpha:\beta$ heterodimer formation and/or surface membrane expression (30), and this might contribute to the suppressed response to the dominant epitopes. It is also possible that effective generation of specific regulator (suppressor) cells in bm12 could inhibit the T cell responses to dominant epitopes, or that T cells responding to another AChR epitope(s) in bm12 suppress the response to peptides $\alpha146$-162 or $\alpha182$-198. The preservation in bm12 of lymphocyte proliferative response to the other peptide recognized by B6 T cells (i.e., peptide $\alpha111$-126) indicates that the bm12 possesses T cells responding to epitopes within 111-126, and also that the Ia molecule of bm12 is capable of binding and presenting this peptide to bm12 T cells.

Our data on the reduced bm12 lymphocyte response to peptide $\alpha146$-162 agree with the results of Infante et al. (31) and Bellone et al. (32), where the latter used a related peptide $\alpha150$-169. However, Infante et al. failed to observe any difference between B6 and bm12 T-cell responses to peptide $\alpha182$-198, while Bellone et al. failed to induce B6 splenic CD4+ cells to proliferate to the immunodominant epitopes of AChR-α subunit when immunized three times with T-AChR. However, Bellone, et al. could induce B6 CD4+ cells to proliferate to related peptides $\alpha150$-169 and $\alpha181$-200, when primed with the respective peptides (32).

We failed to induce EAMG by immunizing EAMG susceptible B6 (H-2b) mice with either peptide $\alpha146$-162 or $\alpha182$-198. However, peptide $\alpha146$-162 and $\alpha182$-198 had the capacity to stimulate T cells as well as B cells. But the antibodies produced after immunization with either of these peptides were not overtly pathogenic (antigenic modulation/complement activation/inhibiting acetylcholine binding). Epitope-specific neonatal tolerance has been achieved (33) with synthetic peptides. Therefore, we decided to further investigate the role of epitopes within peptide $\alpha146$-162 by inducing neonatal tolerance to this peptide. When mice were neonatally tolerized with peptide $\alpha146$-162, a marked reduction in the incidence of clinical MG and anti-AChR antibody response was observed, while tolerization with peptide $\alpha182$-198 of the human sequence had insignificant effects on these parameters. This finding is the first indirect evidence to suggest the involvement of the T-AChR region $\alpha146$-162 in the development of clinical EAMG in mice. This region within the AChR-α subunit may be involved in the generation of specific helper T cells which could provide help for the production of pathogenic antibodies against other B cell epitopes of the AChR molecule through T-T or T-B cell intersite influences (epitope linkages) (34). Therefore, other epitopes residing in the AChR-α,β, c, and d subunits could also be involved in EAMG pathogenesis. It is relevant to point out here that peptide $\alpha146$-162 of the human AChR-α subunit is frequently recognized by human autoimmune T cell lines from MG patients (15) and may well be an interspecies pathogenic T cell epitope. This study does not rule out the possible pathogenic role of other epitopes within AChR molecule.

The mechanism involved in the induction of tolerance to autoantigens implicated in autoimmune diseases is still obscure. Whether or not tolerance to autoantigens implicated in autoimmune diseases deletes T cells bearing particular TCR genes is not known. A number of experimental models have suggested activation of antigen-specific, or non-specific suppressor T cells subsequent to neonatal induction of tolerance to antigen (35). It is of immense importance to study the mechanism involved in the development of tolerance to autoantigens in order to develop specific therapy for autoimmune diseases. Neonatal tolerance to autoantigens, like myelin basic protein (MBP), and its dominant T cell determinants has been achieved in the animal model of experimental autoimmune encephalomyelitis (EAE) (36,37). Antigen or peptide specific tolerance was achieved by neonatally injecting MBP or its encephalotogenic determinants emulsified in incomplete Freunds adjuvant. We have achieved epitope specific tolerance in EAMG by neonatal injection

72

of soluble AChR or peptide α146-162. Neonatal tolerance to MBP-1-9NAC peptide decreased the incidence of 1-9NAC peptide induced EAE. However, neonatal tolerance to 1-9NAC peptide failed to prevent EAE induced by the whole MBP. This suggested that there is at least one additional disease-inducing determinant on MBP, other than 1-9NAC, which cross reacts with the autologous protein at the T cell level. It is also very likely that other determinants · in the AChR molecule may be involved in the pathogenesis. However, peptide α146-162 appears to be one of the dominant regions recognized by T cells and is most probably disease inducing in mice with I-Ab genotype (C57BL6). Prevention of clinical EAMG by neonatal tolerance to AChR or peptide α146-162 could have been achieved by, 1) clonal deletion or anergy of AChR or peptide reactive autoimmune clones involved in the pathogenesis, 2) positive selection (suppressor cells), of cells which were able to suppress an autoimmune response to AChR when adult immunized with AChR. At present we do not know how long tolerance to T-AChR peptide α146-162 could last. The immunological mechanisms involved in the suppression of EAMG development by neonatal tolerization with soluble AChR or with peptide α146-162 are being investigated. Passive tolerance to α146-162 is being attempted in order to ameliorate ongoing EAMG in mice.

ACKNOWLEDGEMENTS

The secretarial assistance of Charlene Hoff and Rosa Lopez is appreciated. This work was supported by a grant (#NS26280) from the National Institute of Health, in part by the Welch Foundation due to the award to MZA of the Robert A. Welch Chair of Chemistry, and a grant from the Muscular Dystrophy Association and the Sealy Smith Endowment funds to PC.

REFERENCES

1. S.H. Appel, R.R. Almon, and N.L. Levy. Acetylcholine receptor antibodies in myasthenia gravis. N. Engl. J. Med. 293:760 (1975).
2. A.G. Engel. Myasthenia gravis and myasthenic syndromes. Ann. Neurol. 16:519(1984).
3. J. Safwenberg, L. Hammerstrom, J.B. Lindbluom, G. Matell, E. Moller, P.O. Osterman, and S.I.E. Smith. HLA-A, -B, and -D antigens in male patients with myasthenic gravis. Tissue Antigens 12:136(1978).
4. J. Bell, S. Smoot, C. Newly, K. Toyka, L. Rassenti, K. Smith, R. Hohlfeld, H. McDevitt, and L. Steinman. HLA-DQ beta-subunit polymorphism linked to MG. Lancet 1058(1988).
5. P. Christadoss, V.A. Lennon, C.J. Krco, E.H. Lambert, and C.S. David. Genetic control of autoimmunity to acetylcholine receptors: role of Ia molecules. Ann NY Acad Sci. 377:258(1981).
6. P. Christadoss. Immunogenetics of experimental autoimmune Myasthenia gravis. Critical Reviews in Immunology 9:247(1989).
7. P. Christadoss, J.M. Lindstrom, R.W. Melvold, and N. Talal. Mutation at I-Aß chain prevents experimental autoimmune Myasthenia gravis. Immuno-genetics 21:33(1985).
8. P. Christadoss, C.S. David, and S. Keve. I-Aαk transgene pairs with I-Aßb gene and protects C57BL10 mice from developing autoimmune myasthenia gravis. Clinical Immunology and Immunopathology 62:235(1992).
9. P. Christadoss, C.S. David, M. Shenoy, and S. Keve. Eαk transgene in B10 mice suppresses the development of myasthenia gravis. Immunogenetics 31:241(1990).

10. R. Hohlfeld, K.V. Toyka, K. Heininger, H. Grosse-Wilde, and I. Kalies. Autoimmune human T lymphocytes specific for acetylcholine receptor. Nature 310:244(1984).

11. R. Hohlfeld, I. Kalies, B. Kohleisen, K. Heininger, B.M. Conti-Tronconi, and K.V. Toyka. Myasthenia gravis: stimulation ofantireceptor autoantibodies by autoreactive T-cell lines. Neurology 36:618(1986).

12. P. Christadoss, V.A. Lennon, and C.S. David. Genetic control of experimental autoimmune myasthenia gravis in mice. 1. Lymphocyte proliferative response to acetylcholine receptor is under H-2 linked Ir gene control. J. Immunol. 123:2540(1979).

13. P. Christadoss, V.A. Lennon, C.J. Krco, and C.S. David. Genetic control of experimental autoimmune myasthenia gravis in mice. III. Ia molecules mediate cellular immune responsiveness to acetylcholine receptors. J. Immunol. 138:1141(1991).

14. P. Christadoss, and M.J. Dauphinee. Immunotherapy for myasthenia gravis: a murine model. J. Immunol. 136:2437(1986).

15. M. Oshima, T. Ashizawa, M. Pollack, and M.Z. Atassi. Autoimmune T cell recognition of human acetylcholine receptor: the site of T cell recognition in myasthenia gravis on the extracellular part of the α subunit. Eur. J. Immunol. 20:2563(1990).

16. T. Yokoi, B. Mulac-Jericevic, J. Kurisaki, and M.Z. Atassi. T lymphocyte recognition of acetylcholine receptor: Localization of the full T cell recognition profile on the extracellular part of the α chain of Torpedo californica acetylcholine receptor. Eur.J.Immunol. 17:1697(1987).

17. A.R. Pachner, F.S. Kantor, B. Mulac-Jericevic, and M.Z. Atassi. An immunodominant site of acetylcholine receptor in experimental myasthenia gravis mapped with T lymphocyte clones and synthetic peptides. Immunol. Letters 20:199(1989).

18. P. Christadoss, C.J. Krco, V.A. Lennon, and C.S. David. Genetic control of experimental autoimmune myasthenia gravis in mice. II. Lymphocyte proliferative response to acetylcholine receptor is dependent on Lyt 1+23- cells. J. Immunol. 126:1646(1981).

19. P. Christadoss, J. Lindstrom, and N. Talal. Cellular immune response to acetylcholine receptors in murine experimental autoimmune myasthenia gravis. Inhibition with monoclonal anti-I-A antibodies. Cell. Immunol. 81:1(1983).

20. G. Widera, and R.A. Flavell. The nucleotide sequence of the murine I-Eßb immune response gene: evidence for gene conversion events in class II genes of the major histocompatability complex. EMBO J. 3:1221(1984).

21. M. Dinaro, U. Hammerling, L. Rask, and P.A. Peterson. The Eßb gene may have acted as the donor gene in a gene conversion--like event generating the Aßbm12 mutant. EMBO J. 3:2029(1984).

22. K.R. McIntyre, and J.G. Seidman. Nucleotide sequence of mutant I-Aßbm12 gene is evidence for genetic exchange between mouse immune response genes. Nature 308:551(1984).

23. B. Mulac-Jericevic, and M.Z. Atassi. α-Neurotoxin binding to acetylcholine receptor: localization of the full profile of the cobratoxin-binding regions on the α-chain of Torpedo califonica acetylcholine receptor by a comprehensive synthet strategy. Prot. Chem. 6:365(1987).

24. B. Mulac-Jericevic, J.I. Kurisaki, and M.Z. Atassi. Profile of the continous antigenic regions on the extracellular part of the α chain of an acetylcholine receptor. Proc. Nat. Acad. Sci. USA 84:3633(1987).

25. J. Lindstrom, B. Einarson, and S. Tzartos. Production and assay of antibodies to acetylcholine receptor. Methods Enzymol. 74:432(1981).

26. B. Mulac-Jericevic, T. Manshouri, T. Yokoi, and M.Z. Atassi. Region of α neurotoxin binding on the extracellular part of the α subunit of the human acetylcholine receptor. J. Prot. Chem. 7:173(1988).

27. P. Christadoss, J.M. Lindstrom, N. Talal, C.R. Duvic, A. Kalantri, and M. Shenoy. Immune response gene control of lymphocyte proliferation induced by acetylcholine receptor specific helper factor derived from lymphocytes of myasthenic mice. J. Immunol. 127:1845(1986).

28. A.S. Rosenthal. Determinant selection and macrophage function in genetic control of the immune response. Immunol. Rev. 40:136(1978).

29. B. Benacerraf. A hypothesis to relate the specificity of T lymphocytes and the activity of I region-specific Ir genes in macrophages and B lymphocytes. J. Immunol. 120:1809(1978).

30. F. Ronchese, M.A. Brown, and R.N. Germain. Structure--function analysis of the Aßbm12. J. Immunol. 139:629(1978).

31. A. Infante, P. Thompson, K.A. Krolick, and K.A. Wall. Determined selection in murine experimental autoimmune myasthenia gravis. Effects of the bm12 mutations on T cell recognition of acetylcholine receptor epitopes. J. Immunol. 146:2977(1991).

32. M. Bellone, N. Ostlie, S. Lei, Xiao-Dong Wa, and B.M. Conti-Tronconi. The I-Abm12 mutations, which confers resistance to experimental myasthenia gravis, drastically affects the epitope repertoire of murine CD4+ cells sensitized to nicotin-ic acetylcholine receptors. J. Immunol. 147:1484(1991).

33. C.R. Young, and M.Z. Atassi. T lymphocyte recognition of sperm whale myoglobin specificity of T cell recognition following neonatal tolerance with either myoglobin or synthetic peptides of an antigenic site. J. Immunogen. 10:161(1983).

34. M.Z. Atassi, S. Yokota, S.S. Twining, H. Lehman, and C.S. David. Preliminary communication: Genetic control of the immune response to myoglobin. VI. Inter-site influences in T lymphocyte proliferative response from analysis of cross-reactions of ten myoglobins in terms of substitution in the antigenic sites and in environmental residues of the sites. Mol. Immunol. 18:945(1981).

35. G.J.V. Nossal. Cellular mechanism of immunolical tolerance. Ann Rev Immunol. 1:33(1983).

36. Y. Qin, D. Sun, M. Groto, R. Meyermann, and H. Wekerle. Resistance to experimental autoimmune encephalomyelitis induced by neonatal tolerization to myelin basic protein: clonal elimination vs. regulation of autoaggressive lymphocytes. Eur J Immunol. 19:373(1989).

37. J. Clayton, G.M. Gammon, D.G. Ando, D.H. Kono, L. Hood, and E.E. Sercarz. Peptide-specific prevention of experimental allergic encephalomyelitis. Neonatal tolerance induced to the dominant T-cell determinant of myelin basic protein. J Exp Med 169:1681(1989).

T CELL REACTIVITY TO SELF AND ALLOGENEIC MHC-PEPTIDES

Zhuoru Liu, Paul Harris, and Nicole Suciu-Foca

College of Physicians and Surgeons of
Columbia University
Department of Pathology
630 West 168th Street
New York, New York 10032

INTRODUCTION

Although considerable knowledge about the basis of immune tolerance to self has been recently gained, there is much to be learned about the mechanism of self-non-self discrimination. This is particularly important in the case of MHC antigens since these molecules serve as restriction elements for T cell recognition of endogenous and exogenous peptides. It is presumed, although not yet proven, that T cells capable of reacting to self-MHC molecules are deleted from the repertoire during intra-thymic differentiation. In contrast, T cells recognizing allogeneic MHC molecules and/or peptides derived from their processing are expected to be present, at a relatively high frequency, in the peripheral repertoire.

In previous studies, using 20-meric synthetic peptides derived from the amino acid sequence of the DRB1*0101 chain, we have demonstrated that: 1) DR1 peptides bind with high affinity to both syngeneic (DR1) and allogeneic (DR11) molecules and elicit T cell reactivity to processed DR1 molecules [1,2]; 2) T cells responding to a synthetic DR1 peptide, recognize this peptide in context of syngeneic MHC-class II molecule(s) and exhibit a restricted TCR-VB gene usage; 4) the precursor frequency of T cells reacting to syngeneic and allogeneic HLA-DR is quite similar. This latter observation implies that anti-self MHC reactive T cells are not deleted from the repertoire and that they may, potentially, contribute to autoreactivity. Similarly, T cells recognizing processed forms of allogeneic MHC molecules may contribute to allograft rejection. Since alloreactivity is caused primarily by T cells able of recognizing directly the allogeneic MHC molecule and/or the peptide bound to it, we have investigated the relative contribution of indirect recognition to T cell responses in MLC.

MATERIALS AND METHODS

Peptides: Synthetic peptides corresponding to residues 1-20, 21-42, 43-60 and 61-80 of the first domain of the DRB1*0101 chain were prepared as previously described [1].

Alloreactive T cell lines: PBMCs from an individual carrying the DRB1*01101/1102 allele were stimulated in vitro with irradiated PBMCs carrying the DRB1*0101/0301 allele. Following 10-days of culture, T cells were harvested, washed and restimulated with a cocktail containing equal amounts (1μg/ml) of the four synthetic peptides. Irradiated PBMCs of responder-origin and rIL2 (5 units/ml) were added to the culture. The culture was restimulated under the same conditions at 10-days interval. T cell reactivity to allogeneic stimulating cells and to DR1-peptides was determined in blastogenesis assays as previously described [2,3].

RESULTS AND DISCUSSION

Precursor Frequency of T Cells Involved in Indirect and Direct Recognition. Responding T cells, that have been primed for 10 days in 1^0MLR with DR1 positive stimulating cells were tested at concentrations ranging from 4×10^4 to 10^2/culture for reactivity to DR1 stimulating cells, and to the cocktail of DR1 peptides. This Limiting Dilution Analysis (LDA) showed that the frequency of T cells responding to DR1 positive stimulating cells was 1/328. The frequency of T cells responding to the peptide mixture in the presence of autologous APCs, was 1/43,992. Hence, the precursor frequency of cells involved in indirect recognition is approximately 100 fold lower than that of cells using the direct allorecognition pathway.

Recognition of DR1 Peptides in Context of Syngeneic APCs. LDA studies revealed that allostimulation of T cells with DR1 positive cells results in the generation of clones capable to respond to synthetic DR1 peptides presented by

TABLE 1: TCL Reactivity to DR1 allopeptides

| | | 2^0 Stimulus | | |
| | DR1-peptides | | DR1 PBMC | |
Responding T Cells $1^0$10 Days-MLC	with autologous APC	without autologous APC	with autologous APC	without autologous APC
Precursor frequency	1/43,992			1/328
mean cpm of triplicates	2,195	615	4,141	629

autologous APCs. In order to determine whether such peptides also result from the natural processing of the DR1 molecule, the following experiments were performed. Responding T cells which have been primed in 1°MLR with DR1 positive PBMCs, were restimulated four consecutive times with the DR1 peptide mixture, in cultures containing irradiated autologous PBMCs and rIL-2. The resulting T cell line (TCL) was challenged with: the peptide mixture and with allogeneic, DR1 positive stimulating cells, in cultures with or without APCs autologous to the responder.

Blastogenesis assays showed that in the presence of autologous APCs, the TCL reacted both to the mixture of synthetic DR1 peptides and to allogeneic, DR1-positive stimulating cells. In the absence of autologous APCs, however, the TCL failed to react against DR1 positive stimulating cells. This indicates that the TCL recognizes peptides resulting from the processing of the DR1 molecule by syngeneic APCs. The synthetic DR1 peptides used for in vitro immunization were, therefore, representative of peptides resulting from the natural processing of the DR1 molecule by the responder's APC (Table 1).

The finding that allostimulation results in the generation of T cells capable of indirect recognition of processed allopeptides has several important implications. First, it explains the development of anti-HLA antibodies following allostimulation by pregnancies, transfusions or transplantation. T helper cells recognizing processed allopeptides may provide the lymphokines required for the differentiation and maturation of B cells with Ig receptors for the corresponding HLA alloantigen. Indeed, the developement of anti-HLA antibodies cannot be explained by the direct recognition pathway. Second, the fact that alloreactivity obeys the rules of MHC-restricted T cell recognition, opens new avenues for specific suppression of chronic antibody mediated rejection, which is the major cause of transplant failure. Thus, for example MHC-class II blockade by mAbs or competitive peptides and/or anti-idiotypic suppression of allopeptide specific T helper cells may be attempted.

Finally, indirect recognition of MHC peptides may be responsible for the development of anti-self Ia immunity as observed in syngeneic GVHD and in other autoimmune conditions [4].

REFERENCES

1. Harris, P., Liu, Z., and Suciu-Foca, N. MHC class II binding of peptides derived from HLA-DR1. J. Immunol. 148:2169 (1992).
2. Liu, Z., Braunstein,N., and Suciu-Foca, N. T cell recognition of allo-peptides in context of syngeneic MHC. J.Immunol. 148:35 (1992).
3. Liu, Z., Y. Sun, Y. Xi, P. Harris and N. Suciu-Foca. T cell recognition of self-Human Histocompatibility Leukocyte Antigens (HLA)-DR Peptides on Context of Syngeneic HLA-DR Molecules. J.Exp. Med. 175: 1663 (1992).
4. King, D.W., Reed, E., and Suciu-Foca, N. Complexes of soluble HLA antigens and anti- HLA autoantibodies in human sera: Possible role in maintenance of self-tolerance. J. Immunol.Res. 8:249 (1992).

ANTIRIBOSOMAL ANTIBODIES IN SLE, INFECTION, AND FOLLOWING DELIBERATE IMMUNIZATION

Keith B. Elkon[1], Eloisa Bonfa[2],
Herbert Weissbach[3], and Nathan Brot[3]

[1]The Hospital for Special Surgery
Cornell University Medical Center, New York, NY
[2]University of Sao Paolo, Sao Paulo, Brazil
[3]Roche Institute of Molecular Biology
Roche Research Center
Nutley, NJ 07110
Tel No. (201) 235-4746

INTRODUCTION

It has long been known that patients with systemic lupus erythematosus (SLE) and related diseases have, in addition to antinuclear antibodies (ANA), autoantibodies directed against cytoplasmic constituents[1,2]. In SLE, these antibodies are predominantly directed against ribosomes[3-5]. In 1985, two groups of investigators independently identified the ribosomal P proteins as the major protein antigens recognized by anti-ribosomal antibodies in SLE[6,7]. In addition, evidence for antibodies that bound to ribosomal RNA (rRNA) was presented[6,7]. Although antibodies against other ribosomal proteins, S10[8], L5[9], and L12[10] have been detected, these are present in a very small number of patients with SLE.

As in the case of all autoantibodies in SLE, the mechanism of induction is poorly understood. However recent studies of anti-P induction in experimental animals[11] and in infectious diseases[12-13] offer potential clues to understanding anti-P production in spontaneous autoimmune disease.

ANTI-RIBOSOMAL P PROTEIN ANTIBODIES

a.   The P Proteins

The P (phospho) proteins, P0, P1, and P2, also known as the A (alanine-rich) proteins, are 3 proteins located on the large (60S) subunit of eukaryotic ribosomes. Since these proteins were identified independently in different species, they have been given different names (L7/L12 in E. coli, YP in yeast, eL12 in A. salina and rpA in Drosophila). For simplicity, all of these eukaryotic proteins will be referred to as P proteins. The apparent molecular weights of human P0, P1, and P2 on SDS-polyacrylamide gels are 38, 19, and 17 kD, respectively[6]. Since the sizes of the P proteins, as predicted by their CDNAS[14] are smaller than those indicated above, these proteins migrate aberrantly on polyacrylamide gels. The abnormal migration may be due to phosphorylation or the unusual amino acid compositions of these proteins. Hasler et al.[15] have recently reported that a ribosome-bound protein kinase which relatively selectively phosphorylates the P proteins has properties similar to casein kinase II. Although P0, P1, and P2 share only minimal sequence homology with their prokaryotic homologues L10 and L7/L12[16], the P proteins are highly conserved in eukaryotes. This is not uniformly true amongst ribosomal proteins and suggests an important function for these proteins (see below).

Although the topography of the P proteins on the ribosomes has not been mapped by immune electron microscopy, it seems most likely, from studies of the *E. coli* homologues L10 and L7/L12[17], that the P proteins are located in the highly exposed region on the stalk of the 60S subunit (**Fig. 1**). The detection of P0, P1, and P2 in a low molecular weight complex following dissociation of ribosomes[18] as well as direct cross-linking studies[19] suggest that homodimers of P1 and of P2 are bound to P0 (Fig. 1). The most highly conserved portions of the P proteins are their amino-(N) and carboxyl-(C) termini[20]. However, only the C-terminus is common to P0, P1, and P2.

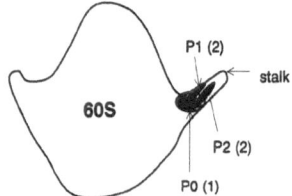

Figure 1. Proposed topography of the P proteins
P1 and P2 exist as homodimers bound at their amino termini to P0 on the stalk of the large (60S) ribosomal subunit. The topography is based on immune electron microscopy of the E. coli homologues as well as cross-linking studies of mammalian P proteins[19]. The numbers in parenthesis refer to the number of copies of each protein on the ribosome.

The exact functions of the P proteins on the ribosome are not fully understood. However, P1/P2, like L7/L12, are thought to be required for all 3 phases of protein synthesis viz. initiation, translocation, and termination[21-23]. Selective elution of P1 and P2 from ribosomes results in the total loss of GTPase activity and protein synthesis[21,23]. Similarly, monoclonal[24] and polyclonal[25] anti-P sera inhibit GTPase activity and protein synthesis in general.

b.   Epitope Recognition by Anti-P Antibodies

Using affinity purification of anti-P0 and anti-P1/P2 antibodies from nitrocellulose, it was demonstrated that anti-P0 antibodies cross-react with P1/P2 and vice versa[6]. Since direct sequencing revealed that the C-terminal 22 amino acids of

82

P1 and P2 from Artemia salina (brine shrimp) proteins are identical[26], we synthesized N- and C-terminal peptides of P2 (referred to as N-22 and C-22 respectively) and tested SLE antisera for reactivity. As predicted, the SLE anti-P antibodies bound to the C- but not the N-terminal peptides[18]. Surprisingly, the C-terminal peptide completely absorbed anti-P binding to P0, P1, and P2 on immunoblots indicating that anti-P antibodies bind to a single linear (sequential) epitope on all 3 P proteins. However, since the P protein complex is disrupted and the proteins are partially denatured on immunoblots, we cannot exclude the possibility that antibodies to conformational epitopes exist in these sera. The inability of the C-22 peptide to absorb anti-P binding to the P proteins on the intact ribosome support this assertion[18]. Interestingly, the immunodominant epitope recognized by lupus anti-P sera and the single phosphorylation site on P2 both map to the conserved C-terminus of the antigen[15].

Fine specificity epitope mapping revealed that the smallest epitope recognized by all anti-P sera comprised 11 amino acids but that many sera bound to heptapeptides (**Fig. 2**)[20]. Human anti-P sera showed similar binding to human and A. salina C-22 peptides whereas MRL/lpr mouse anti-P antibodies demonstrated higher binding to self compared to A. salina C-22[11].

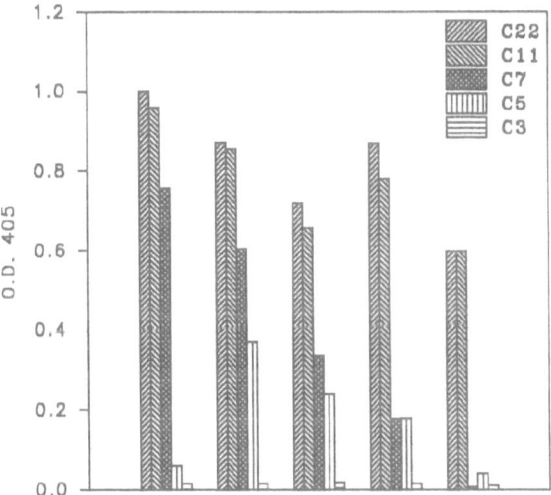

Figure 2. Fine specificity of human anti-P autoantibodies using synthetic peptide antigens. Five different representative SLE anti-P sera were tested for binding to synthetic peptides conjugates by ELISA. C 3...C 22 refer to peptides with homology to the carboxyl (C) -terminal amino acids of 3..22 residues of human P2[14]. Modified from [20].

c.   Clinical and Serological Associations of Anti-P

Anti-P antibodies were initially detected in 10-20% of SLE patients[7,27]. Immunoblot analysis revealed that anti-P is highly specific for SLE[27]. The specificity of anti-P has subsequently been confirmed in several studies[28-30] and is similar to the marker antibodies, anti-dsDNA, and anti-Sm. Anti-P antibodies may be present in sera from patients who satisfy 4 or more criteria for SLE but who do not have anti-

DNA antibodies "ANA positive"[27]. For this reason, they are useful additional serological markers of SLE. Since anti-P and anti-Sm both appear to be good serological markers of SLE, an association between these two autoantibodies would be expected. Elkon et al.[31] observed a higher frequency and level of anti-P antibodies in patients with anti-Sm. Using less sensitive tests, a negative association between anti-P and anti-La was reported[32].

Retrospective clinical evaluation of SLE patients initially failed to reveal significant differences in the clinical associations between patients with and without anti-P antibodies[27]. However, it was noted that several patients with lupus psychosis had anti-P. Using the synthetic C-terminal peptide as antigen and a sensitive radioimmunoassay, Bonfa et al.[32] reported that the majority of patients with lupus psychosis had anti-P and that, in two patients studied longitudinally, the levels of anti-P were elevated prior to and during the exacerbation. Individual case reports[33] and a large, multicenter study[28] observed similar findings although Schneebaum et al. emphasized the strong relationship between anti-P and affective psychosis[28]. Van Dam et al., in a study conducted by questionnaire, did not find an association between anti-P and lupus psychosis[29]. Although some discrepancies are, most likely explained by the retrospective nature of all of these studies, other differences may be due to the criteria used to define lupus psychosis, the tests for detection of anti-P and the time at which serum was drawn. In the study reported by Bonfa et al.[32], patients had clinical manifestations for at least 2 weeks, required hospital admission and most of the patients had other signs of lupus activity.

The elevation in anti-P levels observed in lupus psychosis are potentially important from the point of view of etiology and diagnosis of CNS lupus. Although a direct role for anti-P antibodies in the induction of lupus psychosis remains to be proven, two recent findings are compatible with a pathogenic role for this antibody. In a comparison of anti-P titers / mg IgG in 8 paired serum/CSF samples from patients with neuropsychiatric lupus, a several hundred-fold depletion of anti-P in the CSF was observed - suggesting anti-P absorption or deposition in the CNS[28]. Koren et al.[34], using immunofluorescence, confocal microscopy and electron microscopy demonstrated expression of a P protein epitope *on the surface* of human neuroblastoma cells. Although the presence of a similar epitope on normal neurones remains to be demonstrated, direct binding of anti-P to neuronal cells could explain some forms of neuropsychiatric disease in SLE. As discussed previously[32], CNS injury would only occur in those patients with disruption of the blood brain barrier.

Since the majority of anti-P positive patients will not have CNS disease at the time the antibody is detected, and, since not all patients with lupus psychosis have anti-P[32], a single positive test is of limited value. In contrast, rising levels of anti-P in the context of a behavioral disorder, supports the diagnosis of lupus psychosis. Recent studies by Sato et al.[30] suggest that anti-P levels may also reflect other forms of clinical activity. These authors detected anti-P in 42% of SLE patients with active disease (nephritis, lupus psychosis, serositis, thrombocytopenia, pneumonitis) but in only 8% of patients without active disease.

Anti-28S rRNA Antibodies

Approximately 30-60% of sera obtained from patients with SLE and some other autoimmune diseases bind to single and/or double-stranded RNA[35-37]. In most cases, no sequence or conformational specificity of these antibodies are evident. Evidence for antibodies directed against ribosomal r(RNA) was also obtained many years ago using relatively insensitive methods for detection[3]. Recently, Uchiumi et al. identified a single SLE patient serum that selectively immunoprecipitated 28S rRNA[38]. Moreover, using cloned fragments of human and mouse rRNA, the epitope recognized by the anti-28S antibodies was mapped to a 59 nt fragment, nt 1943-2002[38]. This region of 28S rRNA has 100% sequence identity with yeast rRNA[39] and is thought to represent the GTPase center of the ribosome[40,41]. GTP cleavage is an essential step for the activity of initiation, elongation, and termination factors on the ribosome[42].

Since the P proteins in eukaryotes and L7/L12 in E. coli are also required for GTPase activity[21,22,23] and since autoantibodies are frequently directed against several antigens in a RNP particle[42], we determined whether anti-28S sera were present in a higher frequency in patients with anti-P. Approximately 75% of anti-P positive sera contained antibodies to the 28S rRNA fragment synthesized by in vitro transcription whereas only 8% of SLE sera without anti-P contained these antibodies[43] (Fig. 3). This binding was specific for the 1943-2002 nt fragment since anti-P sera did not show the same high frequency of binding to nt 1943-2002 transcribed in the antisense orientation or to a control fragment (nt 4430-4528) of 28S rRNA. Antibody RNase protection studies revealed that most anti-28S antibodies tested also bound to nt 1943-2002 although some sera protected a slightly smaller fragment within the same region[43]. Using several site-directed mutants of the rRNA antigen, we observed that recognition was remarkably sensitive to changes in secondary structure. Overall, these studies support the idea that autoantibodies in SLE are directed against structurally-related components of RNP complexes most likely reflecting immunogenicity of these complexes (see below).

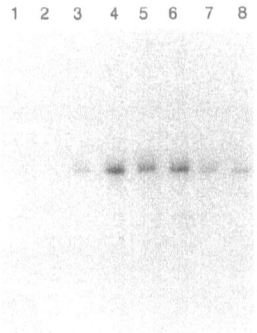

Figure 3. Anti-28S rRNA autoantibodies in SLE. A fragment of 28S rRNA (nucleotides 1922-2020) was amplified from Hela genomic DNA by the polymerase chain reaction. The corresponding RNA was transcribed in vitro in the presence of $[\alpha\text{-}^{32}P]CTP$ and incubated with IgG bound to Sepharose-protein A beads. Immunoprecipitation was performed with 24 anti-P positive (examples are shown in lanes 1,3-8) and 24 anti-P negative (example, lane 2) SLE sera. Following immunoprecipitation, the RNA was isolated by phenol/chloroform extraction, resolved on a 7 M urea polyacrylamide gel and subjected to autoradiography. Experimental details are provided in[43].

Antiribosomal Antibodies (ARA) in Murine Lupus

Several inbred mouse strains spontaneously develop a disease very similar to human SLE (reviewed in 44). Amongst these strains, the MRL mouse has previously been noted to produce antibodies against ds-DNA[44] cardiolipin[45], Sm[46], RNP, and histones[47]. We tested MRL mice for ARA and observed that approximately 10% of MRL/lpr mice aged 3 months or over had anti-P and 10% had anti-S10 antibodies[8].

Remarkably, the MRL antisera recognized the same epitope as human anti-P sera viz. the C-terminus[48]. ARA were not, however, detected in the serum of NZB/W F1 mice nor in a small number of BXSB mice[48].

Detection of a similar profile of ARA in human SLE and MRL mice suggests a similar mechanism for induction of autoantibodies. These similarities could reside in the B or T cell repertoire, the initiating stimulus or in defects in regulation. Investigation of these mechanisms in MRL mice may provide important clues to the etiology of SLE in humans.

Experimental Induction of Anti-Ribosomal Antibodies

As discussed above, most of the evidence suggests that the antigens themselves drive autoantibody production (reviewed in 49). However, this evidence is indirect and studies with non-protein antigens such as DNA[50] and cardiolipin[51] have failed to induce autoantibodies by immunization with the native antigens. To determine whether immunization with intact ribosomes would induce ARA in autoimmune and non-autoimmune strains of mice, we immunized young MRL (prior to the onset of autoimmunity) and H2-matched (C3H) mice with mouse ribosomes[11]. Immunization by several different protocols failed to induce ARA production. This finding suggests that high levels of self antigen are insufficient to break tolerance even in an autoimmune strain of mouse[11]. In contrast, immunization with xenogeneic ribosomes (from A.salina) did induce ARA against several foreign ribosomal proteins. Anti-P antibodies were prominent amongst the induced ARA (**Fig. 4**). Both MRL/lpr and MRL/+ mice produced higher levels of anti-P than the normal strains indicating enhanced reactivity of MRL B cells[52].

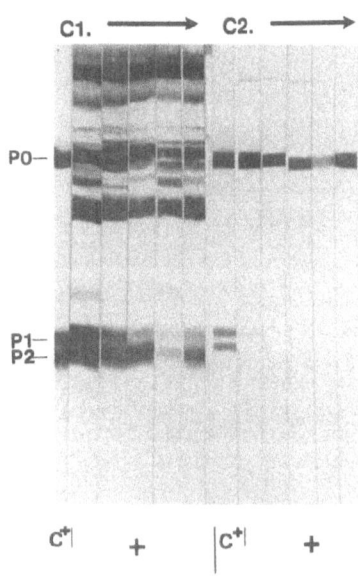

Figure 4. Experimental induction of anti-P antibodies in mice. MRL/++ mice were immunized three times with A. salina ribosomes. Sera were then diluted 1/500 and tested for antiribosomal antibodies and autoantibodies by immunoblotting. The nitrocellulose contained either total A. salina (panel C1) or total mouse (panel C2) ribosomal proteins. C+ refers to sera used as a positive control. The positions of P0, P1 and P2 are indicated. Modified from[11].

When sera from mice immunized with Artemia ribosomes were tested for reactivity with self (mouse) ribosomes, only autoantibodies against P0 were detected. This suggests that tolerance to the P proteins, or P0 in particular, is either more difficult to maintain or that, because of the very high level of conservation of the C-termini, cross-reactions are inevitable. Since 5 copies of the C-termini of the P proteins are present on the ribosome (see Fig. 1), we determined whether multivalency was required for anti-P antibody production. Selective elution of P1 and P2 dimers from Artemia ribosomes prior to immunization resulted in the complete loss of immunogenicity against P0. This finding suggests either that the multivalency of the P proteins or the presence of P1/P2 is uniquely required for immunogenicity[11]. Although we cannot, at present distinguish between these two possibilities, we favor the idea that a multivalency determines immunogenicity since multivalency has been shown to be an important factor in antigen selection in graft versus host disease (GVHD)[52]. It is of interest that autoantibody specificities apparently identical to anti-P and anti-Sm have been detected in GVHD produced in B10.A (4R) X B10.A (2R) F1 hybrids[53].

Anti-P Antibodies in Chaga's Disease

In an "experiment of nature", humans infected with the protozoan hemoflagellate, T. cruzi, are immunized with the T. cruzi antigens, including ribosomes. Although all patients develop antibodies against T. cruzi ribosomes, patients who develop Chaga's disease (a cardiomyopathy thought to be due to an autoimmune response to cardiac tissue), have significantly higher levels of antibodies against the T. cruzi protein, P2 (JL5)[12]. Analogous to the immunized mice, these patients also develop anti-P autoantibodies that bind to human P2 on Western blots[12] and to the human C-22 synthetic peptide in an ELISA[13]. Since the titers of these autoantibodies are lower than in SLE and may require affinity purification to be identified[12], it seems likely that these are cross-reactive antibodies similar to those induced in experimental mice. Whether or not these antibodies play a role in tissue injury, elevation in patients with autoimmune cardiomyopathy may prove to be useful as a marker of an ongoing deleterious immune response.

Summary and Conclusions

ARA occur in approximately 10% of randomly selected SLE patients but in up to 40% of patients with active disease. Anti-P antibodies appear to be a highly specific diagnostic marker for SLE since they are rarely detected in other multisystem autoimmune disorders. ARA are most frequently directed against the P proteins and the shared conserved C-terminus of the P proteins is immunodominant in almost all sera tested. Anti-P antibodies increase in titer in patients with active disease and have been reported to be detected more frequently in patients with severe behavioral disturbances. This may be particularly true of patients with affective disorders. The clinical utility of serological tests for anti-P in central nervous system lupus must await large, prospective studies.

Other ARA antibodies have been detected in patients with SLE. These antibodies include anti-28S rRNA, anti-S10, and anti-L12. In all cases, the frequency with which these antibodies are detected is increased in sera containing anti-P. The P proteins and the 28S rRNA epitope play essential, but as yet undefined, roles in GTPase activity on the ribosome. The L12 protein is the mammalian homologue of the E. coli and yeast proteins known to bind to the 28S rRNA epitope. These

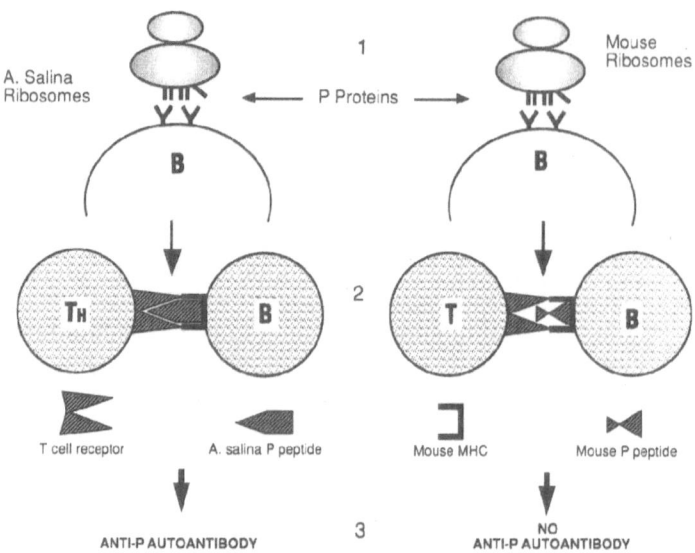

Figure 5. Hypothetical model of anti-P autoantibody induction. 1. The multivalent P protein complex on the 60S ribosomal subunit of either mouse or A. salina ribosomes cross-links surface Ig on resting B cells. The cells become partially activated. 2. According to the model proposed by Lanzavecchia[55], B cells present P protein peptides to T cells. Mice of the appropriate MHC type have T cells that recognize the A. salina peptide-MHC complex (left panel) and provide the necessary second signal for antibody production 3. In the case of mice immunized with mouse ribosomes (right panel), autoreactive T cells are either not present or T suppressor cells actively suppress further antibody production by the activated B cell. Reproduced from 11.

findings indicate that some SLE patients produce autoantibodies against multiple components of a functionally related domain of the ribosome. This, in turn, supports the notion that the ribosome initiates and/or maintains autoantibody production.

Despite the evidence supporting an antigen driven immune response, attempts to induce anti-P antibodies by immunization with *autologous* ribosomes in the autoimmune strain of mouse, MRL, have been unsuccessful. It therefore seems likely that the ribosomal components must be altered in some way to break tolerance or that other abnormalities of the immune system are necessary for autoantibody production. Immunization with *foreign* ribosomes induce anti-P autoantibodies in mice and in apparently normal humans infected with the hemoflaggelate, T. cruzi. The ability of the P proteins to break tolerance in these situations is, most likely, explained by the provision of a T cell epitope (the foreign P protein) together with the multivalency of the P proteins on the ribosome (which activate autoreactive B cells). We therefore propose (**Fig. 5**) a two-signal model for autoantibody production similar to that suggested for T-B collaboration in the normal immune response[54] and also in the GVHD model of lupus[52]. The reasons for T cell activation and for the activation rather than tolerization of B cells by multivalent antigens remain important questions for further studies.

Acknowledgements: This work was supported, in part, by the National Institutes of Health, USA. We gratefully acknowledge help and collaboration from Drs. W. Danho, A. Gharavi, I. Wool, T. Uchiumi, R.A. Eisenberg, and S.G. Reed. The expert secretarial assistance from Ms. V. Te Eng Fo is appreciated.

REFERENCES

1. G.L. Asherson, Antibodies against nuclear and cytoplasmic cell constituents in systemic lupus erythematosus and other diseases, *Br J Exp Pathol* 40:209 (1959).
2. H.R.G. Deicher, H.R. Holman, and H.G. Kunkel, Anticytoplasmic factors in the sera of patients with systemic lupus erythematosus and certain other diseases, *Arthritis Rheum* 3:1 (1960).
3. E.W. Lamon and J.C. Bennett, Antibodies to ribosomal ribonucleic acid (rRNA) in patients with systemic lupus erythematosus (SLE), *Immunology* 19:439 (1970).
4. P.H. Schur, L.A. Moroz, and H.G. Kunkel, Precipitating antibodies to ribosomes in the serum of patients with systemic lupus erythematosus, *Immunochemistry* 4:447 (1967).
5. B.C. Sturgill and R.R. Carpenter, Precipitating antibodies to ribosomes in the serum of patients with systemic lupus erythematosus, *Arthritis Rheum* 8:213 (1965).
6. K.B. Elkon, A.P. Parnassa, and C.L. Foster, Lupus autoantibodies target the ribosomal P proteins, *J Exp Med* 162:459-471 (1985).
7. A.-M. Francoeur, C.L. Peebles, K.J. Heckman, J.C. Lee, and E.M. Tan, Identification of ribosomal protein antigens. *J Immunol* 135:2378-2384 (1985).
8. E. Bonfa, A.P. Parnassa, D.D. Rhoads, D.J. Roufa, I.G. Wool, and K.B. Elkon, Antiribosomal S10 antibodies in humans and MRL/lpr mice with systemic lupus erythematosus, *Arthritis and Rheum* 32:1252-1261 (1989).
9. J.A. Steitz, C. Berg, J.P. Hendrick, H. La Branche-Chabot, R. Metspalu, J. Rinke, and T. Yario, A 5S rRNA complex is a precursor to ribosome assembly in mammalian cells, *J Cell Biol* 106:545 (1988).
10. T. Sato, T. Uchiumi, R. Kominami, and M. Arakawa, Autoantibodies specific for the 20 kDa ribosomal large subunit protein L12, *Biochem Biophys Res Comm* 172:496 (1990).
11. J.J. Hines, H. Weissbach, N. Brot, and K.B. Elkon, Anti-P autoantibody production requires P1/P2 as immunogens but is not driven by exogenous self antigen in MRL mice. *J Immunol* 146:3386 (1991).
12. G. Levitus, M.H. Joskowicz, M.H.V. Van Regenmortel, and M.J. Levin, Humoral autoimmune response to ribosomal P proteins in chronic Chagas' heart disease, *Clin. Exp. Immunol.* 85:413 (1991).
13. Y.A.W. Skeiky, D.R. Benson, M. Parsons, K.B. Elkon, and S.G. Reed, Cloning and expression of Trypanosoma cruzi ribosomal protein P0 and epitope analysis of anti-P0 autoantibodies in Chagas' disease patients, *J Exp Med* 176:201 (1992).
14. B.E. Rich and J.A. Steitz, Human acidic ribosomal phosphoproteins P0, P2, and P2: Analysis of cDNA clones, in vitro synthesis, and assembly, *Mol Cell Biol* 7:4065 (1987).
15. P. Hasler, N. Brot, H. Weissbach, A.P. Parnassa, and K.B. Elkon, Ribosomal P proteins P0, P1, and P2 are phosphorylated by casein kinase II at their conserved carboxyl-termini, *J Biol Chem* 266:13,815 (1991).
16. A. Lin, B. Wittman-Leibold, J. McNally, and I.G. Wool, The primary structure of the acidic phosphoprotein P2, *J Biol Chem* 257:9189 (1982).
17. G. Stoffler and M. Stoffler-Meilicke, Immunoelectron microscopy on Escherichia coli ribosomes, *in:* Structure, Function and Genetics of Ribosomes, B. Hardesty, G. Kramer, eds., Springer-Verlag, New York, p 28 (1985).

18. K.B. Elkon, S. Skelly, A.P. Parnassa, W. Moller, H. Weissbach, and N. Brot, The identification and synthesis of a ribosomal protein antigenic determinant in systemic lupus erythematosus, *Proc Natl Acad, USA* 83:7419-7423 (1986).

19. T. Uchiumi, A.J. Wahba, and R.R. Traut, Topography and stochiometry of acidic proteins in large ribosomal subunits from Artemia salina as determined by cross-linking, *Proc Natl Acad Sci, USA* 84:5580 (1987).

20. K.B. Elkon, E. Bonfa, R. Llovet, W. Danho, H. Weissbach, and N. Brot, The properties of the ribosomal P2 protein autoantigen are similar to those of foreign protein antigens, *Proc Natl Acad Sci, USA* 85:5186 (1988).

21. W.P. MacConnell and N.O. Kaplan, The activity of the acidic phosphoproteins from the 80S rat ribosomes, *J Biol Chem* 257:5359 (1982).

22. F. Sanchez-Madrid, R. Reyes, P. Conde, and J.F. Ballesta, Acidic ribosomal proteins from eukaryotic cells. Effect on ribosomal functions, *Eur J Biochem* 98:409 (1979).

23. A.J. Van Agthoven, J.A. Maasen, and W. Moller, Structure and function of an acidic protein from 60S ribosomes and its involvement in elongation factor-2 dependent GTP hydrolysis, *Biochem Biophys Res Comm* 77:989-998 (1977).

24. T. Uchiumi, R.R. Traut, and R. Kominami, Monoclonal antibodies against acidic phosphoproteins P0, P1, and P2 of eukaryotic ribosomes as functional probes, *J Biol Chem* 265:89-95 (1990).

25. D.W. Stacey, S. Skelly, T. Watson, K. Elkon, H. Weissbach, and N. Brot, The inhibition of protein synthesis by IgG containing antiribosome P autoantibodies from systemic lupus erythematosus patients, *Arch Biochem Biophys* 267:398-403 (1988).

26. R. Amons, W. Pluijms, J. Kriek, and W. Moller, The primary structure of protein eL12'/eL12'-P from the large subunit of Artemia salina ribosomes, *FEBS Lett* 146:143 (1982).

27. E. Bonfa and K.B Elkon, Clinical and serological associations of the anti-ribosomal P protein antibody, *Arthritis Rheum* 29:981-985 (1986).

28. A.B. Schneebaum, J.D. Singleton, S.G. West, J.K. Blodgett, L.G. Allen, J.C. Cheronis, and B.L. Kotzin, Association of psychiatric manifestations with antibodies to ribosomal P proteins in systemic lupus erythematosus, *Am J Med* 90:54-62, (1991).

29. A. Van Dam, H. Nossent, J. de Jong, J. Meilof, E.J. ter Borg, A.J.G. Swaak, and R.J.T. Smeenk, Diagnostic value of antibodies against ribosomal phosphoproteins. A cross sectional and longitudinal study, *J Rheumatol* 18:1026 (1991).

30. T. Sato, T. Uchiumi, T. Ozawa, M. Kikuchi, M. Nakano, R. Kominami, M. Arakawa, Autoantibodies against ribosomal proteins found with high frequency in systemic lupus erythematosus patients with active disease, *J Rheumatol* 18:1681 (1991).

31. K.B. Elkon, E. Bonfa, R. Llovet, and R. Eisenberg, Association between anti-Sm and antiribosomal P protein autoantibodies in human SLE and MRL/lpr mice, *J Immunol* 143:1549 (1989).

32. E. Bonfa, S.J. Golombek, L.D. Kaufman, S. Skelly, H. Weissbach, N. Brot, and K.B. Elkon, Association between lupus psychosis and anti-ribosomal P protein antibodies: measurement of antibody using a synthetic peptide antigen, *N Engl J Med* 317: 265-271 (1987).

33. A.P. van Dam, R.H.W.M. Derksen, F. Gmelig Meyling, et al, A prospective

study on antiribosomal P proteins in two cases of familial lupus and recurrent psychosis, *Ann Rheum Dis* 49:779-782 (1990).

34. E. Koren, M.W. Reichlin, M. Koscec, R.D. Fugate, and M. Reichlin, Autoantibodies to the ribosomal P proteins react with a plasma membrane-related target on human cells, *J Clin Invest* 89:1236 (1992).

35. M.R. Attias, R.A. Sylvester, and N. Talal, Filter radioimmunoassay for antibodies to Reovirus RNA in systemic lupus erythematosus, *Arthritis Rheum* 16:719-725 (1973).

36. D. Eilat, A.D. Steinberg, and A.N. Schechter, The reaction of SLE antibodies with native, single-stranded RNA: radioassay and binding specificities, *J Immunol* 120:550-557 (1978).

37. P.H. Schur and M. Monroe, Antibodies to ribonucleic acid in systemic lupus erythematosus, *Proc Natl Acad Sci, USA* 63:1108-1112 (1969).

38. T. Uchiumi, R.R. Traut, K.B. Elkon, and R. Kominami, A human autoantibody specific for a unique conserved region of 28S ribosomal RNA inhibits the interaction of elongation factors 1 alpha and 2 with ribosomes, *J Biol Chem* 266: 2054-2062 (1991).

39. G.M. Veldman, J. Klootwijk, V.C.H.F. de Regt, R.J. Planta, C. Branlant, A. Krol, and J.P. Ebel, The primary and secondary structure of yeast 26 rRNA, *Nucleic Acids Res* 9: 6935-6952, (1981).

40. J. Dijk, R.A. Garret, and R. Muller, Studies on the binding of the ribosomal protein complex L7/12-L10 and protein L11 to the 5' one third of 23S RNA: a functional center of the 50S subunit, *Nucleic Acids Res* 6: 2717-2729 (1979).

41. T.A.L. El-Baradi, V.C.H.F. de Regt, S.W.C. Einerhand, J. Teixido, R.J. Planta, J.P.G. Ballesta, and H.A. Rave HA, Ribosomal proteins EL 11 from *Escherichia coli* and L15 from *Saccharomyces cerevisiae* bind to the same site in both yeast 26S and mouse 28S rRNA, *J Mol Biol* 195 909-917 (1987).

42. H. Weissbach, Soluble factors in protein synthesis. *in:* Ribosomes: Structure, function, and genetics, G.R. Craven, J. Davies, K. Davis, L. Kahan, N. Nomura, eds., Baltimore University Park Press, pp 377-412, (1980).

43. J.-L. Chu, N. Brot, H. Weissbach, and K. Elkon, Lupus antiribosomal P antisera contain antibodies to a small fragment of 28S rRNA located in the proposed ribosomal GTPase center, *J Exp Med* 174:507 (1991).

44. A.N. Theofilopolous and F.J. Dixon, Murine models of systemic lupus erythematosus, *Adv Immunol* 37:269 (1985).

45. A.E. Gharavi, R.C. Mellors, and K.B. Elkon, IgG anticardiolipin in murine lupus, *Clin Exp Immunol* 78:233-238 (1989).

46. R.A. Eisenberg, E.M. Tan, and F.J. Dixon, Presence of anti-Sm reactivity in autoimmune mouse strains, *J Exp Med* 147:582 (1978).

47. O. Costa and J.C. Monier, Antihistone antibodies detected by micro-ELISA and immunoblotting in mice with lupus-like syndrome (MRL/l, MRL/n, and NZB strains), *Clin Immunol Immunopathol* 40:276 (1986).

48. E. Bonfa, A. Marshak-Rothstein, H. Weissbach, N. Brot, K.B. Elkon, The frequency and epitope recognition of anti-ribosome P antibodies from humans with SLE and MRL/lpr mice are similar, *J Immunol* 140:3434 (1988).

49. A.E. Gharavi, J.L. Chu, and K.B. Elkon, Autoantibodies in systemic lupus erythematosus are not due to random polyclonal B cell activation, *Arthritis Rheum* 31:1337-1345 (1988).

50. B.D. Stollar, The antigenic potential and specificity of nucleic acids,

nucleoproteins, and their modified derivatives, *Arthritis Rheum* 24:100 (1981).

51.   M.P. Madaio, S. Hodder, S. Schwartz, and B.D. Stollar, Responsiveness of autoimmune and normal mice to nucleic acid antigens, *J Immunol* 132:872 (1984).

52.   E.S. Gleichman, T. Pals, A.G. Rolink, T. Radaszkiewicz, and H. Gleichman, Graft-versus-host reactions: clues to the etiology of a spectrum of immunological diseases, *Immunology Today* 5:324 (1984).

53.   A.P. van Dam, J.F. Meilof, H.G. van den Brink, and R.J.T. Smeenk, Fine specificities of anti-nuclear antibodies in murine models of graft-versus-host disease, *Clin Exp Immunol* 81:31 (1990).

54.   P. Bretscher, and M. Cohn M, A theory of self-nonself discrimination, *Science* 169:1042 (1970).

55.   A. Lanzavecchia, Antigen-specific interactions between T and B cells, *Nature* 314:536 (1985).

CROSS-REACTIONS OF ANTI-IMMUNOGLOBULIN SERA WITH SYNTHETIC T-CELL RECEPTOR β PEPTIDES: MAPPING ON A 3-DIMENSION MODEL

H. Kaymaz[1], F. Dedeoglu[1], S.F. Schluter[1],
A.B. Edmundson[2], and J.J. Marchalonis[1]

[1] Microbiology and Immunology
University of Arizona, Tucson, AZ
[2] Harrington Cancer Center, Amarillo, TX

ABSTRACT

The derived amino acid sequences of human T-cell receptor β chain shows significant homology to λ light chains of immunoglobulins in its variable, joining, and constant regions. We assessed the cross-reactivity between Tcr β chains and immunoglobulin light chains by determining the capacity of rabbit antisera to human or murine immunoglobulins to react to a synthesized set of nested, overlapping 16-mer peptides corresponding to the VDJC sequence of the Tcr β chain YT35. The observed reactivities were consistent with homologies to λ and κ light chains, the strongest reactivity being with a peptide that corresponds to the "switch peptide" of light chains, as assessed by ELISA binding and competitive inhibitions assays. Other regions reactive with anti-light chain sera corresponded to CDR1 and Fr3 segments of the variable domain and a segment of the constant region predicted to loop out of the tight globular structure. The peptide immunochemical results, together with the identification of specific regions of sequence correspondence between Tcr β and the characterized λ light chain Mcg, allowed us to develop a 3-dimensional model of the β chain consistent with its role in antigen recognition.

INTRODUCTION

That T-cell receptor β chains are related to immunoglobulins can be clearly seen both in their gene arrangements and protein sequences. Detailed sequence comparisons and phylogenetic analyses of protein sequences (1) reveal that Tcr β chains are highly homologous to immunoglobulin light chain (2), particularly λ chains. Regions corresponding to variable (V), joining segment (J) and constant (C) domains can be identified in Tcr β chain. The close relationship between T-cell receptors and immunoglobulins was also suggested in early studies showing serological cross-reactions between anti-immunoglobulin sera and T-cell surface molecules (3.7). In this report, we extend these studies by directly demonstrating antigenic cross-reactions between Tcr β chains and immunoglobulin and mapping the determinants recognized by anti-light chain sera. We also compare the determinants recognized by anti-human λ sera on Tcr β chain and the human myeloma λ chain protein Mcg.

Immunobiology of Proteins and Peptides VII
Edited by M.Z. Atassi, Plenum Press, New York, 1994

Cross-reactivity was assessed by synthesizing a set of overlapping peptides comprising the immunoglobulin-like portion of Tcr β chain sequence. We screened rabbit sera against human immunoglobulins on these peptides and found a set of determinants corresponding to CDR1, Fr3, and J segment regions of the V domain. One other determinant was found in the constant region. The mapping of these determinants buttresses a model we have proposed for the three dimensional structure of the β chain deduced from its similarity to the human λ chain Mcg (8).

MATERIALS AND METHODS

The gene sequence of Tcr β chain YT35 was used to design a series of 22 synthetic 16-mer peptides that overlapped by 5 residues. Sequences of the individual peptides were published elsewhere (9). These peptides were synthesized using an Applied Biosystems Peptide synthesizer at the University of Arizona Biotechnology Center. Commercial rabbit antisera specific for human λ, κ or γ chains were purchased from DAKO Corporation (Carpenteria, CA). Evaluation of antibody reactions were done by indirect and competitive ELISA using previously described conditions (10). For affinity purification of peptide specific antibodies, activated CH Sepharose 4B (Pharmacia, Uppsala, Sweden) was used for coupling the peptide according to instructions of the manufacturer. Mcg is a completely characterized (8) λ myeloma light chain (Vλ5). GUN is a E-Bence-Jones protein with VκIII sequences (11).

RESULTS AND DISCUSSION

We screened several rabbit anti-immunoglobulin sera for their capacity to react with human Tcr β peptides. Essentially, the same regions were recognized by all of the reactive sera. In Figure 1, the reactivities of anti-human λ, anti-human κ and anti-human γ sera are shown along with rabbit serum control. Antisera against human γ heavy chains showed no reactivity against β peptides. However, both anti-human λ and anti-human κ chain antibodies reacted strongly with peptides $\beta3$, $\beta8$, $\beta11$ and $\beta17$. RaHλ also reacted with peptide $\beta21$, but titration showed that this binding was substantially weaker than the others.

It was possible that the cross-reactions of the T-cell receptor peptides with anti-immunoglobulin sera occurred because of the contamination of Ig preparations with T-cell receptors. To exclude this possibility, we affinity purified the reactive sera through peptide-conjugated sepharose columns and tested the isolated peptide-specific antibodies on purified human λ and κ proteins. As is shown in Figure 2, the peptide specific antibodies obtained from the anti-λ sera reacted strongly with the human λ myeloma protein Mcg, but their activity on the κ Bence-Jones protein GUN was much weaker. This was true for all four reactive peptides $\beta3$, $\beta8$, $\beta11$ and $\beta17$. The opposite was true for specific antibodies from the anti-human κ serum. They reacted strongly with κ protein GUN but not with Mcg. The reactions could be inhibited with free protein (not shown). These results demonstrate that the peptide reactivities are the result of true cross-reactions.

A comparison of the reactivity of the anti-human λ sera on both β and λ chain peptides is shown in Figure 3. As expected, the reactivity of the anti-λ sera is much stronger with λ peptides. However, the corresponding regions recognized on the β chain, namely $\beta3$, $\beta8$, B11 and $\beta17$, are also reactive on the λ chain. These common antigenic regions correspond to the CDR1, Fr3, and J segments of the variable region and a segment of the constant region.

We have constructed a 3-dimensional model for the immunoglobulin portion of the Tcr β chain based on the known structure of the Mcg λ chain (8). The amino acid sequences of the β and λ were first aligned. This was straightforward since the two

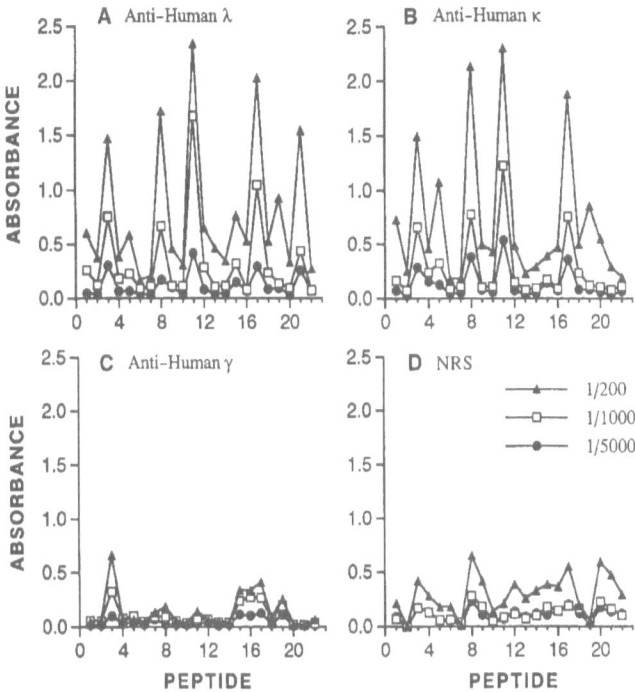

Figure 1. Reactivities of rabbit anti-immunoglobulin sera or normal rabbit serum to the set of Tcr β peptides. Serum were tested at a dilution of (Δ) 1/200, ($\square$) 1/1000, ($\bullet$) 1/5000. [A] antiserum to human λ; [B] antiserum to human κ; [C] antiserum to human γ heavy chain; [D] normal rabbit serum. Reactions were read at 405 nm after 30 minutes incubation with substrate.

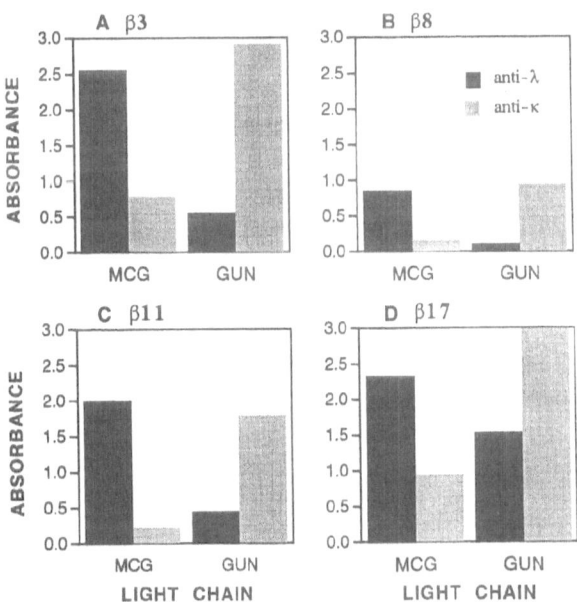

Figure 2. Reactivity of affinity purified antibodies obtained from anti-human λ and from anti-human κ by ELISA on λ myeloma protein Mcg, and κ Bence-Jones protein GUN. Affinity purified antibodies were tested at 2.5 μg/ml.

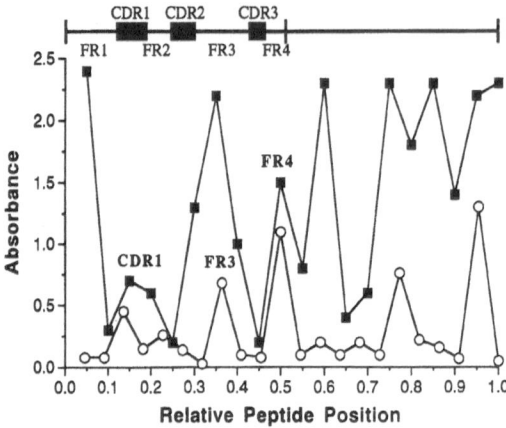

Figure 3. ELISA binding of rabbit antiserum to human λ chain on peptides of the λ chain Mcg (---■---) and the β chain YT35 (---o---) are shown on a normalized X-axis giving the relative peptide position. Upper diagram shows the regions of Mcg.

sequences are more than 30% identical in both the V and C regions. Residues known to be important in the λ chain structures are identical in β chain. The known structural features of the λ chain could be then be correlated with the β chain sequences and a model constructed. This is shown in Figure 4. Only a few modifications had to be made to accommodate this structure. Alterations due to gaps and insertions of the β chain sequence correspond to loop structure. Modifications include a shortening of the first hypervariable loop, a lengthening of the third hypervariable loop, and an extension of the third framework region. In the constant region the defining features of the immunoglobulin fold are retained, but two sets of additional residues cause two large loops to billow out from the globular domain structure into the space between the V and C domains.

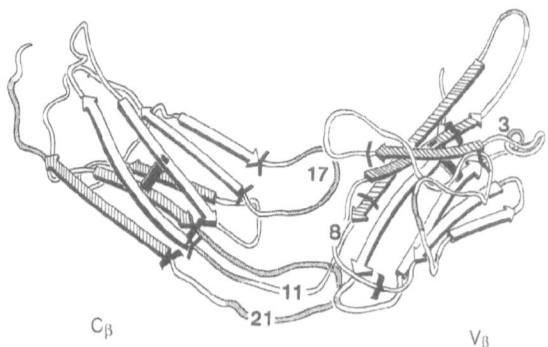

Figure 4. Model of T-cell receptor β chain based on the sequence alignment of YT35 β chain with λ light chain Mcg. Alterations in β chain model are indicated by dotted sections. Segments cross-reactive with anti-Ig sera are marked by < > for peptides #3, #11 and #17; { } for peptides #8 and #21.

The positions of the anti-λ reactive β peptides are indicated on the model in Figure 4. In most cases, the recognized determinants tend to be associated with loops rather than with β sheet structure. These same structures also comprise the antigenic regions recognized by anti-λ sera on the λ myeloma protein Mcg. The correlation of these structures and their antigenic cross-reactivity, together with the high degree of sequence identity between λ and β chains engenders confidence that the proposed model is reasonable.

The peptides reactive with anti-λ have recently been shown to also react with natural antibodies (12). Peptide $\beta 3$ corresponds to CDR1 and can be considered an idiotype determinant. The $\beta 8$ region has been implicated in the activation of particular T-cell subsets by "superantigens" (13). Considered together, these results raise the interesting possibility that antibodies and T-cell receptors raised against these determinants could be important in immune regulation and may be useful in immune-intervention in autoimmune disease.

ACKNOWLEDGEMENTS: This research was supported in part by Grants from the National Cancer Institute, CA42049 (JJM), Grant #CA19616 (ABE). We would like to thank Ms. Diana Humphreys for her help in the preparation of this manuscript.

REFERENCES

1.   J.J. Marchalonis, and S.F. Schluter, Evolution of variable and constant domains and joining segments of rearranging immunoglobulins. *The FASEB Journal*, 3:2469 (1989).

2.   Y. Yanagi, Y. Yoshikai, K. Leggett, S.P. Clark, I. Aleksander, T.W. Mak, A human T-cell specific cDNA clone encodes a protein having extensive homology to immunoglobulin chains, *Nature (Lond)*, 308:145 (1984).

3.   N.M. Hogg, and M.F. Greaves, Antigen-binding Thymus-derived lymphocytes. II. Nature of the Immunoglobulin Determinants, *Immunology* **22**:967 (1972).

4.   N.L. Warner, Membrane immunoglobulins and antigen receptor on B and T lymphocytes, *Adv.Immunol.* **19**, 67 (1974).

5.   R.E. Cone, M. Feldman, J.J. Marchalonis and G.J.V. Nossal, Cytophilic properties of surface immunoglobulin of thymus-derived lymphocytes. *Immunology* 26:49 (1974).

6.   J.J. Marchalonis, Lymphocyte surface immunoglobulins: molecular properties and function as receptors for antigen. *Science* 190:20 (1975).

7.   H. Binz, and H. Wigzell, Shared idiotypic determinants on B and T lymphocytes reactive against the same antigenic determinants, *J. Exp. Med.* 142:197 (1975).

8.   A.B. Edmundson, K.R. Ely, E.E. Abola, M. Schiffer, and N. Pangiotopoulos, Rotational allomerism and divergent evolution of domains in immunoglobulin light chains, *Biochemistry* 14:3953 (1975).

9.   F. Dedeoglu, R.A. Hubbard, S.F. Schluter, and J.J. Marchalonis, T-cell receptors of man and mouse studied with antibodies against synthetic peptides, *Exptl. & Clin. Immunogenetics*, 9:95 (1992).

10. J.J. Marchalonis, F. Dedeoglu, H. Kaymaz, S.F. Schluter, and A.B. Edmundson, Antigenic mapping of a human λ light chain: Correlation with 3-dimensional structure, *J. Protein Chemistry*, 11:129 (1992).

11. C.R. Ross, R.A. Hubbard, S.F. Schluter, A. Diamanduros, A.C. Wang, and J.J. Marchalonis, Antibodies to synthetic peptides corresponding to variable region first framework (Fr1) segments of T cell receptors: Detection of T cell products and cross-reactions with classical immunoglobulins, *Immunologic Research*, 8:81 (1989).

12. J.J. Marchalonis, H. Kaymaz, F. Dedeoglu, S.F. Schluter, D.E. Yocum, and A.B. Edmundson, Human autoantibodies reactive with synthetic autoantigens from T-cell receptor β chain, *Proc. Natl. Acad. Sci. USA*, 89:3325 (1992).

13. A.M. Pullen, J. Bill, R.T. Kubo, P. Marrack, and J.W. Kappler, Analysis of the interaction site for the self-superantigen Mls-1ᵃ on T-cell receptor Vβ, *J. Exp. Med.* 173:1183 (1991).

STRESS PROTEINS IN AUTOIMMUNITY

John Winfield and Wael Jarjour

Thurston Arthritis Research Center
University of North Carolina at Chapel Hill
Chapel Hill, North Carolina 27599, USA

INTRODUCTION[1]

Interest in the idea that stress proteins (also called heat shock proteins) contribute to the development and persistence of autoimmunity derives from five principal observations. First, stress proteins are immunodominant antigens of all prokaryotic infectious microorganisms, many of which have been implicated as inducers of arthritis and autoimmune disease in genetically susceptible individuals. Second, the T-cell repertoire, especially that of T-cells bearing $\gamma\delta$ TCRs, appears to be biased toward recognition of microbial stress proteins (and, possibly, host stress proteins as well) as part of an evolutionarily conserved mechanism for host defense against infection and for resolution of tissue inflammation. Third, stress proteins are highly conserved, as would be expected in view of their fundamentally important roles in cell biology. During infection, which entails "stress" for both the microorganism and the host, there is increased synthesis and altered expression of extremely similar sets of autologous and foreign stress proteins. Fourth, the inflammatory response associated with infection up-regulates HLA glycoproteins that present antigen peptides to T-cell receptors. This may amplify recognition signals for conserved stress protein peptides to a degree sufficient for stimulation of autoreactive T-cells that have escaped deletion in the thymus. The various "permissive" genetic effects that underlie autoimmune disease may allow an autoreactive immune response to stress proteins, or to other self-constituents that cross-react with stress proteins, to persist or recur in susceptible individuals. Finally, two animal models of human autoimmune disease, adjuvant arthritis in the rat and spontaneous type-1 diabetes in non-obese mice, have implicated a stress protein, hsp60, as the critical pathogenetic element.

Research in this area during the past two years has moved from an initial focus on molecular mimicry in arthritis to assessments of the roles of stress proteins in basic immune

---

[1]This topic was reviewed previously in **J. Winfield and W. Jarjour, Do stress proteins play a role in arthritis and autoimmunity?**, *Immunological Reviews 37:193 (1991)*, which cites earlier published work in detail. The present brief emphasizes progress in this field during the past two years.

mechanisms and in the etiopathogenesis of many other autoimmune diseases. In this brief review, some of the emerging concepts will be highlighted. Emphasis will be on: a) $\gamma\delta$ cells and the stress protein, hsp60, and their potential roles in inflammation; b) reactivity of T-cells with stress proteins in rheumatoid arthritis (RA) and juvenile chronic arthritis (JCA); and c) antibodies and autoantibodies to stress proteins. Information concerning two animal models and mammalian stress proteins and their role in cell biology will be presented as background.

ANIMAL MODELS THAT IMPLICATE STRESS PROTEINS

Initial support for an etiological relationship between microbial stress proteins and arthritis was obtained by Irun Cohen and his associates in the adjuvant arthritis model in Lewis rats.[1] A T-cell clone (A2b) derived from *M. tuberculosis* antigen-containing cultures of lymph node cells of rats primed with Freund's complete adjuvant, transferred arthritis when injected intravenously into irradiated, syngeneic naive rats. The A2b clone exhibited immunologic reactivity *in vitro* and *in vivo* with *M. tuberculosis* and cartilage proteoglycan. Using clone A2b to screen mycobacterial antigens expressed in *E. coli*, the stress protein hsp60 was identified as a critical antigen in the adjuvant arthritis model, and a mycobacterial hsp60 nonapeptide (amino acid position 180-188) having some sequence similarity with the link protein of rat proteoglycan was delineated as the specific target. The obvious conclusion here was that a mycobacterial stress protein had induced, via molecular mimicry, a specific autoimmune attack against cartilage in the joint. The potential relevance of these findings to arthritis in man was underscored when synovial fluid T-cells from patients with rheumatoid arthritis and other types of inflammatory arthritis were shown to proliferate in response to recombinant mycobacterial hsp60 and antigen preparations from mycobacteria and gram-negative bacteria.[2-5]

Several years ago, Dana Elias (also in Irun Cohen's group) reported that immunity to hsp60 was involved in the spontaneous diabetes that develops in non-obese (NOD) mice.[1,6] Insulitis appears at age 1 month in these mice. At 2-4 months, hsp60 cross-reactive antigen and hsp60-reactive T-cells are detectable in the circulation, followed about two weeks later by the appearance of anti-hsp60 antibodies in the circulation. By 4-6 months, marked reduction in islet β cells and overt hyperglycemia is apparent. A T-cell clone (C9) has been derived from NOD mice that recognizes an epitope on a 24-amino acid synthetic peptide of human hsp60 (p227). Administration of the C9 clone to 1-month old NOD mice leads to the accelerated development of diabetes. Vaccination of NOD mice with this clone, or with peptide p277, induces resistance to diabetes.

Both of these models have been useful for delineating basic immune mechanisms in autoimmune disease and for developing new approaches to immunotherapy. While the potential applicability of the hsp60/adjuvant arthritis paradigm to some forms of inflammatory arthritis in man remains an open question, a role for hsp60 in early human diabetes is unlikely. Recent evidence indicates that a 64 kD antigen implicated in human diabetes, which had been thought to be hsp60, is glutamic acid decarboxylase.[7]

MAMMALIAN STRESS PROTEINS AND THE STRESS RESPONSE[2]

The Stress Response

Cells of all organisms from *E. coli* to man respond to adverse changes in their local environment by increasing the transcription of genes encoding stress proteins, which provide protection from the stress event - the heat shock or stress response. A variety of stimuli other than heat also can induce the synthesis of stress proteins, *e.g.,* various drugs, hormones, prostaglandins, cytokines, heavy metals and many other toxic agents, and threats to the cell such as hypoxia, glucose starvation, and virus or bacterial infection. When a cell is subjected to heat shock, it becomes unresponsive to most other stimuli and exhibits a variety or morphological and metabolic changes. Certain stress proteins, especially the highly inducible 72 kD member of the hsp70 family, migrate between different cell compartments, *e.g.* from cytoplasm to nucleus/nucleolus and then back to cytoplasm, during sublethal cellular stress in a process that confers protection to the cell upon subsequent stress (acquired thermotolerance).

Classification and Function of Mammalian Heat Shock Proteins

There are two major classes of mammalian heat shock proteins: heat shock proteins (HSPs) and glucose-regulated proteins (GRPs), which are induced when cells are deprived of glucose or oxygen. HSPs have estimated molecular masses of 110, 90, 73, 72, 58, 47, 32, 28, and 8 kD. The three known GRPs have masses of 100 (also termed grp94), 78, and 75 kD. Certain HSPs and GRPs are closely related structurally and can be grouped into families that share similar functions. For example, members of the hsp70 family (proteins of 72 and 73 kD) and hsp90 exhibit considerable sequence homology with grp78 (also referred to a BiP or Ig heavy chain binding protein) and grp100, respectively. **hsp110** is a nucleolar/cytoplasmic phosphoprotein possibly involved in rRNA transcription. **hsp90**, a heavily phosphorylated rod-like structure with 12 isoforms, is the most abundant stress protein in mammalian cells. Located in the cytosol, it binds to and regulates many important proteins, including retrovirus encoded oncogene proteins and the nontransformed 8S/9S form of the steroid hormone receptor. Synthesis of hsp90 is increased in T-cells following stimulation with mitogens, suggesting a possible role in T-cell activation. **grp100** is compartmentalized in the endoplasmic reticulum, the Golgi complex, and, possibly, in the plasma membrane, and may have functions analogous to those of hsp90. The **hsp70 family** (mammalian homologs of DnaK in *E. coli* and the ~70 kD protein in mycobacteria) consists of at least four members that function as chaperonins to facilitate protein folding events in the translocation of polypeptides across membranes and in their assembly into multimeric structures. Characterized members of the hsp70 family include: **hsp73**, a constitutively expressed protein; **hsp72**, a highly inducible form that is virtually indistinguishable from hsp73; **grp75**, a recently identified protein that functions as a chaperonin in mitochondria in a fashion analogous to that of the fourth member, **grp78**, in the endoplasmic reticulum. hsp72 is especially important in the response of the cell to stress and in the induction of thermotolerance. The only **groEL-related stress protein** identified thus far is hsp58 (also called P1 or, in this review, **hsp60**), a constitutively expressed protein located inside mitochondria as "donut"-shaped oligomers. Like groEL homologs in bacteria and plants, mammalian hsp60 is a chaperonin involved in the import of proteins into mitochondria and their subsequent re-folding. Other groEL-related proteins may exist in mammalian cells, although all of the 8-12 additional genes identified thus far appear to be non-functional

---

[2]Citations for this section can be found in **W. Welch** *et al.,* **Response of mammalian cells to metabolic stress: changes in cell physiology and structure/function of stress proteins,** *Current Topics in Microbiology and Immunology* **167:31 (1991).**

pseudogenes. The **small mammalian stress proteins** include histone H2B, ubiquitin, hsp28, hsp32, and hsp47. Ubiquitin is essential for chromosome organization through its interaction with nucleosome histones H2A and H2B and also is important in the ATP-dependent proteolytic degradation of unstable cellular proteins. hsp28 may be involved in the control of cell growth. hsp32 recently was identified as heme oxygenase, which plays a role in defense against oxidative reactions catalyzed by iron and other heavy metals. hsp32 is induced by exposure to ultraviolet light, various tumor promoters, and chemical carcinogens. hsp47 is a novel membrane glycoprotein that binds collagen and fetuin (α-fetoprotein).

STRESS PROTEINS AND THE IMMUNE SYSTEM

Regardless of the ultimate significance of stress proteins in the genesis of autoimmunity, they clearly are important elements in many aspects of the immune system and its function. grp78 (BiP) and a hsp70-related protein appear to function as chaperonins in the expression and secretion of immunoglobulin molecules by B cells and in the handling and presentation of antigen peptides by antigen-presenting cells, respectively. Other putative functions of stress proteins include tumor antigenicity, lymphocyte homing, resistance to target cell lysis, regulation of protooncogene transcription, lymphocyte and macrophage activation, $\gamma\delta$ T-cell selection, and immune surveillance. Genes for hsp70 have been located in the MHC class III region, and RFLPs of class III hsp70 genes are found in several autoimmune diseases, including type 1 diabetes and thyroiditis,[8,9] and SLE (Jarjour and Winfield, unpublished observations).

$\gamma\delta$ T-Cells and hsp60

While hsp60 certainly is not the only, or even the major, activator of mycobacteria-reactive $\gamma\delta$ T-cells,[10-12] considerable evidence suggests that the T-cell repertoire is biased toward recognition of this stress protein. First, ~20% of the T-cell response in adult mice following immunization with *M. tuberculosis* is directed against mycobacterial hsp60.[13] Second, a high proportion of murine neonatal $\gamma\delta$ thymocyte hybridomas recognize mycobacterial antigens, including hsp60 and a conserved seventeen-amino acid synthetic peptide representing amino acids 180-196 of *M. leprae* hsp60.[14,15] Such $\gamma\delta$ thymocyte hybridomas exhibit an unusual, TCR-dependent "self-stimulatory" activity that involves recognition of this conserved synthetic hsp60 peptide. More recent observations by Rebecca O'Brien in Willi Born's group[16] indicate that as many as 10-20% of mature $\gamma\delta$ T-cells in normal spleen and lymph node of adult mice recognize mycobacterial hsp60 and this p180-196 peptide as well. All of the neonatal $\gamma\delta$ thymocyte hybridomas express $V_\gamma 1$ and $V_\delta 6$ TCRs; many of the adult $\gamma\delta$ cells also express $V_\gamma 1$ and $V_\delta 6$, but with extensive junctional diversity, suggesting that cells in this subset are produced throughout life and are not simply descendants of neonatal cells. Reactivity with hsp60 is not limited to $\gamma\delta$ T-cells, however. A substantial proportion of $V_\beta 11$-bearing $\alpha\beta$ T-cells in adult, neonatally thymectomized BALB/c mice are self-reactive and recognize mycobacterial hsp60.[17] Third, brief heat shock of murine T-cells or syngeneic bystander cells induces selective proliferation of $\gamma\delta$ cells,[18] a phenomenon that is enhanced by pre-exposure of the cells to mycobacterial antigens. Finally, limiting dilution analysis of human cord blood cells reveals a high proportion of T-cells that proliferate to mycobacterial hsp60.[19] Taken together, these observations suggest a) that the proportion of T-cells, especially $\gamma\delta$ cells, that recognize hsp60 at birth is high, and b) that this high proportion of hsp60-reactive T-cells is maintained in adults.

It has been proposed that the large subset of $\gamma\delta$ cells that recognize conserved epitopes of autologous stress protein peptides represent a "rapid-response" first line of defense against infection.[20] Consistent with this schema are: a) the dramatic expansion of $\gamma\delta$ T-cells that occurs when peripheral blood mononuclear cells from normal individuals are cultured with *M. tuberculosis* or other mycobacteria and bacteria;[21] b) the rapid accumulation of $V_\gamma 1$, $V_\delta 6$ TCR-bearing T-cells that specifically recognize hsp60 when *Listeria monocytogenes* is injected intraperitoneally into mice;[22] and c) the recent demonstration of the human hsp60 reactivity of a T-cell clone derived from a *Yersinia*-specific T-cell line.[23] Born *et al.* further speculate that autologous stress proteins are "presented" by "stressed" cells to special subsets of $\gamma\delta$ cells. $\gamma\delta$ cells respond by mediating cellular activation, immune surveillance, and regulation of lymphocyte growth and differentiation. Investigations of Doherty[24] with a human influenza A-induced pneumonitis in mice extends Born's hypothesis. Large numbers of $\gamma\delta$ T-cells with $V_\gamma 1, V_\delta 6$ TCRs accumulate in parallel with macrophages expressing large amounts of hsp60, but this occurs late during resolution of the pneumonitis and is dependent upon the $\alpha\beta$ T-cell response that mediates clearance of the virus. Doherty proposes that any virus-specific $\alpha\beta$ T-cell response promotes a local hsp60$^+$ macrophage/$\gamma\delta$ T-cell circuit as a normal "mop-up" protective mechanism. It is reasonable to speculate that inflammatory tissue injury of other types, *e.g.*, joint inflammation in reactive arthritis or rheumatoid arthritis could activate this mechanism as well. A likely enhancer of $\gamma\delta$ T-cell recognition of hsp60 peptides at sites of local inflammation is the up-regulation by stress or cytokines of MHC class Ib glycoproteins (Qa-1 in the mouse), which have been implicated in the presentation of hsp60 peptides to $\gamma\delta$ cells.[25-27]

T-CELL REACTIVITY WITH STRESS PROTEINS IN ARTHRITIS

In a mechanistic sense, all of the ingredients for recognition of autologous stress protein peptides by T-cells in inflamed joints are in place. Stress proteins clearly are expressed by cells in the synovium,[28,29] which is not unexpected given the capacity of inflammatory mediators to induce surface-localized forms of stress proteins or stress protein peptides, *e.g.*, hsp60 on murine bone marrow macrophages following stimulation with IFN_γ.[30] Nor should it be surprising that autoreactive T-cells that recognize conserved stress protein peptides can be found in synovial fluid. The existence of such cells in healthy individuals has been demonstrated convincingly by Munk.[31] The recent experiments of Benichou[32], together with the data presented by Nicole Suciu-Foca at this meeting on recognition of autologous HLA DR1 peptides, prove nicely that tolerance for self-peptides is not complete, which is good because absolute tolerance would severely attenuate the ability to the T-cell repertoire to see foreign shapes. Despite all of this, T-cell reactivity with human hsp60 or conserved hsp60 peptides has been quite difficult to document convincingly in RA.

Adult Rheumatoid Arthritis

On the negative side, many investigators have demonstrated an increased synovial T-cell proliferative response to microbial antigens, including hsp60,[2-5,33-34] but two recent studies that used limiting dilution analysis to estimate the frequency of mycobacterial and hsp60-reactive T-cells in rheumatoid joints could not demonstrate increased numbers.[35,36] Indeed, in one of these studies, no *M. bovis* hsp60-reactive T-cells were found in the synovial fluid.[35] Res *et al.*, in a follow-up of their earlier observation that synovial fluid

T-cells from patients with early RA proliferate in response to an *E. coli* lysate containing recombinant hsp60 from *M. bovis*,[2] now find that contaminating *E. coli* antigens, rather than mycobacterial hsp60, are recognized primarily[37]. Similar observations have been made by Life *et al.*[38] Res also demonstrated that the proliferative response of pleural exudate mononuclear cells to these antigens is similar to that of synovial fluid cells. This latter observation suggests that reactivity of synovial fluid cells to mycobacterial antigens may be a general response of chronic inflammation (which is consistent with the "local hsp60$^+$ macrophage/$\gamma\delta$ T-cell circuit" model of Doherty), rather than an "organ-specific" phenomenon confined to joints. Finally, De Graeff-Meeder and colleagues[28] could not demonstrate reactivity of synovial fluid T-cells from patients with RA to HPLC-purified recombinant human hsp60.

On the other hand, two interesting studies from Norway provide important new information that supports either hsp60 induction of arthritis or the Doherty model.[39,40] In the first study, synovial membrane and synovial fluid T-cells (mostly TCR$\alpha\beta$, DR-restricted) were shown to lyse autologous macrophages that had been pulsed with BCG.[39] Two synovial fluid T-cell clones also lysed targets that had been pulsed with human recombinant hsp60. The second study cloned out synovial fluid T-cells from two very early, very active patients with RA (without doubt, the best kind of patient to study) using *M. leprae* hsp 60, and then screened the clones against several hsp60 synthetic peptides chosen for maximum homology with human hsp60.[40] A clone from each patient recognized a conserved epitope in these peptides. One clone was DQ-restricted and the other was DR-restricted. The clones used different V_β gene segments, but the same D_β and J_β elements. Both exhibited cytotoxicity against autologous antigen-pulsed macrophages. Taken together, these observations provide additional support for the concept of molecular mimicry between microbial stress proteins and self-constituents in the induction of joint inflammation in RA. But as emphasized by the authors, an alternate interpretation of the data is that T-cell reactivity with autologous hsp60 might be a normal part of the inflammatory response, rather than its cause. They speculate that hsp60-reactive cytotoxic T-cells could remove stressed or damaged cells expressing hsp60, perhaps even contributing to the resolution of the joint inflammation. In any case, these reports extend previous observations demonstrating that healthy individuals have T-cells in their circulation that recognize strictly autologous hsp60 epitopes.[41,42]

An altogether new type of molecular mimicry is suggested by reports from two different laboratories that demonstrate cross-reaction of microbial stress proteins with HLA molecules. Anderson and colleagues[43] discovered that an autoreactive T-cell clone from a patient with tuberculoid leprosy recognized a peptide from the third hypervariable region of the β_3 chain of HLA DRB11501, which is associated with this form of leprosy. More recently, Albani[44] in Dennis Carson's group in La Jolla used a rabbit antiserum to the third hypervariable region of HLA DRB10401 to demonstrate immunologic cross-reactivity with recombinant dnaJ, an *E. coli* stress protein that shares an 11-amino acid sequence homology with the DRB10401 susceptibility sequence for RA. An antiserum to recombinant dnaJ recognized the homologous peptide, the intact DRB10401 molecule, and, most importantly, specifically stained DRB10401 homozygous B cells. This provocative observation raises entirely new questions regarding gram-negative bacteria and their stress proteins in the etiology of RA and, as discussed below, already is bearing fruit with respect to the role of stress proteins in juvenile chronic arthritis.

The concept that T-cells with $\gamma\delta$ receptors play a pivotal role in the immune response to stress proteins in the joint also is controversial. In support of the original observations that $\gamma\delta$ T-cells in the joint are especially reactive with mycobacteria and mycobacterial hsp60, [45-48] it has been reported that a high proportion of the synovial fluid T-cells that proliferate in response to mycobacterial hsp60 express $\gamma\delta$ TCRs (especially

104

$V_\gamma 1$).[34] But preliminary data of Pope et al.[49] suggest that when an acetone precipitable fraction of M. tuberculosis and hsp60-containing fractions thereof are used as antigens, no relationship between the relative proportions of $\gamma\delta$ cells and the increased proliferative response of RA synovial T-cells is observed, and there is no enrichment of $\gamma\delta$ cells after stimulation with mycobacterial antigen.

In summary, recent investigation indicates that the mechanisms underlying persistent inflammation in the joints of patients with RA are far more complex than simple autoreactive attack by $\gamma\delta$ T-cells directed against a conserved hsp60/cartilage cross-reactive epitope. Taken together, however, the accumulated data in murine and human systems is beginning to suggest that the Doherty's "local hsp60+ macrophage/$\gamma\delta$ T-cell circuit" model may obtain as a basis for persistence of joint inflammation, even if stress proteins are not the primary triggers.

Juvenile Chronic Arthritis

Several recent reports strongly support a role for microbial stress proteins in JCA. First, De Graeff-Meeder and colleagues[28] found that synovial fluid T-cells from six patients with short-duration JCA proliferated in response to HPLC-purified recombinant human hsp60. Immunohistochemical staining for hsp60 with ML-30, a monoclonal antibody raised against M. leprae hasp60, revealed a dramatically increased expression of hsp60 in different cell types of the joint. Doubt regarding the specificity of ML-30 staining for human antigen is mitigated by similar results using two monoclonal antibodies prepared against human recombinant hsp60.[29] Second, Albani and colleagues, who earlier described a cross-reaction between E. coli dnaJ and HLA DRB 10401, now find that children with polyarticular and systemic JCA have increased cellular and humoral immunity to dnaJ (unpublished work to be presented at the American College of Rheumatology meetings in Atlanta, October, 1992). Finally, Danieli et al.[50] have, in essence, replicated the entire Irun Cohen adjuvant arthritis paradigm in JCA: T-cell reactivity with the mycobacterial hsp60 amino acid position 180-188 epitope and with a partially homologous peptide from cartilage proteoglycan link protein.

ANTIBODIES AND AUTOANTIBODIES TO STRESS PROTEINS

Several laboratories have detected antibodies to microbial stress proteins in arthritis and other autoimmune diseases.[51-53] In these studies, it is not always possible to determine whether detection of antibodies to microbial stress proteins in patient sera represents a true increase in specific antibodies to agents of putative etiologic significance, or whether their detection simply reflects a polyclonal increase in immunoglobulin levels. Only limited data are available concerning the development of true autoantibodies to stress proteins, i.e., antibodies that react with human antigens. Unconfirmed reports suggest that patients with SLE exhibit autoantibodies to ubiquitin and a charged synthetic octapeptide of ubiquinated histone H2A.[54,55] Using immunoblotting techniques, our laboratory described autoantibodies to hsp90 and to the constitutively expressed 73 kD member of the hsp70 family in a minority of patients with SLE[56,57] but when these initial studies were extended in a more comprehensive survey of 268 patients with various rheumatic diseases, inflammatory bowel disease, and several autoimmune skin diseases,[58] autoantibodies to human hsp60, hsp73, and hsp90 were found to be rather uncommon (20% or less of sera in a given disease category) and of uniformly low titer. Thus, it is clear that patients with autoimmune disease do not develop high affinity/high titer autoantibodies to stress proteins of the kind that are commonly seen to DNA and other nuclear antigens in SLE.

STUDIES IN OTHER DISEASES

Research on stress proteins in other autoimmune (or possibly autoimmune) disorders is roughly five years behind that in arthritis, *i.e.*, is just beginning. Except for type 1 diabetes, where stress proteins have been shown not to be important, few conclusions can be drawn at this time. The sometimes difficult lessons learned from arthritis should be useful in the investigation of other diseases, however. A partial list of some of the diseases being studied includes the following: Alzheimer's disease,[40,59-63] multiple sclerosis,[64-67] thyroid disease,[68-70] inflammatory bowel disease,[71-73] and polymyositis.[74] A few vignettes will provide a sense of ongoing research in these diseases. Increased levels of hsp72 and grp78 have been found in the neuritic plaques and neurofibrillary tangles of Alzheimer's disease brains.[61] Polymorphisms of hsp70 genes in the MHC class III region were discovered in autoimmune thyroid disease and type 1 diabetes.[8,9] Muscle fibers were highly positive for MHC class I antigen and hsp60 in a very interesting patient with an unusual form of polymyositis mediated by $\gamma\delta$ T-cells.[74] Stress proteins and $\gamma\delta$ T-cells have attracted particular attention in multiple sclerosis, and, although only a handful of papers have been published thus far, an international meeting devoted to this topic will be held in Spain in spring, 1993. Limited research on Crohn's disease and ulcerative colitis, perhaps surprisingly, does not support a role for $\gamma\delta$ T-cells and stress proteins.[71,73]

CONCLUSIONS

Following the evolution of this area during the past five years has been like a roller coaster ride, with initial high enthusiasm,[75] then skepticism,[76], and now modest enthusiasm again. Although fascinating and extraordinarily useful heuristically as models, the contribution of hsp60 to the pathogenesis of adjuvant arthritis in rats and to diabetes in non-obese mice probably does not obtain in the same fashion in human disease. Considerable conflicting data concerning $\gamma\delta$ T-cell recognition of hsp60 in arthritis and other types of inflammation still must be resolved. But the accumulated data support concepts that $\gamma\delta$ T-cells recognize bacterial hsp60 in an evolutionarily conserved mechanism of host defense against infection and that macrophages and other types of cells present hsp60 peptides to $\gamma\delta$ cells as part of the normal "mop-up" response to inflammation. Both of these mechanisms may contribute to immunopathology, *e.g.*, in reactive arthritis, as has been delineated convincingly in experimental trachoma.[77] The issue of whether microbial hsp60, dnaJ, or other stress proteins induce joint inflammation in disorders like rheumatoid arthritis and juvenile chronic arthritis has not been resolved. It is not unreasonable to predict that stress proteins actually play various combinations of three roles in different situations: a) microbial trigger of autoimmune disease; b) target of $\gamma\delta$ T-cells as part of the normal resolution of inflammatory tissue injury; and c) inappropriate target of excessive autoreactive T-cell attack in the persistent or recurrent inflammation of chronic disease.

ACKNOWLEDGMENTS

The experimental work from the authors' laboratory cited in this article was supported in part by National Institutes of Health grants AR30863, AR7416, and AR30701. Wael Jarjour is a Fellow of the Terri Gotthelf Lupus Research Institute.

REFERENCES

1. I.R. Cohen, Autoimmunity to chaperonins in the pathogenesis of arthritis and diabetes, *Annu. Rev. Immunol.* 9:567 (1991).

2. P.C. Res, F.C. Breedveld, J.D.A. Van Embden, C.G. Schaar, W. Van Eden, I.R. Cohen, and R.R.P. De Vries, Synovial fluid T cell reactivity against 65 kD heat shock protein of mycobacteria in early rheumatoid arthritis, *Lancet* ii:478 (1988).

3. J.S.H. Gaston, P.F. Life, L. Bailey, and P.A. Bacon, Synovial fluid T cells and 65 kD heat-shock protein, *Lancet* 2:856 (1988).

4. J.S. Gaston, P.F. Life, K. Granfors, R. Merilahti Palo, L. Bailey, S. Consalvey, A. Toivanen, and P.A. Bacon, Synovial T lymphocyte recognition of organisms that trigger reactive arthritis, *Clin. Exp. Immunol.* 76:348 (1989).

5. J.S.H. Gaston, P.F. Life, L.C. Bailey, and P.A. Bacon, In vitro responses to a 65-kilodalton mycobacterial protein by synovial T cells from inflammatory arthritis patients, *J. Immunol.* 143:2494 (1989).

6. D. Elias, D. Markovits, T. Reshef, R. van der Zee, and I.R. Cohen, Induction and therapy of autoimmune diabetes in the non-obese diabetic (NOD/Lt) mouse by a 65-kDa heat shock protein, *Proc. Natl. Acad. Sci. U. S. A* 87:1576 (1990).

7. S. Baekkeskov, H.J. Aanstoot, S. Christgau, A. Reetz, M. Solimena, M. Cascalho, F. Folli, H. Richter Olesen, P. DeCamilli, and P.D. Camilli, Identification of the 64K autoantigen in insulin-dependent diabetes as the GABA-synthesizing enzyme glutamic acid decarboxylase [published erratum appears in Nature 1990 Oct 25;347(6295):782], *Nature* 347:151 (1990).

8. N.J. Caplen, A. Patel, A. Millward, R.D. Campbell, S. Ratanachaiyavong, F.S. Wong, and A.G. Demaine, Complement C4 and heat shock protein 70 (HSP70) genotypes and type I diabetes mellitus, *Immunogenetics* 32:427 (1990).

9. S. Ratanachaiyavong, A.G. Demaine, R.D. Campbell, and A.M. McGregor, Heat shock protein 70 (HSP70) and complement C4 genotypes in patients with hyperthyroid Graves' disease, *Clin. Exp. Immunol.* 84:48 (1991).

10. K. Pfeffer, B. Schoel, H. Gulle, S.H.E. Kaufmann, and H. Wagner, Primary responses of human T cells to mycobacteria: a frequent set of γ/δ T cells are stimulated by protease-resistant ligands, *Eur. J. Immunol.* 20:1175 (1990).

11. D. Kabelitz, A. Bender, S. Schondelmaier, B. Schoel, and S.H.E. Kaufmann, A large fraction of human peripheral blood y/b+ T cells is activated by mycobacterium tuberculosis but not by its 65-kD heat shock protein, *J. Exp. Med.* 171:667 (1990).

12. A systematic molecular analysis of the T cell-stimulating antigens from Mycobacterium leprae with T cell clones of leprosy patients Identification of a novel M. leprae HSP 70 fragment by M leprae-specific T cells, *J. Immunol.* 147:3530 (1991).

13. S.H. Kaufmann, U. Vath, J.E. Thole, J.D. van Embden, and F. Emmrich, Enumeration of T cells reactive with Mycobacterium tuberculosis organisms and specific for the recombinant mycobacterial 64-kDa protein, *Eur. J. Immunol.* 17:351 (1987).

14. R.L. O'Brien, M.P. Happ, A. Dallas, E. Palmer, R. Kubo, and W.K. Born, Stimulation of a major subset of lymphocytes expressing T cell receptor γδ by an antigen derived from *Mycobacterium tuberculosis, Cell* 57:667 (1989).

15. W. Born, L. Hall, A. Dallas, J. Boymel, T. Shinnick, D. Young, P. Brennan, and R. O'Brien, Recognition of a peptide antigen by heat shock-reactive γδ T lymphocytes, *Science* 249:67 (1990).

16. R.L. O'Brien, Y-X. Fu, R. Cranfill, A. Dallas, C. Ellis, C. Reardon, J. Lang, S.R. Carding, R. Kubo, and W. Born, Heat shock protein Hsp60-reactive γδ cells: A large, diversified T-lymphocyte subset with highly focused specificity, *Proc. Natl. Acad. Sci. USA* 89:4348 (1992).

17. A. Iwasaki, Y. Yoshikai, H. Yuuki, H. Takimoto, and K. Nomoto, Self-reactive T cells are activated by the 65-kDa mycobacterial heat-shock protein in neonatally thymectomized mice, *Eur. J. Immunol.* 21:597 (1991).

18. R. Rajasekar, G-K. Sim, and A. Augustin, Self heat shock and γδ T-cell reactivity, *Proc. Natl. Acad. Sci. U. S. A* 87:1767 (1990).

19. H.P. Fischer, C.E.M. Sharrock, and G.S. Panayi, High frequency of cord blood lymphocytes against mycobacterial 65-kDa heat-shock protein, *Eur. J. Immunol.* 22:1667 (1992).

20. W. Born, M.P. Happ, A. Dallas, C. Reardon, R. Kubo, T. Shinnick, P. Brennan, and R. O'Brien, Recognition of heat shock proteins and γ/δ cell function, *Immunol. Today* 11:40 (1990).

21. M.E. Munk, A.J. Gatrill, and S.H.E. Kaufmann, Target cell lysis and IL-2 secretion by γ/δ T lymphocytes after activation with bacteria, *J. Immunol.* 145:2434 (1990).

22. K. Nomoto, and Y. Yoshikai, Heat-shock proteins and immunopathology: regulatory role of heat-shock protein-specific T cells, *Springer. Semin. Immunopathol.* 13:63 (1991).

23. E. Hermann, A.W. Lohse, Zee.R. van der, W. Van Eden, W.J. Mayet, P. Probst, T. Poralla, Buschenfelde.K.H. Meyer zum, and B. Fleischer, Synovial fluid-derived Yersinia-reactive T cells responding to human 65-kDa heat-shock protein and heat-stressed antigen-presenting cells, *Eur. J. Immunol.* 21:2139 (1991).

24. P.C. Doherty, W. Allan, M. Eichelberger, and S.R. Carding, Heat-shock proteins and the γδ T cell response in virus infections: implications for autoimmunity, *Springer. Semin. Immunopathol.* 13:11 (1991).

25. D. Vidovic, M. Roglic, K. McKune, S. Guerder, C. MacKay, and Z. Dembic, Qa-1 restricted recognition of foreign antigen by a γ-δ T-cell hybridoma, *Nature* 340:646 (1989).

26. K. Ito, L. Van Kaer, M. Bonneville, S. Hsu, D.B. Murphy, and S. Tonegawa, Recognition of the product of a novel MHC TL region gene (27b) by a mouse γ-δ T cell receptor, *Cell* 62:549 (1990).

27. F. Imani, and M.J. Soloski, Heat shock proteins can regulate expression of the Tla region-encoded class Ib molecule Qa-1, *Proc. Natl. Acad. Sci. USA* 88:10475 (1991).

28. Meeder.E.R. De Graeff, Zee.R. van der, G.T. Rijkers, H.J. Schuurman, W. Kuis, J.W. Bijlsma, B.J. Zegers, and W. Van Eden, Recognition of human 60 kD heat shock protein by mononuclear cells from patients with juvenile chronic arthritis, *Lancet* 337:1368 (1991).

29. L.ucassen-M-A.van-der-Zee-R.. Boog-C-J.de-Graeff-Meeder-E-R., Two monoclonal antibodies generated against human hsp60 show reactivity with synovial membranes of patients with juvenile chronic arthritis, *J-Exp-Med.* . (1992).

30. S.H.E. Kaufmann, B. Schoel, A. Wand-Württenberger, U. Steinhoff, M.E. Munk, and T. Koga, T-cells, stress proteins, and pathogenesis of mycobacterial infections, *Curr. Top. Microbiol. Immunol.* 155:125 (1990).

31. M.E. Munk, B. Schoel, S. Modrow, R.W. Karr, R.A. Young, and S.H. Kaufmann, T lymphocytes from healthy individuals with specificity to self-epitopes shared by the mycobacterial and human 65-kilodalton heat shock protein, *J. Immunol.* 143:2844 (1989).

32. G. Benichou, P. Takizawa, P. Ho, C. Killion, C. Olson, M. McMillan, and E. Sercarz, Immunogenicity and tolerogenicity of self-major histocompatability complex peptides, *J. Exp. Med.* 172:1341 (1990).

33. R.M. Pope, M.A. Pahlavani, E. LaCour, S. Sambol, and B.V. Desai, Antigenic specificity of rheumatoid synovial fluid lymphocytes, *Arthritis Rheum.* 32, No.11:1371 (1989).

34. K. Söderström, E. Halapi, E. Nilsson, A. Grönberg, J. van Embden, L. Klareskog, and R. Kiessling, Synovial cells responding to a 65-kDa mycobacterial heat shock protein have a high proportion of a TcR$\gamma\delta$ subtype uncommon in peripheral blood, *Scand. J. Immunol.* 32:503 (1990).

35. F.D. Crick, and P.A. Gatenby, Limiting-dilution analysis of T cell reactivity to mycobacterial antigens in peripheral blood and synovium from rheumatoid arthritis patients, *Clin. Exp. Immunol.* 88:424 (1992).

36. H.P. Fischer, C.E. Sharrock, M.J. Colston, and G.S. Panayi, Limiting dilution analysis of proliferative T cell responses to mycobacterial 65-kDa heat-shock protein fails to show significant frequency differences between synovial fluid and peripheral blood of patients with rheumatoid arthritis, *Eur. J. Immunol.* 21:2937 (1991).

37. P. Res, J. Thole, and R. de Vries, Heat-shock proteins and autoimmunity in humans, *Springer. Semin. Immunopathol.* 13:81 (1991).

38. P.F. Life, E.O. Bassey, and J.S. Gaston, T-cell recognition of bacterial heat-shock proteins in inflammatory arthritis, *Immunol. Rev.* 121:113 (1991).

39. S.G. Li, A.J. Quayle, Y. Shen, J. Kjeldsen Kragh, F. Oftung, R.S. Gupta, J.B. Natvig, and O.T. Forre, Mycobacteria and human heat shock protein-specific cytotoxic T lymphocytes in rheumatoid synovial inflammation, *Arthritis Rheum.* 35:270 (1992).

40. N. Perez, J. Sugar, S. Charya, G. Johnson, C. Merril, L. Bierer, D. Perl, V. Haroutunian, and W. Wallace, Increased synthesis and accumulation of heat shock 70 proteins in Alzheimer's disease, *Brain Res. Mol. Brain Res.* 11:249 (1991).

41. T. Koga, A. Wand Wurttenberger, J. DeBruyn, M.E. Munk, B. Schoel, and S.H. Kaufmann, T cells against a bacterial heat shock protein recognize stressed macrophages, *Science* 245:1112 (1989).

42. J.R. Lamb, V. Bal, P. Mendez-Samperlo, A. Mehlert, A. So, J. Rothbard, S. Jindal, R.A. Young, and D.B. Young, Stress proteins may provide a link between the immune response to infection and autoimmunity, *Int. Immunol.* 1:191 (1989).

43. D.C. Anderson, W. van Schooten, M. Barry, A. Janson, T. Buchanan, and R.P. de Vries, A mycobacterium leprae specific human T cell epitope cross-reactive with an HLA DR2 peptide, *Science* 242:259 (1988).

44. S. Albani, J.E. Tuckwell, L. Esparza, D.A. Carson, and J. Roudier, The susceptibility sequence to rheumatoid arthritis is a cross-reactive B cell epitope shared by the Escherichia coli heat shock protein dnaJ and the histocompatibility leukocyte antigen DRB10401 molecule, *J. Clin. Invest.* 89:327 (1992).

45. J. Holoshitz, F. Koning, J.E. Coligan, J. de Bruyn, and S. Strober, Isolation of CD4-CD8- mycobacteria-reactive T lymphocyte clones from rheumatoid arthritis synovial fluid, *Nature* 339:226 (1989).

46. A. De Maria, M. Malnati, A. Moretta, D. Pende, C. Bottino, G. Casorati, F. Cottafava, G. Melioli, M.C. Mingari, N. Migone, S. Romagnani, and L. Moretta, CD3+4-8-WT31- (T cell receptor γ+) cells and other unusual phenotypes are frequently detected among spontaneously interleukin 2-responsive T lymphocytes present in the joint fluid in juvenile rheumatoid arthritis. A clonal analysis, *Eur. J. Immunol.* 17 (12):1815 (1987).

47. F.M. Brennan, M. Londei, A.M. Jackson, T. Hercend, M.B. Brenner, R.N. Maini, and M. Feldmann, T cells expressing γδ chain receptors in rheumatoid arthritis, *J. Autoimmun.* 1:319 (1988).

48. S. Porcelli, M.B. Brenner, J.L. Greenstein, S.P. Balk, C. Terhorst, and P.A. Bleicher, Recognition of cluster of differentiation 1 antigens by human CD4-CD8-cytolytic T lymphocytes, *Nature* 341:447 (1989).

49. R. Pope, J. Lessard, and A. Landay, δ/γ T cell receptor (TCR) positive T lymphocytes: lack of evidence for significant involvement in activation of synovial lymphocytes by mycobacterial antigens, *Arthritis Rheum.* 33, No.9, Supp.:S57 (1990).

50. M. Giovanna Danieli, D. Markovits, A. Gabrielli, A. Corvetta, P.L. Giorgi, R. van der Zee, J.D.A. Van Embden, G. Danieli, and I.R. Cohen, Juvenile rheumatoid arthritis patients manifest immune reactivity to the mycobacterial 65-kDa heat shock protein, to its 180- 188 peptide, and to a partially homologous peptide of the proteoglycan link protein, *Clin. Immunol. Immunopathol.* 64:121 (1992).

51. G. Tsoulfa, G.A. Rook, J.D. van Embden, D.B. Young, A. Mehlert, D.A. Isenberg, F.C. Hay, and P.M. Lydyard, Raised serum IgG and IgA antibodies to mycobacterial antigens in rheumatoid arthritis, *Ann. Rheum. Dis.* 48:118 (1988).

52. L. McLean, J. Archer, M. Cawley, B. Kidd, F. Pegley, and P. Thompson, 65 kDa stress protein (SP65) antibodies in ankylosing spondylitis (AS) and rheumatoid arthritis (RA), *Arthritis Rheum.* 33, No.9 Supplement:S68 (1990). (Abstr.)

53. N. Mensi, D.R. Webb, C.W. Turck, and G.A. Peltz, Characterization of Borrelia burgdorferi proteins reactive with antibodies in synovial fluid of a patient with Lyme Arthritis, *Infect. Immun.* 58:2404 (1990).

54. S. Muller, J.P. Briand, and M.H. Van Regenmortel, Presence of antibodies to ubiquitin during the autoimmune response associated with systemic lupus erythematosus, *Proc. Natl. Acad. Sci. U. S. A* 85:8176 (1988).

55. S. Plau', S. Muller, and M.H.V. Van Regenmortel, A branched, synthetic octapeptide of ubiquitinated histone H2A as target of autoantibodies, *J. Exp. Med.* 169:1607 (1990).

56. S. Minota, S. Koyasu, I. Yahara, and J.B. Winfield, Autoantibodies to the heat-shock protein hsp90 in systemic lupus erythematosus, *J. Clin. Invest.* 81:106 (1988).

57. S. Minota, B. Cameron, W.J. Welch, and J.B. Winfield, Autoantibodies to the constitutive 73-kD member of the hsp70 family of heat shock proteins in systemic lupus erythematosus, *J. Exp. Med.* 168:1475 (1988).

58. W.N. Jarjour, B.D. Jeffries, J.S. Davis,IV, W.J. Welch, T. Mimura, and J.B. Winfield, Autoantibodies to stress proteins: a survey of various rheumatic and inflammatory diseases, *Arthritis Rheum.* 34:1133 (1991).

59. M.E. Cheetham, J.P. Brion, and B.H. Anderton, Human homologues of the bacterial heat-shock protein DnaJ are preferentially expressed in neurons, *Biochem. J.* 284:469 (1992).

60. M.A. Pappolla, R.A. Omar, K.S. Kim, and N.K. Robakis, Immunohistochemical evidence of antioxidant stress in Alzheimer's disease, *Am. J. Pathol.* 140:621 (1992).

61. J.E. Hamos, B. Oblas, D. Pulaski Salo, W.J. Welch, D.G. Bole, and D.A. Drachman, Expression of heat shock proteins in Alzheimer's disease, *Neurology* 41:345 (1991).

62. A. Morandi, B. Los, L. Osofsky, L. Autilio Gambetti, and P. Gambetti, Ubiquitin and heat shock proteins in cultured nervous tissue after different stress conditions, *Prog. Clin. Biol. Res.* 317:819 (1989).

63. J.G. Guillemette, L. Wong, D.R. Crapper McLachlan, and P.N. Lewis, Characterization of messenger RNA from the cerebral cortex of control and Alzheimer-afflicted brain, *J. Neurochem.* 47:987 (1986).

64. K.W. Wucherpfennig, J. Newcombe, H. Li, C. Keddy, M.L. Cuzner, and D.A. Hafler, $\gamma\delta$ T-cell receptor repertoire in acute multiple slerosis lesions, *Proc. Natl. Acad. Sci. USA* 89:4588 (1992).

65. K. Selmaj, C.F. Brosnan, and C.S. Raine, Expression of heat shock protein-65 by oligodendrocytes in vivo and in vitro: implications for multiple sclerosis, *Neurology* 42:795 (1992).

66. M.S. Freedman, T.C. Ruijs, L.K. Selin, and J.P. Antel, Peripheral blood γ-δ T cells lyse fresh human brain-derived oligodendrocytes, *Ann. Neurol.* 30:794 (1991).

67. K. Selmaj, C.F. Brosnan, and C.S. Raine, Colocalization of lymphocytes bearing $\gamma\delta$ T-cell receptor and heat shock protein hsp65+ oligodendrocytes in multiple sclerosis, *Proc. Natl. Acad. Sci. U. S. A* 88:6452 (1991).

68. A.M. McGregor, Heat shock proteins and autoimmune thyroid disease; too hot to handle [editorial; comment], *J. Clin. Endocrinol. Metab.* 74:720 (1992).

69. A.E. Heufelder, B.E. Wenzel, and R.S. Bahn, Cell surface localization of a 72 kilodalton heat shock protein in retroocular fibroblasts from patients with Graves' ophthalmopathy, *J. Clin. Endocrinol. Metab.* 74:732 (1992).

70. A.E. Heufelder, B.E. Wenzel, and R.S. Bahn, Methimazole and propylthiouracil inhibit the oxygen free radical-induced expression of a 72 kilodalton heat shock protein in Graves' retroocular fibroblasts, *J. Clin. Endocrinol. Metab.* 74:737 (1992).

71. D.C. Markesich, E.T. Sawai, J.S. Butel, and D.Y. Graham, Investigations on etiology of Crohn's disease. Humoral immune response to stress (heat shock) proteins, *Dig. Dis. Sci.* 36:454 (1991).

72. A. Elsaghier, C. Prantera, G. Bothamley, E. Wilkins, S. Jindal, and J. Ivanyi, Disease association of antibodies to human and mycobacterial hsp70 and hsp60 stress proteins, *Clin. Exp. Immunol.* 89:305 (1992).

73. L.K. Trejdosiewicz, A. Calabrese, C.J. Smart, D.J. Oakes, P.D. Howdle, J.E. Crabtree, M.S. Losowsky, F. Lancaster, and A.W. Boylston, $\gamma\delta$ T cell receptor-positive cells of the human gastrointestinal mucosa: occurrence and V region gene expression in Heliobacter pylori-associated gastritis, coeliac disease and inflammatory bowel disease, *Clin. Exp. Immunol.* 84:440 (1991).

74. R. Hohlfeld, A.G. Engel, K. Ii, and M.C. Harper, Polymyositis mediated by T lymphocytes that express the γ/δ receptor, *N. Engl. J. Med.* 324:877 (1991).

75. J.B. Winfield, Stress proteins, arthritis, and autoimmunity, *Arthritis Rheum.* 32:1497 (1989).

76. J.B. Winfield, and W. Jarjour, Do stress proteins play a role in arthritis and autoimmunity?, *Immunol. Rev.* 121:193 (1991).

77. R.P. Morrison, R.J. Belland, K. Lyng, and H.D. Caldwell, Chlamydial disease pathogenesis. The 57-kD chlamydial hypersensitivity antigen is a stress response protein, *J. Exp. Med.* 170:1271 (1989).

POLYCLONAL B CELL ACTIVATION AND B CELL CROSS-REACTIVITY DURING AUTOANTIBODY PRODUCTION IN SYSTEMIC LUPUS ERYTHEMATOSUS

Dennis M. Klinman[1], Akira Shirai[2], and Yoshiaki Ishigatsubo[2]

[1]Division of Virology, Center for Biologics Evaluation and Research, Food and Drug Administration, Bethesda, MD 20892
[2]Yokohama City University Medical School Yokohama, Japan

INTRODUCTION

Increased levels of serum autoantibodies are the sine qua non of systemic lupus erythematosus[1]. Several models have been proposed to explain how such autoantibody production is induced and maintained. The first holds that specific autoantigens (or antigens cross-reactive with self) stimulate an anti-self response (i.e., DNA inducing anti-DNA antibodies)[2]. The second proposes that autoantibody production results from a generalized process of antigen-independent polyclonal B cell activation[3,4]. Several years ago, we suggested a synthesis of these ideas: that polyclonal activation initiated autoantibody production while autoantigenic stimulation perpetuated the process[5]. Recent findings suggest that cross-reactive B cells also play an important role in determining the nature of the antibodies produced in animals with systemic autoimmune disease.

B CELLS OF MANY DIFFERENT SPECIFICITIES ARE STIMULATED IN SLE

Autoantibody production has been studied for decades and the literature on this subject is extensive. One cannot help but be impressed by the diversity of self and foreign antigens recognized by antibodies from lupus-prone animals. These include nuclear antigens (DNA, histones, Sm and SnRNP), determinants on cell membranes (erythrocytes, T cells, macrophages and kidney cells), cytoplasmic proteins (myosin and actin), as well as foreign antigens (TNP, ARS and SRBC). Since individual researchers usually study the response to only one or a few determinants, the general nature of this B cell activation has not always been appreciated. However, the literature contains evidence that humans and mice with lupus have increased numbers of B cells reactive with scores of different self and foreign antigens. This is more consistent with a process of polyclonal activation than antigen-specific immune stimulation. Whereas an autoantigen-driven response should be antigenically restricted, a polyclonal stimulus should activate cells of many different specificities, as observed experimentally.

Immunobiology of Proteins and Peptides VII
Edited by M.Z. Atassi, Plenum Press, New York, 1994

It has been suggested that the response of autoimmune animals cannot be polyclonal, since not all B cells reactive with all antigens are stimulated. Yet polyclonal activation is not the same as toti-clonal activation. Homeostatic mechanisms seem to prevent the in vivo stimulation of a majority of lymphocytes, even in a strongly immunostimulatory environment. For example, repeated treatment with polyclonal activators results in the stimulation of only 2% of available B cells, a frequency quite similar to that seen in mice with advanced lupus. While 2% may not seem like a large fraction of the lymphocyte pool, only 0.1% of lymphocytes in normal mice are activated to secrete Ig (Figure 1)[6].

The specificity of the antibodies produced by polyclonally-activated B cells can be influenced by several factors. First, there is wide variation in the frequency with which primary B cells recognize different antigens. Less than 0.05% of B cell precursors react with DNP or fluorescein while 0.5 - 5% of splenic lymphocytes recognize DNA[7-9]. Lymphocytes also differ in their susceptibility to polyclonal activating factors. Memory B cells, for example, are more easily stimulated than primary lymphocytes of the same specificity. Autoreactive lymphocytes which have been actively suppressed in vivo might also be hyper-responsive to immunostimulatory signals once released from such suppression. Such a situation would result in the preferential stimulation of autoreactive B cells in the absence of a specific autoantigenic stimulus. Given differences in precursor frequency and responsivity to (polyclonal) activating factors, variability in the number of B cells recognizing different antigens in an immunostimulatory environment would be expected.

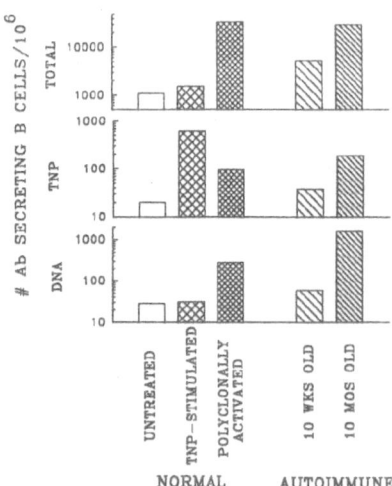

Figure 1. The number of splenic B cells from normal BALB/c mice and lupus-prone NZB/W mice was determined by ELIspot assay. Note that immunization with TNP-KLH results in a repertoire skewed towards anti-TNP production, while treatment with a polyclonal activator (LPS) yield a repertoire similar to that manifest by autoimmune mice.

116

By one month of age, the number of B cells secreting Ig in lupus-prone mice is significantly elevated when compared to normals (Figure 1) 5. This is an important observation, since it indicates that the process(es) responsible for B cell activation in lupus are present shortly after birth.

Studies in my lab have shown that the fraction of B cells reactive with a number of different antigens increases proportionally in young lupus-prone mice (Figure 2)[10]. This is consistent with a process in which the repertoires expressed by young mice are expanded polyclonally.

By comparison, the fraction of B cells reactive with specific autoantigens increases preferentially only in animals with clinically detectable autoimmunity. For example, Eisenberg et al reported that anti-Sm antibodies were preferentially overproduced by adult but not young MRL-lpr/lpr mice[11,12]. In NZB/W mice, Kotzin et al found increased anti-histone antibodies and Hahn et al found increased anti-DNA antibodies only in adult mice with active lupus[13,14].

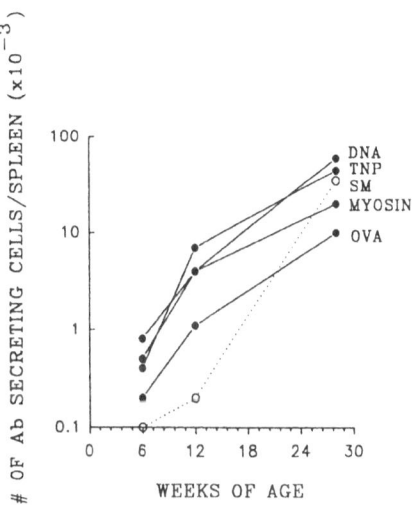

Figure 2. The number of splenic B cells from MRL/lpr mice producing antibodies reactive with these antigens was determined by ELIspot assay. Note the simultaneous and equivalent increase in reactivity between 6 and 12 weeks, followed by the preferential increase in anti-SM reactivity later in disease.

It has been argued that IgG secreting B cells are uniquely responsible for disease pathogenesis, and that the polyclonal activation of IgM producing cells is therefore irrelevant. To examine this possibility, the correlation between early polyclonal activation versus (auto)antigen-specific stimulation on the development of lupus was analyzed in (NZB X NZW)F_1 X NZB backcross mice. Results showed that the presence of polyclonal (IgM) activation early in life was an excellent predictor of the subsequent development of immune-mediated glomerulonephritis (P <.001). In contrast, young animals expressing repertoires biased towards the production of anti-DNA antibodies were at no greater risk for the development of disease than were non-anti-DNA producing littermates[10].

Table 1. Early polyclonal activation predicts autoimmune disease

Repertoire expressed at 10 weeks of age	% of (NZB x NZW)F_1 x NZB mice surviving at 7 months
Normal	95
Polyclonally activated[1]	42
Skewed towards anti-DNA[2]	94

[1]Approximately 50% of mice expressed repertoires showing a polyclonal increase in the number of Ig secreting B cells.
[2]Approximately 10% of mice expressed a repertoire in which the fraction of B cells secreting anti-DNA antibodies was significantly increased over littermate controls.

In older mice, by comparison, increased IgG anti-DNA production was significantly correlated with the presence of active lupus nephritis. Thus, IgG anti-DNA antibodies appear to contribute to end-organ damage in SLE, but may not be involved in the initiation of the disease process.

EFFECT OF POLYCLONAL ACTIVATORS ON THE B CELL REPERTOIRE

Studies have been conducted to determine whether an autoimmune-like B cell repertoire can be induced by treating normal mice with polyclonal activators[15]. In the early 1980's, it was shown that autoantibodies were produced when normal animals were repeatedly stimulated with LPS[16,17]. Similar treatment hastened the development of autoimmune disease in lupus-prone mice. More recent studies showed that polyclonal activators could induce normal DBA/2 mice to express repertoires quantitatively and qualitatively similar to those found in autoimmune NZB mice (Figure 1)[6]. Similarly, introduction of the BCL2 transgene into normal animals (expression of BCL2 results in polyclonal B cell activation) resulted in the production of antibodies against a diverse array of autoantigens (including Sm and DNA) and the development of immune-complex mediated glomerulonephritis[18]. These findings indicate that polyclonal B cell activation can result in a pattern of autoantibody production and disease similar to that found in SLE.

The factor(s) or process(es) responsible for inducing polyclonal activation in lupus are currently unknown. Possibilities include the overproduction of stimulatory cytokines/endotoxins, an increased responsiveness to such agents, or a defect in immune suppression (reviewed in 15). Factors capable of inducing lymphocyte maturation and/or proliferation are present at increased concentrations in mice with systemic autoimmunity. For example, significantly elevated levels of IFN_{gamma} are present in the serum of MRL/lpr mice, a finding consistent with the overproduction of IgG2a

118

by mice of that strain[19-21]. It also appears that B cells from autoimmune animals are hyper-responsive to immunostimulatory cytokines which are present in vivo[22,23].

By comparison, attempts to demonstrate that autoantigens are present at elevated levels (or in unusually immunogenic forms) in autoimmune mice have generally failed. As a rule, autoantigens are poor immunogens and must be extensively modified to induce an autoimmune response and there is little evidence that such modifications occur in vivo. It is also noteworthy that in cases where immune responses have been elicited by (modified) autoantigens, they are often directed against epitopes which are not recognized by naturally occurring autoantibodies.

ROLE OF AUTOANTIGEN IN THE PERPETUATION OF AUTOANTIBODY PRODUCTION

While polyclonal activation appears to play a dominant role in initiating autoantibody production, it is clear that autoantigen (or an antigen cross-reactive with self) contributes to the perpetuation of this process. For example, Eisenberg et al. showed that the concentration of anti-Sm autoantibodies in MRL/lpr-lpr mice increased with age and was associated with isotype switching (Figure 2)[24]. Moreover, they demonstrated that anti-Sm antibodies could be induced by exogenously administered Sm and that Sm was present in the serum of high-responder mice. Our own work confirmed that the proportion of B cells producing anti-Sm antibodies rose disproportionately in **adult** MRL-lpr/lpr mice, as would be expected for an antigen-driven response[25]. Yet young mice of the same strain expressed repertoires characterized by polyclonal activation, reflected by the simultaneous and proportional proliferation of B cells reactive with a number of self and foreign antigens[25].

Evidence that antigen plays a role in the selection of autoreactive clones is also derived from studies of the genes that encode such autoantibodies. First, analyses of anti-DNA secreting hybridomas has shown that the ratio of replacement to silent nucleotide mutations is higher than expected, arguing that positive selection has occurred. Second, such replacements tend to cluster in the complementarity defining regions of the Ig molecule. Finally, there is some (albeit inconclusive) evidence that clonal diversity in the anti-DNA response declines with age, consistent with antigen-driven selection of only a subset of B cells[26,27].

B CELL CROSS-REACTIVITY AND AUTOANTIBODY PRODUCTION

Sera from normal individuals contain cross-reactive antibodies capable of binding to a diverse array of foreign and self antigens[28,29]. These 'natural' antibodies are characteristically of the IgM heavy chain class and tend to be of low affinity[30,31]. It has been postulated that they protect the host from pathogens during the period between infection and the development of a more specific immune response. Yet the stimulation of cross-reactive B cells is strictly regulated in vivo, perhaps reflecting their potential to contribute to the development of autoimmune disease.

Our lab used a novel chamber ELIspot assay to investigate the cross-reactivity of B cells secreting Ig under physiologic conditions[32]. Results indicate that up to one-quarter of spontaneously activated IgM secreting lymphocytes are cross-reactive. By comparison, immunization induced the activation of IgG secreting cells that cross-reacted at much lower frequencies[32]. These findings are consistent with the view that the antigen-driven selection of somatically mutated B cells yields clones of high affinity and low cross-reactivity[33].

Table 2. Cross-reactivity of anti-DNA secreting human B cells

B cell source	Isotype	% Cross-reactivity with		
		TNP	DNA	Ova
Normal	IgM	15	100	13
Normal	IgG	4	100	5
Autoimmune	IgM	18	100	13
Autoimmune	IgG	17	100	13

The cross-reactivity of <u>in vivo</u> activated B cells from 17 patients with active systemic lupus erythematosus was compared to that of B cells from 13 normal volunteers. Note that in normals but not lupus patients, a significantly smaller fraction of IgG versus IgM anti-DNA producing cells are cross-reactive.

It was therefore surprising to observe that IgG autoantibody secreting cells from humans with lupus cross-reacted significantly more frequently than normal. This was not due to a generalized defect in the selection or maturation of the IgG secreting cell repertoire, however, since the cross-reactivity of B cells from lupus patients producing IgG against non-autoantigens was no different than normals (Table 2).

Presumably, cross-reactive IgG secreting lymphocytes normally would be tolerized or suppressed <u>in vivo</u>, perhaps due to their ability to bind self antigens and promote autoimmune disease. The immunoregulatory defect that permits the stimulation of autoreactive B cells in lupus patients also appears to allow the activation of cross-reactive clones. Supporting this contention, preliminary studies of murine AIDS and graft-versus-host disease show that other conditions which elicit autoantibody production are also associated with the increased use of cross-reactive B cells. The observation that cross-reactive B cells may be used more commonly in autoimmune states raises a serious issue with respect to the interpretation of many previous studies. Historically, researchers compared the concentration of a specific serum antibody to the total concentration of serum Ig. For example, IgG anti-DNA levels may have risen eight fold while serum IgG levels rose only two-fold, leading the investigator to conclude that a specific increase in serum anti-DNA had been detected. However, such a comparison presumes that the nature of the serum Ig pool had not changed. If the two-fold increase in serum Ig was due to an eight-fold rise in the production of cross-reactive antibodies, then the concentration of IgG antibodies reactive with a variety of antigens, including DNA, might increase by much more than two-fold. This concept is modeled in Figure 3.

120

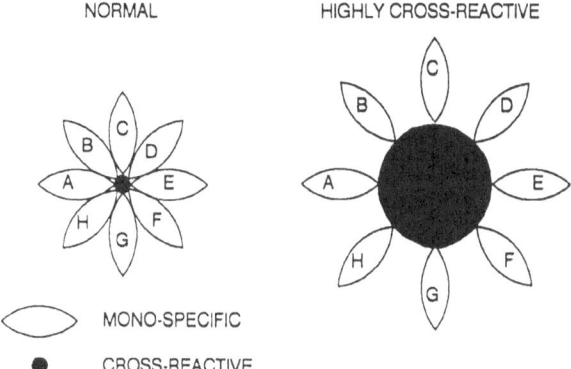

Effect of Increased B Cell Cross Reactivity
of the Antigen-Binding Repertoire

NORMAL   HIGHLY CROSS-REACTIVE

MONO-SPECIFIC

CROSS-REACTIVE

Figure 3. In normal animals, a relatively small fraction of B cells are cross-reactive. In autoimmune animals, this fraction is significantly increased. In the latter situation, the sum of mono-reactive and cross-reactive B cells recognizing any given antigen would appear to be increased our to proportion to the total number of Ig secreting cells.

CONCLUSION

Considerable insight into the immune abnormalities that culminate in the development of systemic autoimmune disease has been obtained by long-term studies of mice and humans with lupus. While the active disease state is characterized by the specific overproduction of pathogenic autoantibodies, the pre-clinical state is characterized by a process of polyclonal B cell activation. We believe that this polyclonal activation compromises the host's ability to tolerize/suppress autoreactive lymphocytes. Immune-mediated destruction of target cells establishes a positive feedback loop, in which the release of self antigens by injured cells induces further autoantibody production. Throughout this process, the immunostimulatory milieu created by polyclonal activating factor(s) facilitates the stimulation of self-reactive and cross-reactive B cells. We conclude that the therapy of patients with SLE should be initiated as early as possible and be directed towards reducing polyclonal activation (rather than at specific responses to self antigens).

REFERENCES

1. A.N. Theofilopoulos and F.J. Dixon. Etiopathogenesis of murine systemic lupus erythematosus. Immunol.Rev. 55:179 (1981).
2. J.A. Hardin. The lupus autoantigens and the pathogenesis of SLE. Arthritis Rheum. 29:457 (1986).
3. S.P. Izui, P.J. McConahey and F.J. Dixon. Increased spontaneous polyclonal activation of B lymphocytes in mice with spontaneous autoimmune disease. J.Immunol. 121:2213 (1978).
4. D.M. Klinman and A.D. Steinberg. Systemic autoimmune

121

disease arises from polyclonal B cell activation. J.Exp.Med. 165:1755 (1987).

5. D.M. Klinman, Y. Ishigatsubo and A.D. Steinberg. Acquisition and maturation of expressed B cell repertoires in normal and autoimmune mice. J.Immunol. 141:801 (1988).

6. D.M. Klinman. Regulation of B cell activation in autoimmune mice. Clin.Immunol.and Immunopath. 530:250 (1989).

7. N.R. Klinman, R.L. Riley, M.R. Stone, D. Wylie and D. Zharhary. The specificity repertoire of prereceptor and mature B cells. Ann NY Acad.Sci. 179:130 (1986).

8. D.S. Pisetsky and S.A. Caster. The B-cell repertoire for autoantibodies: frequency of precursor cells for anti-DNA antibodies. Cell.Immunol. 15:294 (1982).

9. N.H. Sigal and N.R. Klinman. The B-cell clonotype repertoire. Adv.Immunol. 26:255 (1978).

10. D.M. Klinman. Polyclonal B cell activation in lupus-prone mice precedes and predicts the development of autoimmune disease. J Clin Invest 86:1249 (1990).

11. R.A. Eisenberg, S.Y. Craven and P.L. Cohen. The stochastic control of anti-Sm autoantibodies in MRL/Mp-lpr/lpr mice. J.Clin.Invest. 80:691 (1987).

12. R.A. Eisenberg, S.Y. Craven and P.L. Cohen. Isotype progression and clonality of anti-Sm autoantibodies in MRL/Mp-lpr/lpr mice. J.Immunol. 139:728 (1987).

13. D.G. Ando, F.M. Ebling and B.H. Hahn. Detection of native and denatured DNA antibody forming cells by the enzyme-linked immunospot assay. A clinical study of (NZB x NZW) F1 mice. Arthritis Rheum. 29:1139 (1986).

14. M. Gioud, B.L. Kotzin, F.L. Rubin, F.G. Joslin and E.M. Tan. In vivo and in vitro production of anti-histone antibodies in NZB/NZW mice. J.Immunol. 131:269 (1983).

15. A.D. Steinberg, S. Baron and N. Talal. The pathogenesis of autoimmunity in New Zealand mice. I. Induction of anti-nucleic acid antibodies by polyinosinic. polycytidylic acid. Proc.Natl.Acad.Sci.USA 63:1102 (1969).

16. L. Hang, M.T. Aguado, F.J. Dixon and A.N. Theofilopolous. Induction of severe autoimmune disease in normal mice by simultaneous action of multiple immunostimulators. J.Exp.Med. 161:423 (1985).

17. K.H. Slack, L. Hang, L. Barkley, R.J. Fulton, L.A. D'Hoostelaere, A. Robinson and F.J. Dixon. Isotypes of spontaneous and mitogen induce autoantibodies in SLE prone mice. J.Immunol. 132:1271 (1984).

18. A. Strasser, S. Whittingham, D.L. Vaux, M.L. Bath and A. Harris. Enforced BCL2 expression in B-lymphoid cells prolongs antibody responses and elicits autoimmune disease. Proc.Natl.Acad.Sci. 88:8661 (1991).

19. G.J. Prud'homme, C.L. Park, T.M. Fieser, R. Kofler, F.J. Dixon and A.N. Theofilopoulos. Identification of a B cell differentiation factor (s) spontaneously produced by proliferating T cells in murine lupus strains of the lpr/lpr genotype. J.Exp.Med. 157:730 (1983).

20. L. Murray and C. Martens. The abnormal lymphocytes in lpr mice transcribe Interferon-gamma and tumor necrosis factor-alpha genes spontaneously in vivo. Eur.J.Immunol. 19:563 (1989).

21. S. Umland, R. Lee, M. Howard and C. Martens. Expression of lymphokine genes in splenic lymphocytes of autoimmune mice. Mol.Immunol. 26:649 (1989).

22. L.R. Herron, R.L. Coffman, M.W. Bond and B.L. Kotzin. Increased autoantibody production by NZB/NZW B cells in

response to Il-5. J.Immunol. 141:842 (1988).

23. D.M. Klinman. IgG1 and IgG2a production by autoimmune B cells treated in vitro with IL-4 and IFN-gamma. J.Immunol. 144:2529 (1990).

24. E.W. Shores, R.A. Eisenberg and P.C. Cohen. Role of the Sm antigen in the generation of anti-Sm antibodies in the SLE-prone MRL mouse. J.Immunol. 136:3662 (1987).

25. D.M. Klinman, R.A. Eisenberg and A.D. Steinberg. Development of the autoimmune B cell repertoire in MRL-lpr/lpr mice. J.Immunol. 1440:506 (1990).

26. M. Shlomchik, A. Marshak-Rothstein, C.B. Wolfowicz, T.L. Rothstein and M.G. Weigert. The role of clonal selection and somatic mutation in autoimmunity. Nature 328:805 (1987).

27. M. Shlomchik, M. Mascelli, H. Shan, M.S. Zadic, D. Pisetsky, A. Marshak-Rothstein and M. Weigert. Anti-DNA antibodies from autoimmune mice arise by clonal expansion and somatic mutation. J.Exp.Med. 171:265 (1990).

28. B. Guilbert, G. Dighiero and S. Avrameas. Naturally occurring antibodies against nine common antigens in human sera. J.Immunol. 128:2779 (1982).

29. G. Dighiero, P. Lymberi, J. Mazie, S. Rouyre, G.S. Butler-Browne, R.G. Whalen and S. Avrameas. Murine hybridomas secreting natural monoclonal antibodies reactive with self antigens. J.Immunol. 131:2267 (1983).

30. G. Dighiero, P. Lymberi, D. Holmberg, I. Lundquist, A. Coutinho and S. Avrameas. High frequency of natural autoantibodies in normal newborn mice. J.Immunol. 134:765 (1985).

31. A. Kaushik, A. Lim, P. Poncet, X. Ge and G. Dighiero. Comparative analysis of natural antibody specificites among hybridomas originating from spleen and peritoneal cavity of adult NZB and BALB/c mice. Scand.J.Immunol. 27:461 (1988).

32. D.M. Klinman. Cross-reactivity of IgM secreting B cells from normal BALB/c mice. J.Immunol. In Press (1992).

33. Y. Naparstek, J. Andre-Schwartz, T. Manser, L.J. Wyosocki, L. Breitman, B.D. Stollar, M. Gefter and R.S. Schwartz. A single germline VH gene segment of normal A/J mice encodes autoantibodies characteristic of systemic lupus erythematosus. J.Exp.Med. 164:514 (1986).

AUTOANTIBODY ACTIVITY AND V GENE USAGE BY B-CELL MALIGNANCIES

Guillaume Dighiero

Immunohematologie et Immunopathologie
Institute Pasteur
75724 Paris Cedex 15, France

INTRODUCTION

During the last few years, we have been particularly interested in the study of the origins and significance of natural autoantibodies. With Guilbert and Avrameas[1] at Pasteur Institute, we demonstrated the constant presence of natural autoantibodies directed against a broad variety of autoantigens in normal human serum. These results led us to extrapolate them in order to formulate the hypothesis that natural autoantibodies reacting with a wide variety of autoantigens will be found in normal human serum and will constitute a substantial part of normal circulating Igs. Taken together with those reported from other groups, these results challenge the clonal deletion theory as a general and unique explanation for self-tolerance. However, autoantibodies against very specific antigens defining the polymorphism (e.g. the A and B antigens in the case of subjects belonging repectively to the A or B group) or against polymorphic antigens of the major histocompatibility system have not been described. Therefore, clonal deletion mechanisms cannot, at the present time, be completely excluded for such particular cases. In addition, results from transgenic mice led some authors to conclude that clones specific for self antigens associated to cellular membranes are deleted in bone-marrow, whereas clones specific for soluble self antigens are anergized in peripheral lymphoid organs[2].

The presence of natural autoantibodies in normal serum implies the existence of normal autoreactive B cell clones under physiological conditions. To substantiate this further, we established natural autoantibody secreting hybridomas from the spleen of adult[3] and new-born[4] normal unimmunized BALB/c mice. As it was the case for natural autoantibodies observed in human serum, these hybridomas secreted monoclonal antibodies frequently displaying a polyspecific binding activity. The frequency of this precursor B cell repertoire was impressively high, specially in new-born mice.

Immunobiology of Proteins and Peptides VII
Edited by M.Z. Atassi, Plenum Press, New York, 1994

These autoreactive B cells secrete the so called "NAAB" characterized by their broad and puzzling reactivity, which is mainly directed against very well conserved public epitopes. They fulfill the definition of an autoantibody since they are self-reactive, but they are not self-specific[5].

In 1983[3], while reporting our results with murine hybridomas displaying NAAB activity and derived from BALB/c mice, we postulated: "If correct, this hypothesis would imply that these B cells carry a polyspecific receptor capable of fixing several different antigens. During active immunization, the binding of a given antigen to this receptor would then induce the corresponding B cell by a series of divisions and mutations, which, under the selective pressure of the antigen, would lead to the production of highly specific antibodies for a given epitope of these antigens. In that case, these B cells and their receptors would then be encoded by the germ-line, and the natural polyspecific antibodies would constitute the synthesized products secreted by these cells". This provocative hypothesis assumed that NAAB were the product of the germ-line repertoire and that they constituted a template upon which Ag driven selection and somatic mutation could operate to derive highly specific immune antibodies.

The germ-line origin was substantiated during the following years, since: 1) polyreactive NAAB appear to predominate early during ontogeny[4] even in germ-antigen free animals[6]; 2) NAAB express recurrent idiotypes[7-15] and appear to be highly interconnected[16]; 3) structural studies have demonstrated their germinal origin[17-20] and the fact that these autoantibodies are encoded by structural genetic elements similar to those employed by non-autoimmune autoantibodies[21,22].

The template hypothesis is still awaiting an experimental verification. Only indirect evidence has been obtained as yet. Naparstek[23] have studied a set of antibodies that use the $VHId^{CR}$ gene, employed for the anti-arsonate system. They have found that germ-line-encoded antibodies of this system do not bind arsonate, but display a polyreactive pattern of binding by reacting with DNA and cytoskeleton proteins. Upon immunization with arsonate, the $VHId^{CR}$ gene undergoes somatic mutations allowing it to become a high affinity anti-arsonate antibody. This process is accompanied by loss of NAAB activity. Guilbert[24] studied hybridomas derived from mice previously immunized with several different antigens and reported that, besides hybrids binding to specific immunizing antigen, they may obtain hybrids binding to specific antigen but also to self antigen, suggesting that the latter could be in an intermediate stage between the preimmune and specifically immune repertoire. In our laboratory, Matthes[25], have recently compared the epitopes recognized on tubulin by natural polyspecific autoantibodies and by induced antibodies. They found that whereas NAAB recognized the same or very overlapping epitopes in the central part of both α and ß subunits of tubulin, induced antibodies recognized very different epitopes located either in amino-terminal or carboxyl-terminal parts of the tubulin sub-units. Whether these results indicate a different genetic origin for NAAB and induced antibodies, or extensive somatic mutations occurring in induced antibodies, is an unsolved issue.

An alternative hypothesis to template theory is that these NAAB constitute a phylogenetically very well conserved system which may act as a first ring of defense. In that sense, these probably low affinity widely reacting antibodies could allow some degree of protection. In primitive animals with no possibility of further diversification of the immune system, they could constitute the unique defense system. This primitive system could be conserved during evolution, since it allows some protection; meanwhile, the immune system develops more specific antibodies. Whatever the case, verification of these hypotheses requires careful experimentation at the clonal level[5].

Considerable evidence has accumulated indicating that this autoreactive B repertoire frequently undergoes malignant transformation. This evidence arose from the study of monoclonal immunoglobulins (MIg), chronic lymphocytic leukemia (CLL) and follicular non-Hodgkin lymphomas (FNHL).

ANTIBODY ACTIVITY OF MIg

MIg correspond to normal synthetic products whose counterpart can be found in the heterogenous normal Ig compartment. The antibody-like activity of MIg has been described against a large number of antigens, e.g. bacterial antigens, plasma proteins, tissue antigens and non-biological haptens[26]. However, an impressive and unexpected frequency of activities has been reported against: 1) the Fc fraction of IgG (rheumatoid factor (RF); 2) I blood group antigen (cold agglutinins, [CA]); and 3) cytoskeleton proteins and DNA (polyreactive autoantibodies); 4) anti-myelin associated glycoproteins (MAG).

MIg with RF activity

Since the first report by Kritzman[27], of a monoclonal IgMK paraprotein with anti-IgG activity, an increasing number of cases have been reported, and the frequency of this specificity has been estimated at more than 10% of total IgM, paraproteins[28]). Most monoclonal components with reported RF activity, were found to form a cryoprecipitate. Almost all cases corresponded to IgMK MIg, but rare cases of human monoclonal IgG and IgA with RF activity and cryoprecipitables have been described.

Pioneer work from Kunkel's laboratory demonstrated the presence of cross-reactive idiotypic (CRI) specificities among these MIg[29]. Sixty per cent of these MIg displaying RF activity were found to share a major CRI called Wa; 20% belonged to a less common CRI designated Po, and a few expressed a more rare CRI, named Bla.

During recent years, considerable work, largely emanating from the group of Dennis Carson, has contributed important information concerning this type of MIg, by precisely defining genetic origins in serological and structural terms[30].

These studies were mainly focused on MIg sharing the Wa CRI. It was found that:

a) Almost all Wa + RF share the 17109 CRI related to light chains and the G6 idiotype related to heavy chains.

b) They constantly express the minor subgroup VKIIIb light chain.

c) The VK light chain is derived from a single germinal gene (Hum Kv 325), since most Wa + paraproteins display an identical or nearly identical light chain sequence, as stated by 13 complete light chain sequences.

d) There is strong sequence homology among μ chains expressing the Wa idiotype. Most of them use the VH1 family (80%) and the minority use VH2 and VH3 families.

Although information derived from Po+ RF MIg is less extensive, they appear to constantly use a VK germinal gene (Hum Kv 328) and to use a conserved VH3 sequence[30,31]. More recently, it has been demonstrates that the idiotype Bla was encoded by a gene of the VH4 gene family[32].

MIg with CA activity

CA paraproteins are almost constantly IgMK paraproteins, which usually react with a set of antigenic determinants directed against the Ii system, or compound antigens including Ii (AI, HI, etc..). Their activity is increased by cold, but thermal amplitude is variable.

As for RF cryoglobulins, CRI associated with light and heavy chains have been described. The vast majority of CA employ a VKIII gene and a VH4 gene[33,34]. The amino acid sequence of light chains revealed striking similarities. Interestingly, some CA used the same Hum Kv-325 germinal gene employed by Wa+ RF paraproteins, indicating that antigenic specificity against IgG or anti-I was supplied by heavy chains[31]. Nucleotidic sequence has been recently established for 2 clones secreting cold agglutinins, which were derived from 2 different patients suffering from cold agglutinin disease. Both clones were shown to use a common VH4 germ-line gene (VH4[21,33]). Interestingly, a single base substitution at the CDR1 region of both clones resulted in the replacement of an aspartic acid by a glycin[34], suggesting that aminoacid substitutions in the CDRs could be responsible for the acquisition of cold agglutinin specificity.

MIg with polyreactive activity

Prompted by our results in normal human serum in the early 80's, we screened 612 MIg for the presence of antibody activity directed against cytoskeleton proteins and DNA. Our results indicated that about 6% of all MIg and 10% of IgM paraproteins bound to these antigens, and that most of them displayed a polyreactive pattern of binding comparable to that found in normal human serum, indicating that MIg frequently correspond to expansion of a clone normally producing a NAAB[35,36]. Delaggi[37] also reported the presence of IgM paraproteins binding to intermediate filaments, and Shoenfeld[38] found that more than 10% MIg shared the 16-6 CRI derived from a monoclonal Ig with anti-DNA activity. However, only 25% of these MIg were demonstrated to possess anti-DNA activity. Another anti-idiotypic reagent (F4) was found to be present in 12% of MIg and was found to be strongly associated with IgG isotype and anti-DNA activity[39].

The presence of shared CRI within these polyreactive MIg, favors their germinal origin[25,38]. Yet we do not presently dispose of structural information of these paraproteins although, studies of monoclonal NAAB derived from normal and autoimmune mice and humans suggest a germinal origin. The use of multiple VH and VL genes has been reported for monoclonal NAAB[34]. However, the frequent expression of MIg with anti-DNA activityu of the 16-6 idiotype suggests the frequent usage of the VH3 V26 gene.

MIg with anti-MAG activity

A peripheral neuropathy is observed in about 5% of Waldenström's macroglobulinemia patients. Work from Saito[40] and Brouet[41], succeeded to demonstrate that in a majority of these cases MIg display an antibody activity against a myelin associated glycoprotein (MAG). The epitope recognized correspond to a glycuronyl sulfate group. However, the pathogenic role of the MIg is not definitively established. Brouet[41], reported a recurrent idiotype among 9 MIg with anti-MAG activity. Six out of these 7 MIgs for which studies could be performed were expressing VH3 and the remaining VH2. Interestingly the rare VKIV family was found in 3 cases, VKI in 2 and VKII in 1 and the remaining patient expressed lambda light chain[41].

ANTIBODY ACTIVITY OF THE CD5+ CLL B-LYMPHOCYTES

In our laboratory, we have recently examined the antibody activity of B lymphocytes proliferating in this disease, by fusing B lymphocytes with the non-secreting murine myeloma X-63. We could demonstrate that a high percentage of CLL B lymphocytes are committed for the production of natural autoantibodies[42]. Similar results have been reported by Bröker[43] and by Sthoeger[44].

As for gene expression, Humphries[45] reported that 30% of CLL patients were expressing VH251, which is one of the two germinal members of the VH5 family. Logtenberg[46] found that CLL B lymphocytes express VH4 in 50% of cases, VH5 in 20% and VH6 in 15%. In a recent work, Kipps[47,48] found that a high proportion of B-CLL cells expressing K at the membrane reacted with a murine anti-idiotypic antibody raised against a monoclonal IgM RF, expressing the Wa idiotype. Analysis of Kappa light chain variable region genes expressed by leukemic cells sharing idiotype Wa, enabled these authors to demonstrate that they were employing the germinal unmutated Hum Kv 325 germinal gene. Similar restriction was found for VH genes, since the germinal VH1 51P1 gene, was found to be expressed in 20% of CLL cases[48]. Spatz[49] recently reported the nucleotide sequence of a clone derived from a CLL patient, which was found to bind MAG and denaturated DNA. He was found to express a VKIIIa gene with 96% of homology with the germinal counterpart and a VH3 gene whose germinal counterpart could correspond to the VH3-26 gene.

We have studied VH expression in 40 CD5+ B-CLL and found that VH1 was employed in 17%, VH2 in 8%, VH3 in 36%, VH4 in 17%, VH5 in 8% and VH6 in 14%[50]. Although, our study failed to find the same incidence of the small VH4, VH5 and VH6 families; it shows a clear over-representation of these.

These results confirm that CD5+ B-CLL lymphocytes are frequently committed to the production of natural autoantibodies. The results from Kipps et al. indicating the use by the CLL B-lymphocyte of a restricted set of genes, are confirmed by the high frequency of natural autoantibody activity found by us among CLL B-lymphocytes.

ANTIBODY ACTIVITY OF THE CD5-B LYMPHOCYTE FROM FOLLICULAR NON-HODGKIN LYMPHOMAS (FNHL)

Our results with CLL were in aid to the hypothesis that CD5+ B mostly secrete autoantibodies. However, in a recent work , based upon dectection of mRNA transcript of Ly1 gene among 40 murine hybridomas displaying natural autoantibody activity, we could demonstrate that both Ly1+ and Ly1- B lymphocyte subsets were involved in the production of natural autoantibodies[51]. To gain better insight into this problem, we have recently studied 31 hybridomas obtained in the laboratories of R.A. Miller and R. Levy, from CD5 negative B-cell NHL. Our results indicated that 8 out of the 31 hybridomas displayed rheumatoid factor activity and 2 out of these 11 displayed a multispecific activity[52]. These results obtained with Igs derived from CD5- B-cell tumors strongly support the idea that CD5- B-cells are also involved in the production of natural autoantibodies.

Cleary[53] reported that at the difference of CLL and Acute Lymphoblastic Leukemia, where a bias in expression of VH4,VH5 and VH6 families has been demonstrated, CD5-B cells proliferating in NHL appear to employ VH gene families in a more stochastic way, by privilegiating the multigenic VH3 family. In addition, there is an active somatic mutational process in B-cell follicular NHL, that is rarely observed in B-CLL.

CONCLUSION

As a whole, these results indicate that autoreactive B cell repertoire frequently undergoes malignant transformation. The activation of this repertoire through continuous challenge by self antigens may create propitious conditions for mutations and chromosomal translocation to occur. However, this does not explain the results observed with MIgs and in CLL, where a recurrent usage of a few number of genes has been observed.

The binding specificities of antibody molecules is the consequence of the assembling of V, D and J genes for heavy chains and of V and J genes for light chains. These combination events generate a large diversity of antigen binding specifities, which in adult lymphoid tissue appears to be random. However, Ab specificities are developmentally regulated and there is increasing evidence indicating that the fetal preimmune repertoire employs a limited set of V genes, which frequently encode autoantibodies[54-56]. As these genes are recurrently expressed during B-cell malignancies, it could alternatively be postulated, that this developmentally regulated autoreactive B-cell repertoire has a selective advantage to undergo malignant transformation. The reasons accounting for overexpression of these recurrent genes are not clear as yet. The role of environmental selective influences such as the continuous challenge by autoantigens or through idiotype-antiidiotype interactions needs to be more precisely defined. The study of the expression of these recurrent V genes in pre-B cell malignancies which do not express functional surface Igs, should allow to conclude in this point. In the case that stimulation through antigen receptor could be excluded, the possibility that overexpression of these developmentally regulated recurrent V genes, could be accounted by the existence of putative transcriptional enhancers located upstream of these genes, needs to be further explored[57].

REFERENCES

1. B. Guilbert, G. Dighiero, S. Avrameas. Naturally occurring antibodies against nine common antigens in normal humans. I. Detection, isolation and characterization. J Immunol 128:2779 (1982).

2. C. Goodnow, S. Adelstein, A. Basten. The need for central and peripheral tolerance in the B-cell repertoire. Science 248:1349 (1990).

3. G. Dighiero, P. Lymberi, J.C.Mazié, S. Rouyre, G.S. Butler-Browne, R.G. Whalen, S. Avrameas. Murine hybridomas secreting natural monoclonal antibodies reacting with self antigens. J Immunol 131:2267 (1983).

4. G. Dighiero, P. Lymberi, D. Holmberg, I. Lundquist, A. Coutinho, S. Avrameas. High frequency of natural autoantibodies in normal newborn mice. J Immunol 134:765 (1985).

5. G. Dighiero. Autoreactive B-cell repertoire. in: Molecular Immunobiology of self reactivity. CA Bona and A Kaushik ed., M Dekker Inc., New-York (1991).

6. NA Bos, CG Meeuwsen, H Hooijkaas, R Benner, BS Wostman, JR Pleasants. Early development of Ig secreting cells in young of germ-free BALB/c mice fed a chemically defined ultrafiltered diet. Cell Immunol 105:235 (1987).

7. J. Rauch, E. Murphy, J.B. Roths, B.D. Stollar, R.S. Schwartz. A high frequency idiotypic marker of anti-DNA autoantibodies in MR-lpr/lpr mice. J Immunol 129:236 (1982).

8. F. Tron, C.L. Le Guern, P.A. Cazenave, J.F. Bach. Intrastrain recurrent idiotypes among anti-DNA antibodies of (NZB/NZW)F1 hybrid mice. Eur J Immunol 12:761 (1982).

9. T.N. Marion, A.R. Lawton, J.F. Kearney, D.E. Briles. Anti-DNA autoantibodies in (NZB x NZW)F1 mice are clonally heterogenous, but a majority share a common idiotype. J Immunol 128:668 (1982).

10. D.A. Isenberg, Y.Shoenfeld, M.R. Madaio, J. Rauch, M. Reichlin, B.D. Stollar. Z-DNA specific antibodies in human systemic lupus erythematosus. J Clin Invest 71:314 (1983).

11. P. Lymberi, G. Dighiero, T. Ternynck, S. Avrameas. A high incidence of cross-reactive idiotypes among murine natural autoantibodies. Eur J Immunol 15:702 (1985).

12. A. Morgan, D.A. Isenberg, Y. Naparstek, J. Rauch, D. Duggan, R. Khiroga, N.A. Staines, A. Shattner. Shared idiotypes are expressed on mouse and human anti-DNA autoantibodies. Immunol 56:393 (1985).

13. J.F. Kearney, M. Vakil. Idiotype directed interactions during ontogeny play a major role in the establishment of the adult B cell repertoire. Immunol Rev 94:39 (1986).

14. M. Monestier, A. Manheimer-Lory, B. Bellon, C. Painter, H. Dang, N. Talal, M. Zanatti, R. Schwartz, D. Pisetsky, R. Kuppers, N. Rose, J. Brochier, L. Klareskog, R. Holmdahl, F. Alt, C. Bona. Shared idiotopes and resticted immunoglobulin variable region heavy chain genes characterize murine autoantibodies of various specificities. J Clin Invest 78:753 (1986).

15. T. Matthes, G. Dighiero. Detection of private and recurrent idiotopes on natural antitubulin antibodies by monoclonal anti-idiotopic antibodies. J Immunol 140:148 (1988).

16. D. Holmberg, S. Forsgren, F. Ivars, A. Coutinho. Reactions among IgM antibodies derived from normal, neonatal mice. Eur J Immunol 14:435 (1984).

17. R. Baccala, T.V. Quang TV, Gilbert M, Ternynck T, Avrameas S. Two murine natural polyreactive autoantibodies are encoded by nonmutated germ-line genes. Proc Natl Acad Sci USA 86:4624 (1989).

18. R. Kofler, D.I. Noonars, D.E. Levy, M.C. Wilson, N.P. Moller, F.J. Dixon, A.N. Theofilopoulos. Genetic elements utilized for a murine lupus anti-DNA autoantibody are closely related to those for antibodies to exogenous antigens. J Exp Med 161:805 (1985).

19. W. Trepicchio Jr, A. Maruya, K.J. Barrett. The heavy chain genes of a lupus anti-DNA autoantibody are encoded in the germ line of a nonautoimmune strain of mouse and conserved in strains of mice polymorphic for this gene locus. J Immunol 139:3139 (1987).

20. I. Sanz, P. Casali, J.W. Thomas, A.L. Notkins, J.O. Capra. Nucleotide sequences of eight human natural autoantibody VH regions reveal apparent restricted use of VH families. J Immunol 142:4054 (1989).

21. C. Painter, M. Monestier, B. Brown, C.A. Bona. Functional and molecular studies of V genes expressed in autoantibodies. Immunol Rev 94:75 (1986).

22. C.A. Bona. V genes encoding autoantibodies, molecular and phenotypic characteristic. Ann Rev Immunol 6:327 (1988).

23. Y. Naparstek, J. André-Schwartz, T. Manser, L. Wysochi, L. Breitman, B.D. Stollar, M. Gefter, R.S. Schwartz. A single germline VH gene segment of normal A/J mice encodes autoantibodies characteristic of systemic lupus erythematosus. J Exp Med 164:614 (1986).

24. B. Guilbert, W. Mahana, M. Gilbert, J.C. Mazié, S. Avrameas. Presence of natural autoantibodies in hyperimmunized mice. Immunol 56:401 (1985) .

25. T. Matthes, A. Wolff, P. Soubiran, F. Gros, G. Dighiero. Antitubulin antibodies. II. Natural autoantibodies and induced antibodies recognize different epitopes on the tubulin molecule. J Immunol 141:3135 (1988).

26. M. Seligmann, J.C. Brouet. Antibody activity of human myeloma globulins. Semin Hematol 10: 163 (1973).

27. J. Kritzman, H.G. Kunkel, J. McCarthy, R.C. Mellors. Studies of a Waldenström-type macroglobulin with rheumatoid factor properties. J Lab Clin Med 57: 905 (1961).

28. J.J. Crowley, R.D. Goldfien, R.E. Schrohenlober, H.L. Spiegelberg, G.J. Silverman, R.A. Mageed, R. Jefferis, W.J. Koopman, D.A. Carson, S. Fong. Incidence of three cross-reactive idiotypes on human rheumatoid factor paraproteins. J Immunol 140: 3411 (1988).

29. V. Agnello, F.G. Joslin, H.G. Kunkel. Cross idiotypic specificity among monoclonal IgM anti-gammaglobulins. Scand J Immunol 1: 283 (1972).

30. P.P. Chen, D.L. Robbins, F.R. Jirik, T.J. Kipps, D.A. Carson. Isolation and characterization of a light chain variable region gene for human rheumatoid factors. J Exp Med 166: 1900 (1987).

31. G.J. Silverman, F. Goni, J. Fernandez, P.J. Che, B. Frangione, D.A. Carson. Distinct patterns of heavy chain variable region subgroup use by human monoclonal autoantibodies of different specificity. J Exp Med 168: 2361 (1988).

32. G.J. Silverman, R.E. Schrohenholer, M.A. Achavitti, W.J. Koopman, D.A. Carson. Structural characterization of the second major cross-reactive idiotype group of human rheumatoid factors: Association with the VH4 gene family. Arthritis Rheum 33: 1347 (1990).

33. V. Pascual, K. Victor, D. Lelsz, M.B. Spellerberg., T.J. Hamblin, K.M. Thompson, I. Randen, J. Natvig, J.D. Capra, F.K. Stevenson. Nucleotide sequence analysis of the V regions of two IgM cold agglutinins. Evidence that the VH4-21 gene segment is responsible for the major cross-reactive idiotype. J. Immunol 146: 4385 (1991).

34. V. Pascual, and J.D. Capra. Human immunoglobulin Heavy-chain variable region genes: Organization, polymorphism ane expression. Adv. Immuno 49: 1 (1991).

35. G. Dighiero, B. Guilbert, S. Avrameas. Naturally occuring antibodies against nine common antigens in human sera: II. High incidence of monoclonal Ig exhibiting antibody activity against actin and tubulin and sharing antibody specificities with natural antibodies. J Immunol 128: 2788 (1982).

36. G. Dighiero, B. Guilbert, J.P. Fermant, P. Lymberi, F. Danon, S. Avrameas. Thirty-six human monoclonal immunoglobulins (MIg) with antibody activity against cytoskeleton proteins, thyroglobulin and native DNA. Immunological studies and clinical correlations. Blood 62: 264 (1983).

37. K. Dellagi, J.C. Brouet, F. Danon. Cross idiotypic antigens among monoclonal immunoglobulin M from patients with Waldenström's macroglobulinemia and polyneuropathy. J Clin Invest 64: 1530 (1979).

38. Y. Shoenfeld, H. Amital Teplizki, S. Mendlovic, M. Blank, E. Mozes, D.A. Isenberg. The role of the human anti-DNA idiotype 16/6 in autoimmunity. Clin Immunol Immunopathol 51: 313 (1989).

39. A. Davidson, A. Smith, J. Katz, J.L. Preud'homme, A. Salomon, B. Diamond. A cross-reactive idiotype on anti-DNA antibodies define a H chain determinant present almost exclusively on IgG antibodies. J.Immunol; 143: 174 (1989).

40. T. Saito, W. Sherman, N. Latov. Specificity and idiotype of M-proteins that react with MAG in patients with neuropathy. J Immunol 130: 2496 (1983).

41. J.C. Brouet, K. Dellagi, M.C. Gendron, A. Chevalier, C. Schmitt, E. Mihaesco. Expression of a public idiotype by human monoclonal IgM directed to myelin-associated glycoprotein and characterization of the variability subgroup of their heavy and light chains. J Exp Med 170: 1551 (1989).

42. L. Borche, A. Lim, J.L. Binet, G. Dighiero. Evidence that chronic lymphocytic leukemia B lymphocytes are frequently commited to productions of natural autoantibodies. Blood 76: 562 (1990).

43. B.M. Bröker, A. Klajman, P. Youinou, J. Jouquan, C.P. Worman, J. Murphy, L. Mackenzie,

132

R. Quartey-Papafio, M. Blaschek, P. Collins, S. Lal, P.M. Lydyard. Chronic Lymphocytic Leukemia (CLL) cells secrete multispecific autoantibodies. J Autoimmun 1: 469 (1988).

44. Z.M. Sthoeger, M. Wakai, D.B. Tse, V.P. Vinciguerra, S.L. Allen, D.R. Budman, S.M. Lichtman, P. Schulman, L.R. Weiselberg, N. Chiorazzi. Production of autoantibodies by CD5-expressing B lymphocytes from patients with chroniclymphocytic leukemia. J Exp Med 169: 255 (1989).

45. C.G. Humphries, A. Shen, W.A. Kuziel, J.D. Capra, F.R. Blattner, P.W. Tucker. A new human immunoglobulin VH family preferentially rearranged in immature B-cell tumors. Nature 331: 446-449 (1988).

46. T. Logtenberg, F.M. Young, J.H. Van Es, F.H. Gmelig-Meyling, F.W. Alt. Autoantibodies encoded by the most JH proximal human immunoglobulin heavy chain variable region gene. J Exp Med 170: 1347 (1989).

47. T.J. Kipps, E. Tomhave, P.P. Chen, D.A. Carson. Autoantibody associated K light chain variable region gene expressed in chronic lymphocytic leukemiawithlittle orno somaticmutation, implications for etiology and immunotherapy. J Exp Med 167: 840 (1988).

48. T.J. Kipps, E. Tomhave, L.F. Pratt, S. Duffy, P.P. Chen, D.A. Carson. Developmentally restricted immunoglobulin heavy chain variable region gene expressed at high frequency in chronic lympocytic leukemia. Proc Natn Acad Sci USA 86: 5913 (1989).

49. L.A. Spatz, K.K. Wong, M. Williams, R. Desai, J. Golier, J.E. Berman, F.W. Alt, N. Latov. Cloning and sequence analysis of the VH and VL regions of anti-myelin/DNA antibody from a patient with peripheral neuropathy and chronic lymphocytic leukemia. J Immunol 144: 2821 (1990).

50. R. Mayer, T. Logtenberg, J. Strauchen, A. Dimitriu-Bona, L. Mayer, S. Mechanic, N. Chiorazzi, L. Borche, G. Dighiero, A. Mannheimer-Loryh, B. Diamond, F.W. Alt, C. Bona. CD5 and immunoglobulin V gene expression in B-cell lymphomas and Chronic Lymphocytic Leukemia. Blood 75: 1518 (1990).

51. A. Kaushik, R. Mayer, V. Fidanza, H. Zaghouani, A. Lim, C. Bona, G. Dighiero. Ly1 and V-gene expression among hybridomas secreting natural autoantibody. Journal of Autoimmunity; 3: 687 (1990).

52. G. Dighiero, S. Hart, A. Lim, L. Borche, R. Levy, R.A. Miller. Autoantibody activity of immunoglobulins isolated from B-cell follicular lymphomas. Blood 78:581 (1991).

53. M.L. Cleary, T.C. Meeker, S. Levy, E. Lee, M. Trela, J. Sklar, R. Levy. Clustering of extensive somatic mutations in the variable region of an immunoglobulin heavy chain gene from a human B-cell . Cell 44: 97 (1986).

54. G.D. Yancopoulos, S.V. Desiderio, M. Paskind, J.F. Kearney, D. Baltimore, F. Alt. Preferential utilization of the most JH-proximal VH gene segments in pre-B cell lines. Nature 311:727 (1984).

55. H.W. Schroeder, J.L. Hilson, R.M. Perlmutter. Early restriction of the human antibody repertoire. Science 238:791 (1987).

56. G. Dighiero, A. Lim, M.P. Lembezat, A. Kaushik, L. Andrade, A. Freitas. Comparative usage of VH gene family usage by newborn Xid and non Xid mice, newborn NZB and adult NZB and by splenic and peritoneal cavity B cell compartments. Eur J Immunol 18:1979 (1989).

57. P.P. Chen, N.J. Olsen, P.M. Yang, R.W. Soto-Gil, T. Olee, K.A. Siminovitch, D.A. Carson. From human autoantibodies to the fetal antibody repertoire to B cell malignancy: It's a small world after all. Int. Rev Immunol 5:239 (1990).

NATURALLY OCCURRING HUMAN AUTOANTIBODIES
TO DEFINED T-CELL RECEPTOR AND LIGHT CHAIN PEPTIDES

John J. Marchalonis[++], Hulya Kaymaz[++], Samuel F. Schluter[++], and David E. Yocum[*]

[++]Microbiology and Immunology Department
[*] Rheumatology/Allergy, Department of Medicine
University of Arizona, Tucson, AZ 85724

ABSTRACT

We used synthetic peptides duplicating the structures of a human λ light chain (Mcg), and a human T-cell receptor (Tcr) α and a Tcr β chain predicted from gene sequence to determine the presence and loci of activity of natural human autoantibodies directed against these antigen recognition molecules. We report that normal individuals and patients suffering from autoimmune diseases have antibodies directed against regions of λ light chains and Tcr β chains corresponding to the first complementarity determining region and the third framework region of the variable domain and to constant region determinants. The levels of IgM natural antibodies particularly against the CDR1 peptides tend to be higher in RA patients than in normals or SLE patients. Although polyclonal IgG immunoglobulins from healthy individuals did not show detectable reactivity to Tcr α peptides, such reactivity was found in the IgM immunoglobulins of RA patients, thereby showing that Tcr α peptides can be autoantigenic in man. The levels of IgM autoantibodies to Vβ CDR1 peptides tend to decrease with age. By contrast, there was a marked increase in IgG natural autoantibodies to certain CDR1 sequences with advancing age. We suggest that the natural antibodies to defined regions of immunoglobulins and T-cell receptors are part of a physiological network for the regulation of the immune response.

INTRODUCTION

Clinically healthy individuals can produce antibodies directed against immunoglobulins (1, 2) and discrete peptide determinants of heavy chains (3) and of T-cell receptor β chain (4). The functions of these auotantibodies is currently unknown, but it has been shown that their levels can vary in autoimmune diseases (5) and in aging (6). The present study provides new information on the human autoantibody response to T-cell receptor α/β peptides. Furthermore, we use the nested overlapping 16-mer peptides modelling the completely characterized human λ light chain Mcg to compare the sites of autoantibody reactivities with the epitopes already detected using xenoantisera (7).

Immunobiology of Proteins and Peptides VII
Edited by M.Z. Atassi, Plenum Press, New York, 1994

METHODS AND MATERIALS

Peptides and Proteins

The protein sequence of human λ light chain Mcg (8), and the derived protein sequences of human Tcr β chain YT35 (9) and α chain PY14 (10) were used to construct three series of synthetic overlapping peptides consisting of 20, 22 and 19 peptides respectively. These peptides, 16-mers in length and overlapping by 5 residues were synthesized by the University of Arizona Biotechnology Center using an Applied Biosystems Peptide Synthesizer. Purity was determined by amino acid composition and sequence analysis. The complete sets of λ Mcg (7) and Tcr β (11) peptides have been published elsewhere. Table 1 lists the synthetic Tcr α/β and λ peptides used extensively in this study.

Ovalbumin is purchased from Sigma Chemical Company (St. Louis, MO). Mcg is a human λ myeloma protein of known sequence and three dimensional structure (obtained from Dr. Allen B. Edmundson, Harrington Cancer Center, Amarillo, TX).

Antibodies

Commercial rabbit antisera specific for human λ and κ light chains and for γ heavy chain were purchased from DAKO Corporation (Carpenteria, CA). Rabbit sera against Tcr β chain CDR1 and Fr1 peptides were prepared by immunizing rabbits as previously described (11, 12). The pooled human IgG preparation was purchased from Baxter Healthcare Corporation (Glendale, CA). Human sera were from healthy individuals, age range 20-90, 14 rheumatoid arthritis patients (RA), age range 20-75, and 8 systemic lupus erythematosus (SLE) patients, age range 27-74. The human sera were obtained from the Divisions of Rheumatology and Gerontology, Department of Medicine, College of Medicine, University of Arizona.

Table 1. Extensively used synthetic peptides

$\alpha1$	Q S V T Q L G S H V S V S E G A	N-terminus ($V\alpha$)
$\alpha7$	K K S E T S F H L T K P S A H M	Fr3 ($V\alpha$)
$\alpha11$	R L S T R P N I Q N P D P A V Y	$J\alpha/C\alpha$ "switch"
$\alpha17$	S A V A W S N K S D F A C A N A	$C\alpha$
$\beta1$	D A G V I Q S P R H E V T E M G	N-terminus ($V\beta$)
$\beta3$	C K P I S G H N S L F W Y R Q T	CDR1/Fr2 ($V\beta$)
$\beta8$	K I Q P S E P R D S A V Y F C A	Fr3 ($V\beta$)
$\beta17$	Q P L K E Q P A L N D S R Y C L	C-region "loop"
Mcg1	Q S A L T Q P P S A S G S L G Q	N-terminus ($V\lambda$)
Mcg3	T G T S S D V G G Y N Y V S W Y	CDR1/Fr2 ($V\lambda$)
Mcg8	S G L Q A E D E A D Y Y C S S Y	Fr3/CDR3 ($V\lambda$)
Mcg13	K A T L V C L I S D F Y P G A V	$C\lambda$
Mcg17	A A S S Y L S L T P E Q W K S H	$C\lambda$

Immunological Assays

Rabbit anti-human IgM and anti-human IgG were from DAKO Corporation. Goat anti-rabbit IgG and rabbit anti-human IgG peroxidase immunoconjuates were from Jackson Immunoresearch Laboratories (West Grove, PA). Affinity purifications were performed using peptide conjugated Sepharose columns following procedures

136

described elsewhere (7, 12). Standard ELISA assays were performed as previously described (7). Competitive ELISAs were done by preincubating the antibodies with various quantities of homologous and heterologous antigens before adding the mixture into wells coated with homologous antigens and then following the standard ELISA procedure.

RESULTS

Antigenic Analysis of Light Chain λ Mcg.

Figure 1 compares the binding to sequential overlapping Mcg peptides of a commercial rabbit antiserum specific for human λ chains and of a human polyclonal IgG preparation. The results with the anti-λ sera have been published in detail

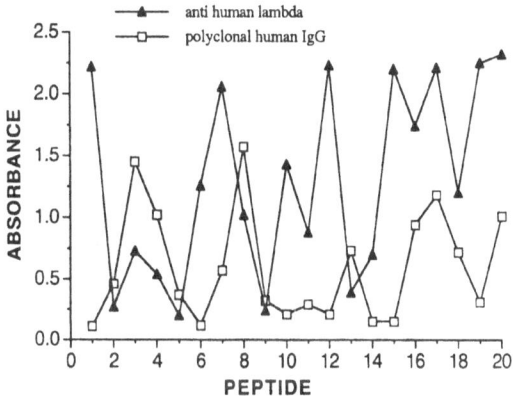

Figure 1. Binding in ELISA of rabbit anti-serum to human light chain (Δ) and polyclonal human IgG (□) to synthetic overlapping 16-mer peptides duplicating the sequence of the human λ Mcg.

elsewhere (7), and are presented here to serve as a reference for the polyclonal normal human IgG. The normal human IgG immunoglobulin binds predominantly to three peptide regions: one defined by peptides #3/4, another at peptide #8, and a third in the region of peptides #16/17. Lesser reactivity is shown to peptides #13 and #20. The first region is a minor determinant in the reaction with the rabbit antibody, but the rabbit antibody similarly reacts strongly to the second region (peptides #6 through #8) and the third region (peptides #16 and #17). The natural antibody shows no reaction for the N-terminal peptide #1, although this is strongly recognized by the xenoantiserum. In addition, peptides #10, #11 and #15 are strongly reactive with the xenoantiserum, but these are not bound by the normal human immunoglobulin. The three major autoantigenic regions correspond to CDR1, Fr3, and an exposed portion of the constant region, respectively. It was possible to affinity purify human autoantibodies to peptides Mcg3, Mcg8 and Mcg17 from the human IgG preparation and to show that the affinity purified antibodies bound to intact Mcg as well as the peptide. The bindings of affinity purified anti-Mcg3 and Mcg8 were specifically inhibitable by free peptide. These antibodies also react with intact Mcg and the reaction is inhibitable by free Mcg.

Comparison of Binding of Rabbit Anti-Immunoglobulins and Polyclonal Human Immunoglobulin to T-cell Receptor β Chain Peptides

Figure 2 illustrates and compares the reaction in ELISA of rabbit antiserum specific for human λ chain (A), rabbit antiserum specific for human κ chains (B), and rabbit antiserum specific for human γ heavy chain (C), with the binding of polyclonal normal human IgG (D) to the nested set of overlapping peptides duplicating the immunoglobulin domain structure of the T-cell receptor β chain YT35. The antibody to heavy chain is essentially negative under these conditions, possibly showing minimal reactivity to peptide β3. The rabbit anti-light chain sera (λ and κ) both react strongly with peptides β3, β8, β11, and β17. The major reactivity of the anti-light chain sera tends to be against peptide β11, which corresponds to the "switch peptide" comprised of the J segment and the beginning of the constant region. The polyclonal IgG immunoglobulin reacts with peptides β3, β8 and β17. These regions correspond to the first complementarity determining region of Vβ, the third framework of Vβ, and to a large loop that billows out of the proposed β sheet structure of the constant region (4). Studies reported elsewhere show that these natural antibodies reactive with T-cell

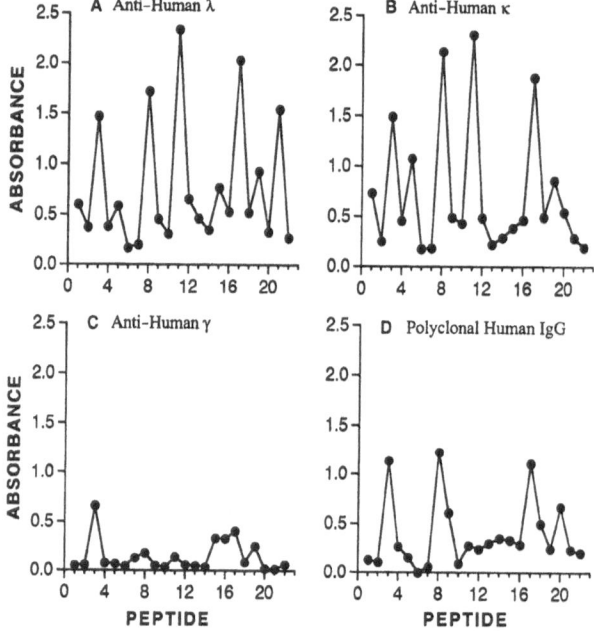

Figure 2. Binding in ELISA of rabbit anti-sera to human λ light chain (A), rabbit anti-serum to human κ light chain (B), rabbit anti-serum to human γ heavy chain (C) and polyclonal normal human IgG immunoglobulin (D) to synthetic overlapping 16-mer peptides modelling the immunoglobulin domain structure of human T-cell receptor β chain.

receptor peptides are present in small quantity in human sera, are inhibitable by peptide, and can be affinity purified (4).

Figure 3 illustrates the binding of polyclonal human IgG to a selected set of synthetic peptides generated for Tcr α chain (PY14), Tcr β chain (YT35) and Mcg sequence. Ovalbumin was chosen as a positive control because we have found that approximately 70% of healthy individuals ranging in age from 20 to 90 have high titer IgG antibodies to this common environmental antigen. The selection of the Tcr and light chain peptides was based upon either their previous determination of reactivity with autoantibodies (β and λ) or their corresponding location to the detected regions (α). A blank consisting of binding to an uncoated plate is always carried out as a negative control to compensate for stickiness or the presence of aggregates. The commercial human intravenous immunoglobulin preparation, which includes sera from more than

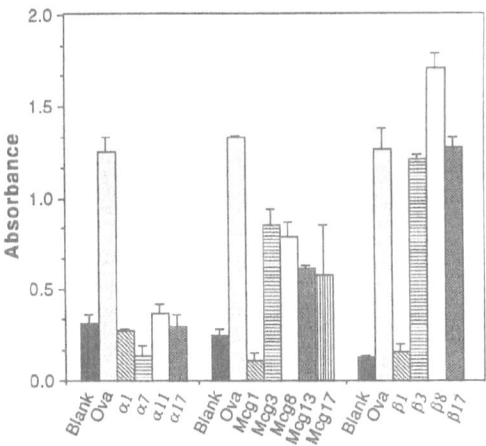

Figure 3. Binding in ELISA of polyclonal human IgG immunoglobulin (Baxter Gammagard) to ovalbumin and to synthetic peptides from human Tcr α and β chains and the Mcg light chain. The IgG preparation was tested at a concentration of 250 ng per ml.

five thousand normal individuals, shows strong reactivity to ovalbumin and to Tcr β peptides #3, #8 and #17. It shows moderate reactivity to Mcg peptides #3, #8, #13, and #17. However, the reactivity to the α peptides is not significantly different than that shown to the uncoated plate.

Binding of Antibodies of Rheumatoid Arthritis Patients to Tcr α and β Peptides

Rheumatoid arthritis patients produce readily detectable levels of IgM autoantibodies, termed rheumatoid factor, that bind to determinants on human γ heavy chain (1, 13), and similar antibodies occur in healthy individuals following immunization (1). We have previously found that individuals with rheumatoid arthritis tend to have higher levels of antibodies against T-cell receptor β peptides than do normal individuals or patients suffering from systemic lupus erythematosus (5). Since similar antibodies can occur in healthy individuals, we have no reason to believe that the anti-T-cell receptor or immunoglobulin antibodies occurring in rheumatoid arthritis patients are fundamentally any different than those of normals. We tested the α peptides and the characterized β peptides for their reactivities against sera from a

Patient	UNC	$\alpha 1$	$\alpha 7$	$\alpha 11$	$\alpha 17$	$\beta 1$	$\beta 3$	$\beta 8$	$\beta 17$
7	97	96	93	133	183	95	388	385	291
8	65	94	125	288	582	75	489	257	183
11	20	59	20	82	196	20	294	67	85
28	110	94	114	247	216	86	422	297	279
29	140	125	144	372	588	99	729	383	365
30	200	167	131	200	213	20	294	ND	337
32	367	288	144	589	649	320	>6,400	4,303	4,200

Geometric Mean Titers 105   117   95   230   323   69   612   385   351

Figure 4. Titers of the IgM natural antibodies to synthetic Tcr peptides in the sera of rheumatoid arthritis patients.

139

panel of rheumatoid arthritis patients. Figure 4 gives the titers of natural IgM antibodies to synthetic Tcr peptides. The response to the β peptides is the expected one; namely, the N-terminal or the $\beta 1$ peptide is unreactive, but there is strong reactivity to the $\beta 3$ (CDR1), $\beta 8$ (Fr3) and $\beta 17$ (C-region loop) peptides. None of the RA patients show IgM reactivity to peptides $\alpha 1$ and $\alpha 7$. However, four patients show IgM reactivity to peptide $\alpha 11$, and six react with peptide $\alpha 17$. In general, the IgM reactivity to these peptides is higher than the IgG reactivity, and these results indicate that people can respond to Tcr α peptides corresponding in position to some of the reactive peptides of Tcr β chain. Figure 5 gives the titers of the IgG natural antibodies to the same set of synthetic peptides. It is noteworthy that only one patient gave a significant response to a T-cell receptor α peptide (patient 32, peptide $\alpha 11$) whereas a number of reactivities were obtained to peptides $\beta 8$ and $\beta 17$.

Patient	UNC	$\alpha 1$	$\alpha 7$	$\alpha 11$	$\alpha 17$	$\beta 1$	$\beta 3$	$\beta 8$	$\beta 17$
8	20	20	20	20	20	20	20	286	186
11	78	20	20	20	20	20	20	325	165
28	20	20	20	20	20	ND	20	ND	ND
29	20	20	20	20	20	20	20	210	82
30	77	20	20	58	97	20	50	ND	261
32	261	123	61	790	300	141	575	1509	919
Geometric Mean Titers	48	27	24	73	41	30	41	414	227

Figure 5. Titers of IgG natural antibodies to synthetic Tcr peptides in the sera of rheumatoid arthritis patients

Figure 6 illustrates group comparisons of normal healthy individuals in the age range 20-40, normal healthy individuals in the age range 70-90 (elderly), and rheumatoid arthritis and SLE patients for the capacity of their IgM and IgG antibodies to bind the sets of α and Mcg peptides described above. The lower figure depicting the IgG response illustrates the high level binding to ovalbumin shown by normal young and old individuals and also by SLE patients and emphasizes the decreased reactivity of RA patients who tend to be anergic to a number of antigens (14). In the IgM compartment, RA patients tend to show the highest level of reactivity even though normals tend to respond to the same antigens. The pattern of reactions with Mcg13 is interesting because normals, RA patients and SLE patients give higher responses than do the elderly individuals. The RA response is significantly higher than that of the normals ($p < 0.05$). IgM responses to $\alpha 11$ and $\alpha 17$ are significant but relatively low compared to the strong reactions with the Mcg peptides or the β peptides (Figure 4). Essentially no IgG binding is detected against the α peptides or Mcg1, but significant activity is detected in all groups against Mcg3, Mcg8, Mcg13, and Mcg17 with the normal IgG.

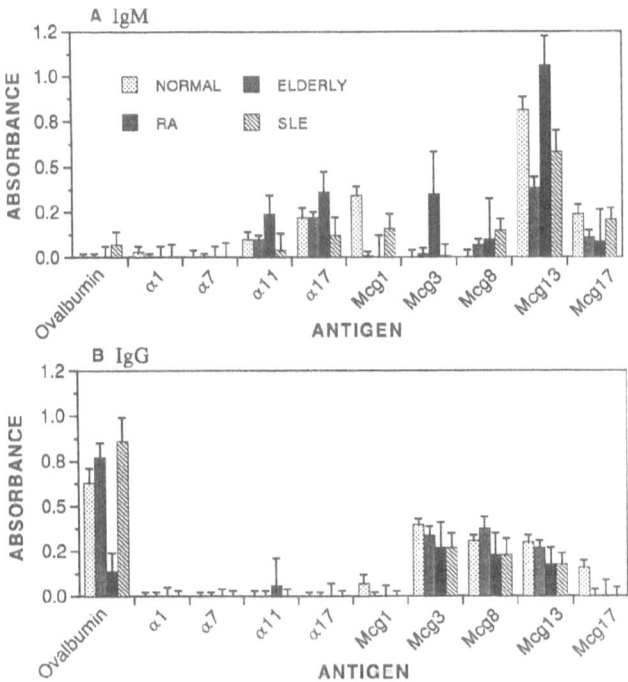

Figure 6. Binding in ELISA of sera of normal young individuals (age range 20-40), healthy elderly individuals (<70 years of age), RA patients and SLE patients to ovalbumin and synthetic peptides from Tcr α chains and λ light chains (Mcg). The sera were all tested at a dilution of 1:200 (4 replicates per sample) and the $\pm$ SE for each group is given. The blank values corresponding to binding to uncoated plates were subtracted from each group.

Figure 7 compares the IgM and IgG reactivities of young (20-40 years of age) with those of healthy elderly (<70 years of age) individuals, to ovalbumin, the $\beta3$ peptide (CDR1) of a V$\beta8.1$ sequence and to the corresponding peptide from a Tcr V$\beta2.1$ gene sequence. With this set of antigens, there is a trend for the IgM response to decrease with age. This is particularly true in the comparison for peptide $\beta3$ where comparison of the actual titers using the Wilcoxon Test (15) gives a p value <0.001. The IgG

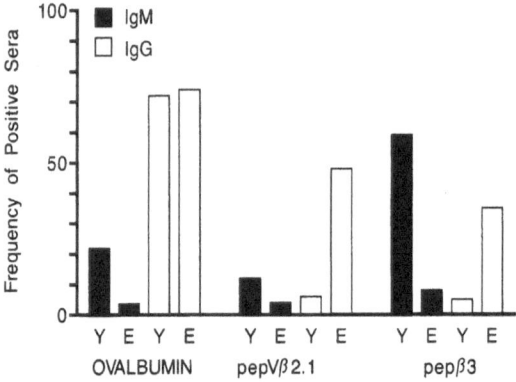

Figure 7. Comparison of IgM and IgG responses of healthy young (Y) and elderly (E) individuals to ovalbumin and to synthetic T-cell receptor CDR1 region peptides $\beta3$ (pep$\beta3$) and V$\beta2.1$ (pepV$\beta2.1$). The data are given as frequency of positive sera per group. The young group consisted of 28 individuals and the elderly group of 29.

levels to ovalbumin are substantially higher than the IgM levels and these remain constant throughout the life of healthy individuals. A trend is shown with both of the Tcr CDR1 peptides for the level of IgG antibodies to increase with age. In the case of the V$\beta2.1$ peptide, group comparisons of the individual titers using the Wilcoxon statistic give a p value <0.02.

DISCUSSION

Salient points germane to clinical autoimmunity and basic immunobiology arise from this study of human natural antibodies. In the first place, normal healthy individuals have the capacity to make IgM and, to a lesser degree, IgG autoantibodies to peptides corresponding to discrete regions of their own T-cell receptors and immunoglobulin light chains. The regions detected in the T-cell receptor β chain correspond to the first complementarity determining region and the third framework region of the variable domain and to an exposed portion of the constant region (4).

Normal polyclonal IgG immunoglobulin contains antibodies reacting with corresponding portions of λ light chain. These antibodies can be affinity purified and react both with the intact light chain and with the peptides used for the affinity purification. It is noteworthy that the reactivity of the affinity purified immunoglobulins is extremely strong relative to the starting material. This most probably can be explained by the removal of competing antigens within the initial preparation. The response to peptides corresponding to the human Tcr α chain PY14 (Vα subgroup I) is considerably less than that to either the β or λ peptides. However, the capacity of humans to respond to their own α peptides is documented here by the reaction of IgM immunoglobulins of rheumatoid arthritis patients with peptides $\alpha11$ ("switch peptide") and $\alpha17$ (constant region).

Levels of IgM antibody directed against the CDR1 peptides tend to decrease with increasing age. In contrast, the IgG reactivity against at least two of these peptides increases substantially with age in healthy individuals. Although we have shown elsewhere that the levels of IgM autoantibodies tend to be higher in RA patients than they are in normals (5), the fact that normal individuals express these antibodies suggests to us that they carry out an immunoregulatory role as opposed to an effector function in destructive autoimmune disease. The CDR1 region is a potentially interesting site for immunoregulation. Based upon molecular modelling approaches, we (4, 16) and Chotia et al., (17) have identified this region as a potentially conserved site for immunoregulation. Although it occupies the position of CDR1 in immunoglobulins, this region shows limited diversity in Tcr β chains by comparison with that of immunoglobulins. Chotia et al., (17) proposed that it might be the site involved in the recognition of MHC products by T-cell receptors in antigen presentation. In contrast, we speculate that the CDR1 region determinants detected here act as public idiotopes (18) that can function in idiotype regulation. We have obtained preliminary evidence supporting this hypothesis by immunizing rabbits with the $\beta3$ peptide. One of the five rabbit Vβ genes sequenced (19) has a CDR1 peptide identical in thirteen out of sixteen positions to our $\beta3$ sequence. Rabbits, like humans, tend to have relatively high levels of autoantibody to this peptide, and immunization with the peptide causes a decrease in level of antibody rather than the expected increase. As a control for this, we immunized with the N-terminal peptide and found that immunization produced the expected result of high levels of specific antibody. Thus, we suggest that the CDR1 region of T-cell receptor β chains presents a set of common regulatory idiotopes. Further studies with immunoglobulin light chains may disclose a similar effect because this region is relatively conserved by comparison with CDR2 and CDR3. The Fr3 region is likewise involved in regulation because V_Ha allotypes contain residues from this region (20) and this region has also been implicated in recognition of superantigens by Tcr β (21). These types of autoimmunoregulation would focus upon selected sets of V regions and the reactivity with the constant region might be analogous to the generation of rheumatoid factors against Cγ to facilitate a shutting off of the IgG response.

ACKNOWLEDGEMENTS: This work was supported in part by Grant CA 42049 from the National Cancer Institute and #I-103 from the Arizona Disease Control Research Commission. We thank Ms. Linda Wilson for expert technical assistance and Ms. Diana Humphreys for her help in the preparation of this manuscript.

REFERENCES

1. Carson, D.A., Chen, P., Fox, R.I., Kipps, T.J., Jirik, F., Goldfine, R.D., Silverman, G., Radoux, V. and Fong, S. Rheumatoid factor and immune networks. Ann. Rev. Immunol. 5:109-106, (1987).
2. Zouali, M., Fine, J.M., Eyquem, A. A human monoclonal IgG 1 with anti-idiotypic activity against anti-human thyroglobulin antibody. J. Immunol. 133: 190-194 (1984).
3. Kaveri, S.V., Kang, C.Y. and Kohler, H. Natural mouse and human antibodies bind to a peptide derived from a germ-line V_H chain. Evidence for evolutionary conserved self-binding locus. J. Immunol. 145:4207-4213 (1990).
4. Marchalonis, J.J., Kaymaz. H., Dedeoglu, F., Schluter, S.F., Yocum, D.E. and Ed Edmundson, A.B. Human autoantibodies reactive with synthetic autoantigens from T-cell receptor β chain. Proc. Natl. Acad, Sci. USA. 89:3325-3329 (1992).

5. Marchalonis, J.J., Kaymaz, H., Schluter, S.F., Yocum, D.E. and Edmundson, A.B. Novel rheumatoid factors binding to T-cell receptor peptide sequences: extension of the concept of anti-immunoglobulin regulation to T-cell receptors. Submitted for publication (1992).

6. Marchalonis, J.J., Schluter, S.F., Wilson, L., Yocum, D.E., Boyer, J.T. and Kay, M.M.B. Natural human antibodies to synthetic peptide autoantigens: correlation with age and autoimmune disease. Gerontology, accepted for publication (1992).

7. Marchalonis, J.J., Dedeoglu, F., Kaymaz, H., Schluter, S.F. and Edmundson, A.B. Antigenic mapping of a human λ light chain: correlation with 3-dimensional structure. J. Protein Chem. 11:129-137 (1991).

8. Edmundson, A.B., Ely, K.R., Abola, E.E., Schiffer, M. and Pangiotopoulos, N. Rotational allomerism and divergent evolution of domains in immunoglobulin light chains. Biochem 14:3953-3961 (1975).

9. Yanagi, Y., Yoshikai, Y., Legget, K. Clark, S.P. Aleksander I., and Mak, T.W. A human T-cell specific cDNA clone encodes a protein having extensive homology to immunoglobulin chains. Nature (Lond.) 308:145-149 (1984).

10. Yanagi, Y., Chan, A., Chen, B. Minden, M. and Mak, T.W. Analysis of cDNA clones specific for human T-cells and the α and β chains of the T-cell receptor heterodimer from a human T-cell line. Proc. Natl. Acad. Sci. 82:3430-3434 (1985).

11. Dedeoglu, F., Hubbard, R.A., Schluter, S.F., and Marchalonis, J.J. T-cell receptors of man and mouse studied with antibodies against synthetic peptides Exptl and Clin. Immunogenetics, in press (1992).

12. Schluter, S.F. and Marchalonis, J.J. Antibodies to synthetic joining segment peptide of the T-cell receptor β-chain: serological cross-reaction between products of T-cell receptor genes, antigen binding T-cell receptors and immunoglobulins. Proc. Natl. Acad. Sci. USA 83:1872-1876 (1986).

13. Sasso, E.H., Barber, C.V. Nardella, F.A., Yount, W.J. and Mannik, M. Antigenic specificities of human monoclonal and polyclonal IgM rheumatoid factors. The $C\gamma2$-$C\gamma3$ interface region contains the major determinants. J. Immunol. 140:3098-3107 (1988).

14. Yocum D.E., Klippel, J.H., Wilder, R.L., Gerber, N.L., Howard, A.A., Wahl, S.M., Lesko, L., Minor, J.R., Preuss, H.G., Yarboro, C., Berkebile, C., and Dougherty, S. Cyclosporin A in severe, treatment-refractory Rheumatoid Arthritis. Annls. of Int. Med. 109:863-869 (1988).

15. Hollander, M., and Wolfe, D.A. Non-parametric Statistical Methods. John Wiley and Sons, New York, (1983).

16. Kaymaz. H., Dedeoglu, F., Schluter, S.F., Edmundson, A.B. and Marchalonis, J.J. Reactions of anti-immunoglobulin sera with synthetic T-cell receptor peptides: implications for the 3-dimensional structure and function of the Tcr β chain. Submitted for publication. (1992).

17. Chothia, C., Bosell, D.R. and Lisk, A.M. The outline structure of the T-cell α/β receptor. EMBO J. 7:3745-3755 (1988).

18. Zanetti, M., Sollazzo, M. and Billetta, R. Functions and structures in a regulatory network for self-reactivity. in Molecular Immunobiology of Self-Reactivity (C.A. Bona and A.K. Kaushik, eds.), Marcell Dekker Inc., New York, pp. 221-238 (1992).

19. Kabat, E.A., Wu, T.T., Perry, H.M., Gottesman, K.S. and Foeller, C. Sequences of proteins of immunological interest, V Edition, Vol. 1. U.S. Department of Health and Human Services, Bethesda, MD (1991).

20. Mage, R.G., Bernstein, K.E., McCartney-Frances, N., Alexander, C.B. Young-

Cooper, G.O., Padlan, E.A., and Cohen, G.H, The structure and genetic basis for expression of normal and latent $V_H a$ allotypes of the rabbit. Mol. Immunol. 21:1067-1081 (1984).

21. Pullen, A.M., Wade, T., Marrack, P., and Kappler, J.W. Identification of the region of T-cell receptor β chain that interacts with the self-superantigen MIs-1ᵃ. Cell 61:1365-1374 (1990).

NATURAL AUTOANTIBODIES

Sylvia K. Chai[1], Liliana Mantovani[1], Marion T. Kasaian[1,2,] and Paolo Casali[1]

[1]The Department of Pathology and Kaplan Comprehensive Cancer Center
New York University School of Medicine
New York, NY 10016

[2]Immulogic Pharmaceutical Corporation
One Kendall Square
Cambridge , MA 02139

INTRODUCTION

It has been recognized for at least half a century that normal individuals display circulating immunoglobulins (Ig) able to bind a variety of foreign antigens, such as bacteria, virus, fungi, as well as self antigens, such as nucleic acids, phospholipids, erythrocytes, serum proteins, cellular components, insulin, and thyroglobulin[1-9]. Because these antibodies arise independently of known and/or deliberate immunization, they have been termed natural antibodies. In contrast to antigen-induced antibodies, which are mainly IgG and monoreactive, a considerable proportion of natural antibodies are IgM and polyreactive, that is they bind several unrelated antigens with different affinities. Natural polyreactive and monoreactive IgG and IgA also exist[7]. Natural antibodies are produced mainly by CD5[+] B cells, the predominant lymphocytes in the neonatal B cell repertoire[10,11]. Because of their broad reactivity for a wide variety of microbial components, natural antibodies play a major role in the primary line of defense against infections. Because of their ability to bind self antigens, they may provide the "templates" for the autoantibodies characteristic of autoimmune conditions, particularly those associated with expansion of CD5[+] B cells, e.g., rheumatoid arthritis[12-14].

THE ANTIGEN-BINDING ACTIVITY OF NATURAL ANTIBODIES

Our studies on natural polyreactive monoclonal antibodies, generated from healthy subjects, have clearly established that the these antibodies bind different antigens in a dose-saturable fashion and with different efficiencies (Fig. 1)[7]. These studies have also shown that the binding of most natural polyreactive antibodies to a given antigen can be cross-inhibited by a variety of different antigens with variable efficiency (Fig. 1).

The ability of polyreactive antibodies to bind different antigens relies on three discrete types of antigen recognition, i.e., recognition of: (i) identical, (ii) similar, and (iii) different epitopes in the context of different antigens (Fig. 2)[5,15,16]. Recognition of identical epitopes, in the context of different antigens, provides the structural correlate for classical immunological "cross reactivity", and is at the basis of autoimmune phenomena secondary to some bacterial infections, e.g. the cross-reaction of an HLA-B27 epitope with that of a *Klebsiella pneumoniae* epitope[17,18,19]. Recognition of structurally similar epitopes

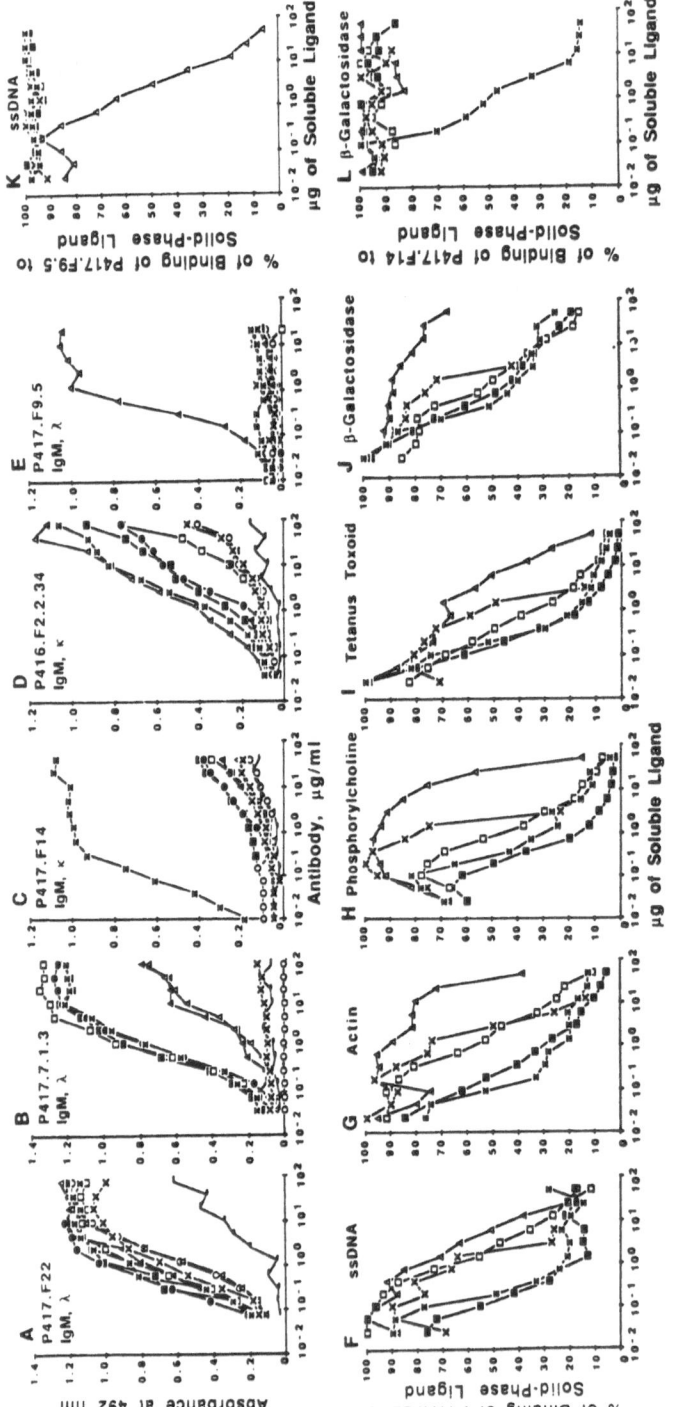

Figure 1. A to E, dose-dependent binding of IgM monoclonal antibodies (mAbs) generated using CD5⁻CD45RA^lo (A to C) and CD5⁺ B cells (D and E) to solid phase ligands. Antigen-binding activity of each mAb is expressed as optical absorbance at 492nm. F to J, dose-dependent inhibition of the binding of mAbs P417.F22 to solid phase ligands by soluble homologous or heterologous ligands. Samples of mAb P417.F22 (0.4µg) were incubated with increasing amounts of soluble ligand. After 18 hours, mixtures were transferred to ELISA plates precoated with ssDNA (F), actin (G), phosphorylcholine (H), tetanus toxoid (I), or ß-galactosidase (J). K and L, dose-dependent inhibition of the binding of mAb P417.F14 (L) to solid phase ssDNA and ß-galactosidase, respectively. In the competitive inhibition experiments, the amount of mAb bound to the solid phase antigen is expressed as a percentage of the binding activity measured in the absence of any soluble ligand (100% of binding activity). The following antigens were used: IgG Fc fragment (o), ssDNA (▲), insulin (●), thyroglobulin (▲), actin (□), phosphorylcholine (x), tetanus toxoid (■), ß-galactosidase (*), and LPS (−)⁷.

148

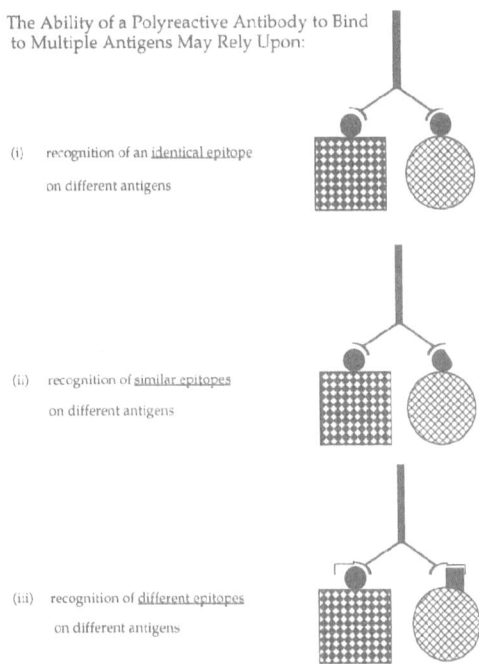

The Ability of a Polyreactive Antibody to Bind
to Multiple Antigens May Rely Upon:

(i)   recognition of an identical epitope
      on different antigens

(ii) recognition of similar epitopes
      on different antigens

(iii) recognition of different epitopes
      on different antigens

Figure 2. The ability of antibodies to bind different antigens can rely on three discrete types of antigen recognition, i.e., recognition of: (i) identical, (ii) similar, and (iii) different epitopes in the context of different antigens. Recognition of identical epitopes (i, full small circle), on different antigens (large hatch-marked square and circle), is at the basis of the classical immunological "cross-reactivity". Recognition of structurally similar epitopes (ii, full and half small circles), on different antigens (large hatch-marked square and circle) relies on the same principle, that is, the binding of different antigens is inherent to structural features of the antigens. The recognition of identical (i) or similar epitopes (ii) in the context of different antigens accounts for the observed polyreactivity of natural antibodies in only a few cases. In contrast with antigen-induced "specific" antibodies, natural polyreactive antibodies in general recognize antigens that are unlikely to share identical or similar epitopes, that is, they are capable of binding different epitopes (iii, full small circle and full square on large hatch-marked square and circle, respectively). This suggests that polyreactivity is a function of features inherent to the binding cleft of the antibody and not a function of structural features shared by different antigens.

relies on the same principle, that is, the binding of different antigens is inherent to structural features of the antigens. The ability of anti-DNA autoantibodies to bind appropriately spaced phosphate residues in the context of a variety of polynucleotides and phospholipids exemplifies this kind of cross-reaction[20]. The recognition of identical or similar epitopes in the context of different antigens accounts for the observed polyreactivity of natural antibodies in only a few cases. In most cases, natural polyreactive antibodies, in contrast to monoreactive high affinity antibodies, have the ability to recognize antigens that are unlikely to share identical or similar epitopes, such as proteins, nucleic acids, phospholipids, and polysaccharides[21,22]. This suggests that polyreactivity is a function of features inherent to the binding cleft of the antibody and not a function of structural features shared by different antigens.

The antigen-combining site of an antibody is lined mainly by three heavy (H) - and three light (L) chain complementarity determining regions (CDRs)[23,24]. Although all six CDRs are involved in antigen binding, the H chain CDR3 seems to play a major role, particularly in the case of large naturally occurring antigen[25]. In a majority of natural polyreactive antibodies, the H chain CDR3 is relatively long[25,26], possibly providing the structural basis for a large antigen-combining site. A broad antigen-combining site would allow polyreactive antibodies to accommodate a variety of dissimilar epitopes. Although a long H chain CDR3 may be critical for polyreactivity, other features of the structures lining the antigen-binding site may provide the correlate for the broad ligand specificity of polyreactive antibodies. For instance, the binding cleft of some murine anti-DNA

Table 6. Features of Human and Murine B-1a, B-1b and B-2 Lymphocyte Subsets.

B cells	surface IgM/IgD ratio	surface CD11b Mac-1	surface CD5 Ly-1	surface CD23 CD23	surface CD45RA B220	Expression of CD5 mRNA Ly-1 mRNA	IgM/IgG producing cell precursors	Frequency of natural antibody-producing cell precursors	Somatic hypermutation	N segment addition
Human:										
B-1a	?	++	yes	?[1]	low to inter[2]	high	high	very high	yes	yes
B-1b	?	++	no	?	low	high	high	very high	yes	no
B-2	low	+	no	?	high	very low	low	low	yes	yes
Mouse:										
B-1a	high	+++	yes	no	low	ND[3]	high	very high	ND	no ?
B-1b	high	+++	no	?	low	ND	high	very high	ND	ND
B-2	low	+	no	yes	high	ND	low	low	yes	yes

[1]inconclusive findings;
[2]intermediate;
[3]not determined.

150

antibodies contains a preponderance of basic arginine residues, enabling the formation of salt bridges with negatively charged antigens, such as DNA, RNA, and oligonucleotides[27,28]. Such an ionic interaction between arginine and nucleic acid phosphate may be an important factor determining the characteristic features of not only naturally occurring antibodies to DNA found in healthy subjects, but also the anti-DNA autoantibodies characteristic of patients with systemic lupus erythematosus (SLE)[27,28].

Consistent with the notion that natural antibodies are generated independently of specific antigenic stimulation, is the germline (unmutated) configuration of the genes encoding their variable (V) regions[5,29]. The expression of unmutated V genes provides a structural correlate for the high degree of idiotypic cross-reactivity observed among a large proportion of natural antibodies[4,12]. Some natural antibodies, encoded in unmutated germline configuration, display low dissociation constants (K_d) (high affinity) for antigen, similar to those of antibodies that have undergone an antigen-driven affinity maturation process[31,32]. High antibody affinity and specificity resulting from an antigen-driven process are due to somatic point mutations in the CDRs of the V regions[33,34]. This raises the issue as to what are the functional differences between natural and antigen-induced antibodies, and whether application of antigen-driven somatic selective pressure to the antibody-producing cell clone yields, as its only result, an increased antibody affinity.

Recently, Foote and Milstein proposed that the antibody-specificity maturation process relies, not only on the antigenic selection of clonotypes displaying high affinity (K_d), but more importantly, on the antigenic selection of clonotypes displaying the ability to bind antigen rapidly (k_{on})[35]. During the maturation of the antibody response to 2-phenyl-5-oxazolone in BALB/c mice, they found that the average K_d values vary little. The antibody repertoire, however, significantly shifts to those clonotypes with high k_{on}. Thus, antibodies are subject to a kinetic selection based on high k_{on}; somatic hypermutation in the Ig V regions would be critical to yield high k_{on} rather than high K_d. High k_{on} values may constitute an essential feature of surface receptors for antigen on memory B cells. It is conceivable that germline-encoded natural antibodies, displaying an affinity for antigen comparable to that of antigen-induced antibodies, may in fact display lower k_{on} values. Nevertheless, as suggested by our recent findings, similar to "conventional" antibodies, natural antibody-producing clonotypes may undergo an antigen-driven affinity maturation, eventually yielding high k_{on} values, as suggested by our recent findings of accumulation of somatic mutations in the CDRs of IgM, IgG, and IgA produced by $CD5^+$ B cells[34,36-39]. Such a process could underlie the generation of highly specific antibodies or autoantibodies.

ANALYSIS OF NATURAL ANTIBODY-PRODUCING CELLS AND THE EXPRESSED Ig V_H GENES

The first indication that natural antibodies are produced by a discrete B cell subset came from studies involving autoimmune NZB mice. These mice naturally develop autoimmune pathology and display an abnormally expanded $Ly-1^+$ B cell population[40]. Ly-1, and its human equivalent, CD5, were first recognized as T cell markers[41,42], and are known to contain a tyrosine phosphorylation site, a feature of many growth factor receptors[42,43]. $Ly-1^+$ B cells from NZB[40] and normal mice[44] spontaneously produce IgM with the features of natural antibodies. $Ly-1^+$ B cells develop from pre-B cells during neonatal life, and are maintained thereafter as self-replenishing IgM^+ B cells, while "conventional" $Ly-1^-$ B cells, develop later in life and are replenished from bone marrow IgM^- progenitors[45]. Some other features which distinguish $Ly-1^+$ B cells from their conventional $Ly-1^-$ counterparts are their unique surface immunoglobulin expression (IgM^{high} and IgD^{low}), their absence from the lymph nodes and bone marrow, and their abundant presence in the peritoneal cavity[44,46].

In humans, the CD5 antigen was first identified on the surface of B lymphocytes from patients with chronic lymphocytic leukemia (CLL)[47-49]. The physiologic equivalent of neoplastic CLL, CD5$^+$ B cells, were identified shortly afterwards in healthy individuals[50,51]. Characteristics such as natural antibody production, specific tissue distribution, and unique ontogenetic development, may distinguish CD5$^+$ from "conventional" B cells in humans, as shown for the murine Ly-1$^+$ equivalents[52-56]. The features of human and murine CD5$^+$ and Ly-1$^+$ B cells, respectively, and those of "conventional" B lymphocytes, as well as those of the recently defined B-1b cell subset (see below), are summarized in Table 1.

In contrast with the vast majority of natural antibodies, whether IgM, IgG or IgA, which are produced by CD5$^+$ B lymphocytes, antigen-induced antibodies, such as IgG to tetanus toxoid, possibly the products of memory B cells, are produced by the CD5$^-$ B lymphocytes[56]. This suggests functions unique to the two B lymphocyte subsets: CD5$^+$ cells are committed to the production of preimmune, natural antibodies, while CD5$^-$ B cells contain the elements that participate in secondary or tertiary high affinity responses. Evidence has accumulated indicating that natural antibody production is not an exclusive property of CD5$^+$ B cells, but may also be a property of some CD5$^-$ B cells[3,56-58]. In order to address this issue, we attempted to segregate, as a discrete subset, natural antibody-producing CD5$^-$ B cells from the total CD5$^-$ B cells. We reasoned that differential expression of CD45RA (CD45RA is the high molecular weight form of the leukocyte common antigen, L-CA[59,60]) could be used (in concert with CD5) to segregate discrete CD5$^-$ B cell fractions committed to the production of natural antibodies. The rationale for this approach was based on experiments in the mouse suggesting that B cells, expressing at low density, the high molecular weight form of the L-CA, produce antibodies with features similar to those of natural antibodies[61-63]. CD5$^-$ B lymphocytes were segregated into those expressing surface CD45RA at low (CD45RA^{lo}) and high (CD45RA^{hi}) density[7,8]. Our experiments showed that like CD5$^+$ B cells, the CD5$^-$CD45RA^{lo} but not the CD5$^-$CD45RA^{hi} B cell subset is highly enriched in IgM-producing cell precursors committed to the production of natural polyreactive and monoreactive antibodies[7]. In contrast, the CD5$^-$CD45RA^{hi} B cell subset is highly enriched in memory B lymphocytes, that is, cells committed to the production of monoreactive, antigen-specific, IgG. Because of their proposed early ontogenic emergence, CD5$^+$ B cells and their murine equivalents (Ly-1$^+$ B cells), have been termed B-1a cells; human CD5$^-$CD45RA^{lo} B cells, like the murine Ly-1 "sister" B cells, have been termed B-1b cells[64]. "Conventional" B lymphocytes identified in the mouse as surface Ly-1$^-$ IgM^{lo}/IgD^{hi}, and in the humans as CD5$^-$CD45RA^{hi} cells, are tentatively termed B-2 cells (Table 1).

B-1b cells account for 2% to 6% of total peripheral B lymphocytes (or 8% of CD5$^-$ cells), and constitute a discrete B cell subset rather than an activation state of B-1a cells[7]. In different individuals, these lymphocytes account for comparable proportions of the total peripheral blood B lymphocytes and in the same individuals, their proportions remain constant over time[7,65]. The high frequency of IgM-producing cell precursors, the prominent commitment to the production of natural antibodies, and their relatively high levels of CD5 mRNA expression suggest that these lymphocytes are related to CD5$^+$ B cells[7,44,66].

It has been suggested that in the early stages of murine development, when natural antibodies are predominant, and when B-1a cells comprise the majority of B lymphocytes, V_H gene expression is nonstochastic[67-71]. In murine fetal B cells, D-proximal V_H gene members (V_H7183 and V_HQ52) are used most frequently, indicating that position-dependent rearrangement mechanisms are operational during early B cell ontogeny[71-73]. Later in the life of the mouse, when "conventional" B-2 cells are predominant, the expression of V_H gene segments in the B cell repertoire normalizes, i.e., members of the largest gene family, V_H558, are the more frequently expressed[30,71-73]. These findings are consistent with the hypothesis that the unique antigen-binding features of natural antibodies result, at least in part, from a biased utilization of certain V_H gene segments. This is further corroborated by the preferential V_HQ52 gene expression by sorted B-1a cells in adult mice[69,70].

In contrast with the findings in the mouse, where the early and adult B cell repertoires differ in V_H gene family utilization, the human early and "mature" B cell repertoires seem to express a similarly biased selection of V_H gene families[74-76]. Human B cells from fetal liver, cord blood, and adult peripheral blood, display comparable overexpression of the V_HIII and V_HIV gene families. This is inconsistent with the hypothesis that B-1a cells (predominant in the fetal and neonatal repertoires) utilize a selection of V_H genes different from that of B-2 cells (which are predominant in the mature B cell repertoire). However, our preliminary experiments, examining different types of B cells sorted from the adult peripheral blood, show that B-1a cells overexpress genes of the V_HIV family, while "conventional" B-2 cells overexpress genes of the V_HIII family. B-1b cells express a selection of V_H genes more similar to that of B-2 than B-1a cells. Since both B-1a and B-1b cells are equally committed to the production of natural antibodies, the above findings suggest that genes of both V_HIV and V_HIII families can be associated with polyreactivity. Further, the preferential utilization of V_HIII genes by B-2 cells suggests that these V_H genes can encode not only polyreactive, but also monoreactive antigen-induced antibodies. Clearly, structures other than the V_H segments play major roles in mediating polyreactivity. Experiments now in progress in our laboratory, aimed at the systematic analysis of the junctional V_H-D-J_H (CDR3) (see previous section) sequence diversity, should define the precise association between the expression of certain Ig genes other than V_H genes and given B cell subsets, and should provide better insight into the structural requirements for polyreactivity.

B-1 CELLS IN AUTOIMMUNE DISEASES

The random processes of V_H-D-J_H and V_L-J_L recombination as well as V_H and V_L gene pairing ensure the potential for the emergence of B clonotypes able to recognize self antigens. Such potential is frequently found among B-1a and B-1b cells. Once generated, these cells may escape normal regulatory mechanisms; elimination of self-reactive B cells takes place in the periphery and may be a relatively inefficient process. Since B-1a and B-1b cells are self-replenishing[44,45,77], they would provide a source of autoantibody-producing cell precursors throughout adult life.

CD5+ B cells are greatly increased in number in several autoimmune diseases, including rheumatoid arthritis[12-14], Sjogren's syndrome[78], and primary antiphospholipid syndrome[79]. In rheumatoid arthritis patients, B-1a cells can account for more than 50% of circulating B lymphocytes and are large, proliferating cells, producing IgM, IgG and IgA rheumatoid factors (RFs)[12,14,36-39]. While most of RFs are polyreactive and low affinity, some are monoreactive and display a high affinity for IgG Fc fragment[14,36,39]. We compared the structural features of mono- and polyreactive RF, derived from both B-1a and B-1b cells[36,37,39]. We found that the assortment of V_H and V_L genes utilized by the high affinity RFs is similar to that utilized by their low affinity polyreactive counterparts, as well as to that of the polyreactive natural antibodies found in healthy subjects[39]. However, in contrast to the germline configuration of the genes encoding the V regions of most low affinity RFs, high affinity RF V genes can contain a number of somatic point mutations[36,37,39]. These mutations cluster in the CDRs and display a high replacement to silent (R/S) ratio, indicative of antigen-driven clonal selection in their generation[36,37,39]. These findings are inconsistent with the notion that B-1 lymphocytes are primordial cellular elements incapable of undergoing antigen-driven selection and rather suggest that these cells may participate in the maturation of a high affinity immune response to self antigens.

As mentioned above, in rheumatoid arthritis patients, high affinity RF autoantibodies are produced predominantly by an expanded surface CD5+ B cell population[36]. In other systemic autoimmune patients, such as those with SLE, CD5+ lymphocytes are not abnormally expanded, and may or may not be the major cell type involved in the production of autoantibodies[58,80]. Previous experiments have suggested that surface CD5-, rather than CD5+ B cells are the major cell type involved in the production of disease-characteristic autoantibodies in SLE patients[13,80]. These autoantibodies are IgG, IgA, and to a lesser extent, IgM, directed to single-stranded and/or double-stranded DNA. Notably, many of these antibodies cross-react with nucleic acids, phospholipids, and other antigens. As in the case of RFs isolated from patients with rheumatoid arthritis, the genes encoding the V regions of these antibodies are not dissimilar to those encoding the V regions of natural antibodies, and they bear the imprints of an antigen-driven selection process, i.e. extensive somatic mutation yielding high R/S ratios in the CDRs[81-84].

Our preliminary results suggest that anti-DNA autoantibodies in SLE patients are produced mainly by (surface CD5$^-$) B-1b cells. In these patients, the B-1b cell population is abnormally elevated, 3- to 5-fold, at the expense of B-2 cells (CD5$^-$CD45RA^{hi})[34]. This reduction in B-2 cells might result from the cytolytic activity of autoantibodies directed to CD45RA; indeed, such antibodies have been detected in the circulation of SLE patients and might be produced by B-1b cells[85,86]. The identification of the B-1b cell subset as an important source of natural autoantibodies in healthy subjects, and as a major contributor to the production of autoantibodies in SLE patients may suggest a role for these cells in autoimmune diseases other than SLE. In cases where the role of B-1a cells and natural autoantibodies cannot be clearly delineated, the examination of B-1b B cell products may prove more revealing.

CONCLUDING REMARKS

Natural antibodies are generated independently of known antigenic stimulation, are mostly IgM, polyreactive, and encoded in germline V genes. Polyreactive IgM natural antibodies are produced by the majority of the (B-1) B cells found in the fetus and neonate. Although endowed with self-reactivity, natural antibodies also bind exogenous antigens. Axenic mice contain a relatively large population of spontaneously generated B cells which produce antibodies able to bind a variety of bacterial antigens[87,88]. This large population of natural autoantibody-producing B cells would have the potential to generate specific, high affinity, IgG and IgA antibodies, subsequent to exogenous antigen exposure.

A central issue related to the understanding of the physiologic and pathologic role of natural antibodies is whether precursors of cells producing natural antibodies, B-1a and B-1b lymphocytes, are capable of undergoing antigen-driven clonal selection, thereby producing antigen-specific, high-affinity, autoantibodies, such as those found in autoimmune patients. We unequivocally showed that both B-1a and B-1b cells can express a hypermutation mechanism similar to that of conventional (B-2) cells. Thus, it can be speculated that application of antigenic pressure to progenitors producing natural low affinity, antibodies could eventually result in the generation of high-affinity, somatically-mutated autoantibodies.

The demonstration that autoantibody-producing B-1 cells originate from a unique lineage would prompt investigations into the characteristic regulatory mechanism of these cells, thereby suggesting a mode for therapeutic intervention in autoimmune patients. For example, the identification of a surface marker exclusive to B-1a and B-1b cells would perhaps make it possible to delete autoantibody-producing B-1a and B-1b cells from the circulation of rheumatoid arthritis or SLE patients using a specific monoclonal antibody. On the other hand, patients suffering from severe bacterial infections could benefit from an enhancement of natural antibody-producing potential of their B-1a and B-1b cells. Finally, in transplant recipients, hyperacute rejection, which has recently been shown to be mediated by natural antibodies[6,89] may be prevented by the use reagents capable of ablating B-1 cells.

Acknowledgments

Most of the work discussed here has been supported by the United States Public Health Service (N.I.H.) Grants AR-40908, and CA-09161. This is publication 17 from the "Jeanette Greenspan Laboratory for Cancer Research". Paolo Casali is a Kaplan Cancer Scholar.

REFERENCES

1. S. Boyden, Natural antibodies and the immune response, *Adv. Immunol.* 5:1 (1965).
2. J.G. Michael, Natural antibodies, *Curr. Top. Microbiol. Immunol.* 48:43 (1969).
3. P. Casali, and A.L. Notkins, CD5$^+$B lymphocytes, polyreactive antibodies and the human B-cell repertoire, *Immunol. Today.* 10:364 (1989a).

4. S. Avrameas, Natural autoantibodies: from "horror autotoxicus" to "gnothi seauton", *Immunol. Today.* 12:154 (1991).

5. P. Riboldi, M.T. Kasaian, L. Mantovani, H. Ikematsu, and P. Casali, Natural antibodies, *in*: "Molecular Pathology of Autoimmunity", C.A. Bona, K. Siminovitch, M. Zanetti, A.N. Theophilopoulos, eds., The Harwood Academic Publishers, New York (1992).

6. M.A. Turman, P. Casali, A.L. Notkins, F.H. Bach, and J.S. Platt, Polyreactive antibodies from $CD5^+$ B cells: antigen specificity and relationship to xenoreactive natural antibodies, *Transplantation.* 52:710 (1991).

7. M.T. Kasaian, H. Ikematsu, and P. Casali, Identification and analysis of a novel human surface $CD5^-$ B lymphocyte subset producing natural antibodies, *J Immunol.* 148:2690 (1992).

8. M.T. Kasaian and P. Casali, Analysis of the human $CD5^-CD45RA^{lo}$ B cell subset, *Ann NY Acad Sci.* 651:59 (1992).

9. M. Nakamura S.E. Burastero, A.L. Notkins, and P. Casali, Human monoclonal rheumatoid factor-like antibodies from CD5 $(Leu-1^+)$ B cells are polyreactive, *J Immunol.* 140:4180 (1988).

10. A.Durandy, L.Thuillier, M.Forveille, and A.Fischer, Phenotypic and functional characteristics of human newborns' B lymphocytes, *J.Immunol.* 144:60 (1990).

11. N.Gadol, and K.A. Ault, Phenotypic and functional characterization of human Leu 1 (CD5) B cells, *Immunol. Rev.* 93:23 (1986).

12. S.E. Burastero, and P. Casali, Characterization of human CD5 (Leu-1, OKT1)$^+$ B lymphocytes and the antibodies they produce, *Contrib Microbiol. Immunol.* 11:231 (1989).

13. P. Casali and A.L. Notkins, Probing the human B-cell repertoire with EBV: Polyreactive antibodies and $CD5^+$ B lymphocytes, *Ann Rev Immunol.* 7:513 (1989).

14. S.E. Burastero, P. Casali, R.L. Wilder, and A.L. Notkins, Monoreactive high affinity and polyreactive low affinity rheumatoid factors are produced by CD5+ B cells from patients with rheumatoid arthritis, *J Exp Med.* 168:1979 (1988).

15. P.Casali, M.T.Kasaian, and G.Haughton, B-1 (CD5 B) cells, *in*: "Autoimmunity," A.Coutinho and M.D.Kazatchkine, ed., John Wiley & Sons, New York (1993). In press.

16. M.Kasaian and P.Casali, Natural antibodies, self recognition and autoimmunity-prone B-1 (CD5 B) cells, *Autoimmunity.* In press (1993).

17. P.L. Schwimmbeck, D.T.Y Yu, and M.B.A. Oldstone, Autoantibodies to HLA B27 in the sera of HLA B27 patients with ankylosing spondylitis and Reiter's syndrome. Molecular mimicry with Klebsiella Pneumoniae as potential mechanism of autoimmune disease, *J. Exp. Med.* 166:173 (1987).

18. R.B. Rayburne and K.M. Williams, Monoclonal antibodies against and HLA-B27-derived peptide reacts with an epitope present on bacterial proteins, *J. Immunol.* 145:2539 (1990).

19. M.W. Cunningham, J.M. McCormack, P.G. Fenderson, M.K. Ho, E.H. Beachey, and J.B. Dale, Human and murine antibodies cross-reactive with streptococcal M protein and myosin recognize the sequence GLN-LYS-SER-LYS-GLN in M protein, *J .Immunol.* 143:2677 (1989).

20. E.M. Lafer, J. Rauch, J.R. Andrzejewski, D. Mudd, B. Furie, R.S. Schwartz, D.B. Stollar, Polyspecific monoclonal lupus autoantibodies reactive with both polynucleotides and phospholipids, *J. Exp. Med.* 153:897 (1981).

21. T. Ternynck, S. Avrameas, Murine natural monoclonal autoantibodies: a study of their polyspecificities and their affinities, *Immunol. Rev.* 94:99 (1986).

22. P. Casali, B.S. Prabhakar, and A.L. Notkins, Characterization of multireactive autoantibodies and identification of LEU-1$^+$ B lymphocytes as cells making antibodies binding multiple self and exogenous molecules, *Intern. Rev. Immunol.* 3:17 (1988).

23. P.M. Colman, Structure of antibody-antigen complexes: Implications for immune recognition, *Adv. Immunol.* 43:99 (1988).

24. E.A. Kabat, Antibody complementarity and antibody structure, *J. Immunol (suppl).* 141:S25 (1988).

25. A.G. Amit, R.A. Mariuzza, S.E. Phillips, and R.J. Poliak, Three-dimensional structure of an antigen-antibody complex at 2.8 A resolution, *Science.* 233:747 (1986).

26. R.L. Stanfield, T.M. Fieser, R.A. Lerner, and I.A. Wilson, Crystal structures of an antibody to a peptide and its complex with peptide antigen at 2.8 A, *Science* 248:712 (1990).

27. D. Eilat, D.M. Webster, and A.R. Rees, V region sequences of anti-DNA and anti-RNA autoantibodies from NZB/NZW F1 mice, *J. Immunol.* 141:1745 (1988).

28. M. Shlomchik, M. Mascelli, H. Shan, M.Z. Radic, D. Pisetsky, A. Marshak-Rothstein, and M. Weigert, Anti-DNA antibodies from autoimmune mice arise by clonal expansion and somatic mutation, *J .Exp. Med.* 171:265 (1990).

29. I. Sanz, P. Casali, J.W. Thomas, A.L. Notkins, and J.D. Capra, Nucleotide sequences of eight human natural antibody V_H regions reveals apparent restricted use of V_H families, *J .Immunol.* 142: 4054 (1989a).

30. G.D.Yancopoulos and F. Alt, Developmentally controlled and tissue-specific expression of unrearranged V_H gene segments, *Cell.* 40:271 (1985).

31. M. Nakamura, S.E. Burastero, Y. Ueki, J.W. Larrick, A.L. Notkins, P. Casali, Probing the normal and autoimmune B cell repertoire with Epstein-Barr virus. Frequency of B cells producing monoreactive high affinity autoantibodies in patients with Hashimoto's disease and systemic lupus erythematosus, *J Immunol.* 141:4165 (1988).

32. Y. Ueki, I. Goldfarb, N. Harindranath, M. Gore, H. Koprowski, A.L. Notkins, A.L., and P. Casali, Clonal analysis of a human antibody response. Quantitation of precursors of antibody-producing cells and generation and characterization of monoclonal IgM, IgG and IgA to rabies virus, *J .Exp .Med.* 171: 19 (1990).

33. H. Ikematsu, M.T. Kasaian, E.W. Schettino, T.H. Steger, P. Casali, Antigen-selected somatic mutations in the V_H regions of natural polyreactive human IgG autoantibodies, In preparation (1992).

34. M.T. Kasaian, H. Ikematsu, J.E. Balow, and P. Casali, Cellular origin and V_H genes of monoreactive and polyreactive IgA and IgG autoantibodies to DNA in patients wtih systemic lupus erythematosus, In preparation (1992).

35. J. Foote, and C. Milstein, Kinetic maturation of an immune response, *Nature* 352:530 (1991).

36. N. Harindranath, I.S. Goldfarb, H. Ikematsu, S.E. Burastero, R.L. Wilder, A.L. Notkins, and P. Casali, Complete sequence of the genes encoding the V_H and V_L regions of low and high affinity monoclonal IgM and IgA1 rheumatoid factors produced by CD5+ B cells from a rheumatoid arthritis patient, *Int. Immunol.* 3:865 (1991).

37. H. Ikematsu, N. Harindranath, and P. Casali, Somatic mutations in the V_H genes of high affinity antibodies to self and foreign antigens produced by human CD5+ and CD5- B cells, *Ann .NY Acad .Sci.* 651:319 (1992).

38. R.C. Williams Jr., C.C. Malone, and P. Casali, Heteroclitic polyclonal and monoclonal anti-Gm(a) and anti-Gm(g) rheumatoid factors react with epitopes induced in Gm (a-), Gm(g-) IgG by interaction with antigen or by nonspecific aggregation, A possible mechanism for the in vivo generation of rheumatoid factors, *J .Immunol.* 149:1817 (1992).

39. L. Mantovani, M.T. Kasaian, R.L. Wilder, and P.Casali, Monoreactive high affinity IgM rheumatoid factor autoantibodies are produced by B-1 cells and utilize somatically mutated V_H and V_L genes, In preparation (1992).

40. K. Hayakawa, R.R. Hardy, M. Honda, L.A. Herzenberg, A.D. Steinberg, and L.A. Herzenberg, Ly-1 B cells: Functionally distinct lymphocytes that secrete IgM autoantibodies, *Proc. Natl. Acad. Sci. USA.* 81:2494 (1984).

41. E.G. Engleman, R. Warnke, R.I. Fox, J. Dilley, C.J. Benike, and R. Levy, Studies of a human T lymphocyte antigen recognized by a monoclonal antibody, *Proc. Nat. Acad. Sci. USA.* 78:1791 (1981).

42. H-J.S. Huang, N.H. Jones, J.L. Strominger, and L.A. Herzenberg, Molecular cloning of Ly-1, a membrane glycoprotein of mouse T lymphocytes and a subset of B cells: Molecular homology to its human counterpart Leu-1/T1 (CD5), *Proc. Nat. Acad. Sci. USA.* 84:204 (1987).

43. N.H. Jones, M.L. Clabby, D.P. Dialynas, H-J.S. Huang, L.A. Herzenberg, and J.L. Strominger, Isolation of complementary DNA clones encoding the human lymphocyte glycoprotein T1/Leu-1. *Nature.* 323:346 (1986).

44. L.A. Herzenberg, A.M. Stall, P.A. Lalor, C. Sidman, W.A. Moore, D.R. Parks, and L.A. Herzenberg, The LY-1 B cell lineage. *Immunol. Rev.* 93:81 (1986).

45. P.A. Lalor, L.A. Herzenberg, S. Adams, A.M. Stall, A.M., Feedback regulation of murine Ly-1 B cell development, *Eur. J. Immunol.* 19:507 (1989).

46. K. Hayakawa, and R.R. Hardy, Normal, autoimmune, and malignant $CD5^+B$ cells: The Ly-1 B lineage?, Ann. Rev. Immunol. 6:197 (1988).

47. L. Boumsell, A. Bernard, V. Lepage, L. Degos, J. Lemerle, J. Dausset, Some chronic lymphocytic leukemia cells bearing surface immunoglobulins share determinants with T cells, *Eur. J. Immunol.* 8:900 (1978).

48. I. Royston, J.A. Majoa, S.M. Baird, G.L. Meserve, J.C. Griffiths, Human T-cell antigens defined by monoclonal antibodies: The 65,000-dalton antigen of T cells (T65) is also found on chronic lymphocytic leukemia cells bearing surface immunoglobulin, *J. Immunol.* 125:725 (1980).

49. L. Boumsell, H. Coppin, D. Pham, B. Raynal, J. Lemerle, J. Dausset, and A. Bernard, An antigen shared by human T cell subsets and B cell chronic lymphocytic leukemic cells: Distribution on normal and malignant cells, *J. Exp. Med.* 152:229 (1988).

50. F. Caligaris-Cappio, M. Gobbi, M. Bofill, and G. Janossy, Infrequent normal B lymphocytes express features of B-chronic lymphocytic leukemia, J. Exp. Med. 155:623 (1982).

51. M. Gobbi, F. Caligaris-Cappio, and G. Janossy, Normal equivalent of cells of B cell malignancies: Analysis with monoclonal antibodies, *Brit. J. Haematol.* 54:393 (1983).

52. N. Gadol, and K.A. Ault, Phenotypic and functional characterization of human Leu 1 (CD5) B cells, *Immunol. Rev.* 93:23 (1986).

53. J.H. Antin, S.G. Emerson, P. Martin, N. Gadol, K.A. Ault, LEU-1^+ (CD5+)B cells: A major lymphoid subpopulation in human fetal spleen: Phenotypic and functional studies, *J. Immunol.* 136:505 (1986).

54. K. Hayakawa, R.R Hardy, and L.A. Herzenberg, Peritoneal Ly-1 B cells: Genetic control, autoantibody production, increased lambda light chain expression, Eur. *J. Immunol.* 16:450 (1986).

55. T.J. Kipps, S. Fong, E. Tomhave, P.P. Chen, R.D. Goldfien, and D.A. Carson, High frequency expression of a conserved kappa variable region gene in chronic lymphocytic leukemia, *Proc. Natl. Acad. Sci. USA.* 84:2916 (1987).

56. P. Casali, S.E. Burastero, M. Nakamura, G. Inghirami, A.I. Notkins, Human lymphocytes making rheumatoid factor and antibody to ssDNA belong to the Leu-1+ B-cell subset, *Science.* 236:77 (1987).

57. K. Hayakawa, R.R. Hardy, D.R. Parks, and L.A. Herzenberg, The "Ly-1 B" cell subpopulation in normal, immunodefective, and autoimmune mice, *J. Exp. Med.* 157:202 (1983).

58. P. Casali, S.E. Burastero, J.E. Balow, and A.L. Notkins, High affinity antibodies to ssDNA are produced by CD5-B cells in systemic lupus erythematosus patients, *J. Immunol.* 143:3476 (1989).

59. K. Hasegawa, H. Nishimura, S. Ogawa, S. Hirose, H. Sato, and T. Shirai, Monoclonal antibodies to epitope of CD45R(B220) inhibit interleukin 4-mediated B cell proliferation and differentiation, *Internat. Immunol.* 2:367 (1990).

60. R.S. Mittler, R.S. Greenfield, B.Z. Schacter, N.F. Richard, and M.K. Hoffmann, Antibodies to the leukocyte antigen (T200) inhibit an early phase in the activation of resting human B cells, J. Immunol. 138:3159 (1987).

61. H. Yakura, F.W. Shen, E. Bourcet, and E.A. Boyse, On the function of Ly-5 in the generation of antigen-driven B cell differentiation. Comparison and contrast with Lyb-2, *J. Exp. Med.* 157:1077 (1983).

62. H. Yakura, I. Kawabata, T. Ashida, and M. Katagiri, Differential regulation by Ly-5 and Lyb-2 of IgG production induced by lipopolysaccharide and B cell stimulatory factor-1 (IL-4), *J. Immunol.* 141:875 (1988).

63. H. Yakura, I. Kawabata, F.W. Shen, and M. Katagiri, Selective inhibition of lipopolysaccharide-induced polyclonal IgG response by monoclonal Ly-5 antibody, *J. Immunol.* 136:2729 (1986).

64. A. Kantor, A new nomenclature for B cells, *Immunol. Today*. 12:388 (1991).

65. T.J. Kipps, and J.H. Vaughn, Genetic influence on the levels of circulating CD5 B lymphocytes, *J. Immunol.* 139:1060 (1987).

66. D.M. Klinman, and K.L. Holmes, Differences in the repertoire expressed by peritoneal and splenic Ly-1 (CD5)+B cells, *J. Immunol.* 144:4520 (1990).

67. G.D. Yancopoulos, S.V. Desiderio, M. Paskind, J.F. Kearney, D. Baltimore, and F.W. Alt, Preferential utilization of the most J_H proximal V_H segments in pre-B cell lines, *Nature*. 311:727 (1984).

68. M.G. Reth, S. Jackson, and F.W. Alt, $V_H D J_H$ formation and D-J_H replacement during pre-B differentiation: Non-random usage of gene segments, *EMBO J.* 5:2131 (1986).

69. H.D. Jeong, and J.M. Teale, Comparison of the fetal and adult functional B cell repertoires by analysis of V_H gene family expression, *J. Exp. Med.* 168:589 (1988).

70. H.D. Jeong, and J.M. Teale, V_H gene repertoire of resting B cells. Preferential use of D-proximal families early in development may be due to distinct B cell subsets, *J. Immunol.* 143:2752 (1989).

71. B.A. Malynn, G.D. Yancopoulos, J.E. Barth, C.A. Bona, and F.W. Alt, Biased expression of J_H-proximal V_H genes occurs in the newly generated repertoire of neonatal and adult mice, *J. Exp. Med.* 171:843 (1990).

72. R.M. Perlmutter, J.F. Kearney, S.P. Chang, L.E. Hood, Developmentally controlled expression of immunoglobulin V_H genes, *Science*. 227:1597 (1985).

73. P. Brodeur, G.E. Osman, J.J. Mackle, and T.M. Lalor, The organization of the mouse Igh-v locus: Dispersion, interdispersion, and the evaluation of V_H gene family clusters, *J. Exp. Med.* 168:2261 (1988).

74. H.W. Schroeder Jr, J.L. Hillson, and R.M. Perlmutter, Early restriction of the human antibody repertoire, *Science*. 238:791(1987).

75. A.M. Cuisinier, V. Guigou, L. Boubli, M. Fougereau, and C. Tonnelle, Preferential expression of V_H5 and V_H6 immunoglobulin genes in early human B cell ontogeny, *Scand. J. Immunol.* 30:493 (1989).

76. P.P. Chen, R.W. Soto-Gil, and D.A. Carson, The early expression of some human autoantibody-associated heavy chain variable region genes is controlled by specific regulatory elements, *Scand. J. Immunol.* 31:673 (1990).

77. L.A.Herzenberg, and A.M.Stall, Conventional and Ly-1 B-cell lineages in normal and μ transgenic mice, *Cold Spring Harbor Sym. Quant. Biol.* 54:219 (1989).

78. M. Dauphinee, Z. Tovar, and N. Talal, B cells expressing CD5 are increased in Sjogren's syndrome, *Arthritis Rheum.* 31:642 (1988).

79. M.C. Velasquillo, J. Alcocer-Varela, D. Alarcon-Segovia, J. Cabiedes, and J. Sanchez-Guerrero, Some patients with primary antiphospholipid syndrome have increased circulating $CD5^+$ B cells that correlate with levels of IgM antiphospholipid, *Clin.. Expl. Rheum.* 9:1 (1991).

80. N. Suzuki, T. Sakane, E.G. Engleman, Anti-DNA antibody production by $CD5^+$ and $CD5^-$ B cells of patients with systemic lupus erythematosus, *J Clin Invest.* 85:238 (1990).

81. A. Mannheimer-Lory, J.B. Katz, M. Pillinger, C. Ghossein, A. Smith, and B. Diamond, Molecular characteristics of antibodies bearing an anti-DNA-associated idiotype, *J. Exp. Med.* 174:1639 (1991).

82. H. Dersimonian, R.S. Schwartz, K.J. Barrett, D.B. Stollar, Relationship of human variable region heavy chain germline genes to genes encoding anti-DNA autoantibodies, *J .Immunol.* 139:2496 (1987).

83. J.H. Van Es, F.H.J. Gmelig-Meyling, W.R.M. Van De Akker, H. Aanstoot, R.H.W.M. Derksen, and T. Logtenberg, Somatic mutations in the variable regions of a human IgG anti-double-stranded DNA autoantibody suggest a role for antigen in the induction of systmic lupus erythematosus, *J .Exp. Med.* 173: 461 (1991).

84. B. Diamond, J.B. Katz, E. Paul, C. Aranow, D. Lustgarten, and M.D. Scharff, The role of somatic mutation in the pathogenic anti-DNA response, *Ann. Rev Immunol.* 10:731 (1992).

85. T.Mimura, P.Fernstein, and J.B.Winfield, Autoantibodies specific for different isoforms of CD45 in systemic lupus erythmatosis, *J Exp. Med.* 172:653 (1990).

86. S.Tanaka, T.Matsuyama, A.D.Steinberg, S.F.Schlossman, and C.Morimoto, Anti-lymphocyte antibodies against CD4+2H4+ cell populations in patients with systemic lupus erythmatosis, *Arth. Rheum.* 32:398 (1989).

87. N.A.Bos, H.Kimura, C.G. Meewsen, H.De Visser, M.P.Hazenberg, B.S.Wostmann, J.R.Pleasants, R.Benner, and D.M.Marcus, Serum immunoglobulin levels and naturally occuring antibodies against carbohydrate antigens in germ-free BALB/c mice fed chemically defined ultrafiltered diet, *Eur.J. Immunol.* 19:2335 (1989).

88. J.R.Underwood, J.S.Pederson, P.J.Chalmers, and B.H.Toh, Hybrids from normal, germ-free, nude and neonatal mice produce monoclonal autoantibodies to eight different intracellular structures, *Ciln.Exp.Immunol.* 60:417 (1985).

89. R.L. Geller, F.H. Bach, M.A. Turman, P. Casali, and J.L. Platt, Natural antibodies bearing a "polyreactive" idiotype are deposited in rejected discordant xenografts, *Transplantation.* In press (1992).

REGULATORY AUTOANTIBODY AND CELLULAR AGING AND REMOVAL

Marguerite M. B. Kay

Regents' Professor
Department of Microbiology and Immunology
and Department of Veterans Affairs
Room 650 LSN
University of Arizona College of Medicine
1501 North Campbell Ave.
Tucson, AZ 85704

INTRODUCTION

Senescent cell antigen (SCA), an aging antigen, was discovered in 1975 (Kay, 1975). It is a protein that appears on old cells and marks them for death. It acts as a specific signal for cellular termination by initiating the binding of IgG autoantibody and subsequent removal by phagocytes (Kay, 1975, 1978,1981, 1984, 1983, 1986, 1988a, 1988b, 1988c, 1990; Kay et al., 1986, 1989, 1982, 1983a, 1988a, 1988b,1991, 1990a, 1990b, 1990c, 1990d; Kay and Lin, 1990; Bennett and Kay, 1981; Singer et al., 1986; Glass et al., 1983, 1985; Bartosz et al., 1982a, 1982b; Khansari et al., 1983; Khansari and Fedenberg, 1983; Walker et al., 1984; Lutz et al., 1984; Petz et al., 1984; Hebbel and Miller, 1984). This appears to be a general physiologic process for removing senescent and damaged cells in mammals and other vertebrates (Kay, 1981). Although the initial studies is done using human erythrocytes as a model, senescent cell antigen was discovered on cells besides erythrocytes in 1981 (Kay, 1981). It occurs on all cells examined (Kay, 1981). The aging antigen itself is generated by the degradation of an important structural and transport membrane molecule, protein band 3 (Kay, 1984). Besides its role in the removal of senescent and damaged cells, senescent cell antigen also appears to be involved in the removal of erythrocytes in clinical hemolytic anemias (Kay et al., 1989, 1990a), sickle cell anemia (Petz et al., 1984; Hebbel and Miller, 1984) and the removal of malaria-infected erythrocytes (Okoye and Bennett, 1985; Friedman et al., 1985). Oxidation generates senescent cell antigen in situ (Kay et al., 1986).

The presence of band 3 related molecules in nonerythroid tissues was first demonstrated in 1983 (Kay et al., 1983b). A protein immunologically related to band 3 was demonstrated in such diverse cell types as isolated neurons, hepatocytes, squamous epithelial cells, alveolar (lung) cells, lymphocytes, neurons, and fibroblasts using an antibody to band 3 that reacts with the transmembrane, anion transport domain of band 3 (Kay *et al.*, 1983b). Since then, band 3 has been described in numerous cell types and tissues including brain, fibroblasts, glial cells, kidney, placental syncytiotrophoblast, eye lens, hepatoma cells, and lymphoid cells (Kay *et al.*, 1991; Kay, 1990; Alper *et al.*, 1989, 1988, Drenckhahn *et al.*, 1984, 1985; Wolpaw and Marint, 1986; Drenckhahn and Merte, 1987; Hazen-Martin *et al.*, 1987, 1986; Demuth *et al.*,1986; Vanderpuye *et al.*, 1988; Verlander *et al.*, 1988; Allen *et al.*, 1987; Brosius *et al.*, 1989; Kudrycki and Shull, 1989). The cytoplasmic amino segment exhibits variation. The transport domain is highly conserved (Figure 1). Band 3 is also present in nuclear (Kay *et al.*, 1983b), Golgi (Kellokumpu *et al.*, 1988), and mitochondrial membranes (Schuster *et al.*, 1986) as well as in cell membranes. Band 3-like proteins in nucleated cells participate in band 3 antibody induced cell surface patching and capping (Kay *et al.*, 1983b). A truncated version of band 3 which lacks the amino terminus has also been described in kidney (Brosius *et al.*, 1989). Band 3 is present in the central nervous system, and differences described in band 3 between young and aging brain tissue (Kay, 1990). One autosomal recessive neurological disease, choreoacanthocytosis, is associated with band 3 abnormalities (Kay *et al.*, 1990a, 1990b, 1990c). The 150 residues of the carboxyl terminus segment of band 3 appear to be altered (Kay *et al.*, 1990a, 1990b, 1990c). In brains from Alzheimer's disease patients, antibodies to aged band 3 label the amyloid core of classical plaques and the microglial cells located in the middle of the plaque in tissue sections and an abnormal band 3 in immunoblots. Band 3 in nonerythoid tissues performs the same functions as it does in erythrocytes. Thus, senescent cell antigen generation from band 3 may be a generalized mechanism for cellular removal.

MECHANISM OF REMOVAL OF SENESCENT CELLS

Neoantigen on Old Cells Triggers Immunoglobulin Binding

The first hint that a "neo-antigen" appeared on senescent cells came from studies showing that IgG autoantibodies selectively bind to old human red blood cells (RBC) aged *in situ* (Kay, 1975; Kay, 1978) during investigations of the mechanism by which macrophages distinguish between senescent and other "self" cells. It is hypothesized that Ig in normal human serum attaches to the surface of senescent RBC until a critical level is reached which results in phagocytosis. Human RBC are used as a model for these studies because of the ready availability of these cells and the ease with which populations of different ages can be separated. In addition, RBC do not synthesize proteins. Therefore, they provide information on cumulative protein aging and damage.

To test this hypothesis, RBC separated by density centrifugation from freshly drawn blood is incubated with specific antibodies to human immunoglobulins (Ig) conjugated to scanning immunoelectron microscopy markers. Old RBC had IgG but no IgA or IgM on their surface as determined by scanning immunoelectron microscopy (Kay, 1975). Young RBC did not have immunoglobulin on their surface. Incubation of old RBC with autologous macrophages resulted in their phagocytosis regardless of whether incubations are performed in medium with serum, autologous Ig-depleted serum or whole serum (Kay, 1975). Young RBC are not phagocytized under any of these conditions. Thus, it appeared that the IgG attached *in situ* to senescent human RBC and rendered them vulnerable to phagocytosis by macrophages. The presence of IgG autoantibodies on the surface of old cells indicates that a new antigen has appeared that is not present on other cells. This is the first indication that a "neo-antigen" appears on senescent cells.Senescent cell IgG is an autoantibody. IgG attached to senescent cells *in situ* is shown to be an autoantibody (Kay, 1978). The antibodies could be dissociated from senescent cells. The dissociated antibodies specifically reattached via the antigen binding (Fab) portion of the IgG molecule to homologous senescent, but not to autologous or allogeneic mature or young cells (Bennett and Kay, 1981). Fab binding is demonstrated by antigen blockade studies, scanning immunoelectron microscopy, and 125I labeled Protein A binding to the Fc region of IgG bound to senescent cells and vesicles (Kay, 1978, 1983, 1986; Kay et al., 1982). Thus, the antibody to the senescent cell antigen is an autoantibody and not a nonspecific or a cytophilic antibody (Kay, 1978). It exhibited specific immunologic binding via the Fab region.

Other investigators have confirmed the presence of IgG on senescent, damaged, and stored red cells (Singer *et al.*, 1986; Glass *et al.*, 1983, 1985; Bartosz *et al.*, 1982a, 1982b; Khansari *et al.*, 1983; Khansari and Fudenberg, 1983; Walker *et al.*, 1984; Lutz *et al.*, 1984; Pet *et al.*, 1984; Hebbel and Miller, 1984; Kay, 1983,1986; Kay *et al.*, 1991a). The amount of IgG on the surface of cells increases with storage. Glass *et al.* (1983) have found a 44% reduction in the mean life span of RBC in old, specific pathogen free rats, as compared to the life span in young rats, as determined by 59Fe pulse-labeling *in vivo*. The proportion of young cells circulating in the blood of old animals was increased. In old rats, young as well as old RBC are heavily labeled with IgG; whereas predominantly old cells carried IgG in young rats. Extending their studies to humans, Glass, *et al.* (Glass *et al.*, 1985) found a significant increase in reticulocytes in healthy elderly humans even though their hematocrits are the same as those of younger individuals. A significant increase in autologous IgG on lower middle aged cells was consistently demonstrated in these elderly individuals. These findings suggest that young cells age prematurely in old individuals. Bartosz *et al.* (1982) have found increased amounts of cell-membrane bound IgG on RBC from patients with Down's syndrome and suggest that accelerated red cell aging occurs in these individuals. Accelerated aging of other cells and systems, including the immune system, has been reported in patients with Down's syndrome. Bosman, *et al.* (Bosman *et al.*, 1991) have found that erythrocytes from patient's with Alzheimer's disease but not multiinfarct dementia show characteristics of accelerated aging.

The results of studies described above demonstrate that Ig was required to initiate phagocytosis of senescent and stored cells and that IgG attaches *in situ* to senescent human RBC (Kay, 1975, 1978).

Erythrocyte Lifespan Studies

These results are confirmed *in vivo* using mice which are bred and maintained in a Maximum Security Barrier devoid of viruses, mycoplasma, and pathogenic bacteria; thus, excluding an exogenous source for the senescent cell antigen (Bennett and Kay, 1981). RBC are labeled *in situ* with 59Fe which labels the newly synthesized hemoglobin in young cells. Red cells are separated on Percoll gradients, 1 or 40 days after radioactive iron injection, into young and old populations, and injected into separate groups of syngeneic mice. Kinetic studies revealed that $\leq 90\%$ of the 59Fe labeled young RBC are removed from the circulation within 45 days. In contrast, $\leq 90\%$ of the 59Fe labeled old RBC are removed within 20 days. The difference in the rate of removal of young and old RBC was statistically significant ($P \leq 0.001$). Kinetic studies on density separated spleen cell populations revealed that the radioactivity decreased in the RBC fraction concomitantly with an increase in radioactivity in the splenic macrophage fraction. The radioactivity was found to be inside macrophages (Bennett and Kay, 1981).

Studies performed in vitro with mouse splenic macrophages and autologous young and old RBC revealed that mouse macrophages phagocytized senescent but not young RBC (P≤0.001). The phagocytosis of middle-aged RBC (~23%) was intermediate between that of young RBC (5%) and old RBC (~50%). This suggested that the appearance of the senescent cell antigen, and, thus, molecular aging of membranes, was a cumulative process.

Cellular Removal in Other Vertebrates

Numerous investigators have confirmed the presence of IgG on old human cells using different methods. In addition, IgG binding to erythrocytes has been shown to be a mechanism of removal of old red cells in SPF mice (Bennett and Kay, 1981), conventionally housed mice (Singer et al., 1986), germ free rats (Glass et al., 1983), cows (Bartosz et al., 1982a), rabbits (Khansari and Fudenberg, 1983), and chickens (Bosman, Harris, and Kay, unpublished). Anion transport by old erythrocytes from chickens was consistent with band 3 aging (Bosman, Harris, and Kay, unpublished).

Role of sialic acid and carbohydrate in recognition and/or removal of senescent cells. Ashwell and coworkers showed that the lifetime of serum glycoproteins could be drastically shortened by removal of external carbohydrates. This involves the disappearance of terminal sialic acid residues and subsequent exposure of penultimate galactose residues. However, sialic acid removal and galactose exposure is not involved in the normal removal of senescent cells (Kay and Bosman, 1985). IgG can not be removed from intact senescent erythrocytes or from their membranes with buffer containing galactose (Kay and Bosman, 1985). Incubation of senescent RBC with galactose does not inhibit their phagocytosis by macrophages indicating that galactose did not displace senescent cell IgG. Absorption of senescent cell IgG with the carbohydrate portion of band 3 did not alter binding to band 3 in immunoblots. The fraction specifically eluted from affinity columns containing the carbohydrate portion of band 3 did not bind to erythrocyte membranes in immunoblots. These results suggest that the IgG binding specifically to senescent erythrocytes was not directed against galactose residues. In addition, synthetic peptide studies described below show that carbohydrate is not required for senescent cell IgG binding because it binds to synthetic peptides which have no carbohydrate attached.

In summary, macrophages recognize old and damaged RBC on the basis of binding of a specific IgG autoantibody to old cells. IgG is required for phagocytosis of old RBC.

IDENTIFICATION OF NEOANTIGEN

Binding of an IgG autoantibody to senescent RBC through immunologic mechanisms indicated that antigenic determinants recognized by these IgG autoantibodies appeared on the membrane surface as RBC aged.

Isolation of Senescent Cell Antigen

As an approach to isolating and identifying this neoantigen, affinity columns are prepared with IgG isolated from old cells. Senescent cell antigen is isolated from sialoglycoprotein mixtures with affinity columns prepared with IgG eluted from senescent cells (Kay, 1981). Material specifically bound by the column is eluted with glycine-HCl buffer, pH 2.3. Both glycoprotein and protein stains of gels of the eluted material revealed a band migrating at a relative molecular weight of 62,000 in the component 4.5 region. These experiments suggested that the 62,000 Mr glycopeptide carried the antigenic determinants recognized by IgG obtained from freshly isolated senescent cells. The 62,000 Mr peptide, but not the remaining sialoglycoprotein mixture from which it is isolated, abolished the phagocytosis-inducing ability of IgG eluted from senescent RBC in the erythrophagocytosis assay (Kay, 1981). This indicated that the 62,000 Mr peptide is the antigen which appeared on the membrane of cells as they aged.

Senescent Cell Antigen is Present on Nucleated Cells

Examination of other somatic cells for the antigen which appears on senescent RBC revealed its presence on lymphocytes, platelets, neutrophils, and cultured human adult liver cells and primary cultures of human embryonic kidney cells as determined by a phagocytosis inhibition assay (Kay, 1981). Senescent cell antigen is isolated from lymphocytes (Kay, 1981; Kay et al., 1982; 1984) with the senescent RBC IgG affinity column. Gel electrophoresis of the material obtained from the column revealed a band migrating at a Mr of 62,000 at the same position as the antigen isolated from senescent RBC (Kay, 1981). This finding confirmed the results obtained with the phagocytosis inhibition assay, indicating that the antigen which appeared on senescent RBC also appeared on other somatic cells.

Appearance of the 62,000 Mr antigen on RBC initiates binding of IgG autoantibodies in situ and phagocytosis of senescent cells by macrophages (Kay, 1975, 1978, 1981, Bennett and Kay, 1981). The antigen is present on stored human lymphocytes, platelets, and neutrophils, and on cultured liver and kidney cells. In addition, IgG autoantibodies in normal serum have been shown to bind to senescent RBC in situ in humans (Kay, 1975, 1978), mice (Bennett and Kay, 1981), rats (Glass et al., 1983), cows (Bartosz et al., 1982a), rabbits (Khansari and Fudenberg, 1983), and chickens (Bosman, Harris, Kay, unpublished). Thus, the immunological mechanism for removing senescent and damaged red cells appears to be a general physiological process for removing cells programmed for death in mammals and, possibly, other vertebrates (Kay, 1975, 1978, 1981).

Identification of Senescent Cell Antigen as a Band 3 Product

Since mature erythrocytes cannot synthesize proteins, senescent cell antigen is probably generated by modification of a preexisting protein of higher molecular weight (Kay, 1986; Kay et al., 1982, 1983a). It is postulated that senescent cell antigen is a component of the 4.5 region that is derived from band 3 (Kay, 1986; Kay et al., 1982, 1983a) based on both extraction and isolation conditions, relative molecular weight, and its characterization as a glycosylated peptide (Kay, 1981).

Experiments designed to test this hypothesis revealed that senescent cell antigen is immunologically related to band 3 and may represent a physiologically significant breakdown product of the parent molecule (Kay, 1986; Kay et al., 1982, 1983a). Both band 3 and senescent cell antigen abolished the phagocytosis inducing ability of IgG eluted from senescent cells; whereas, spectrin, bands 2.1, 4.1, actin, glycophorin A,

PAS staining band 1-4, and desialylated PAS staining bands 1-4 did not. In addition, rabbit antibodies to both purified band 3 and senescent cell antigen, and IgG eluted from senescent cells reacted with band 3 and its breakdown products as determined by immunoautoradiography of RBC membranes indicating that these molecules share common antigenic determinants not possessed by other red cell membrane components (Kay, 1986; Kay et al., 1982, 1983a).

These results confirmed those obtained with the erythrophagocytosis assay by indicating that band 3 carries the antigenic determinants of senescent cell antigen. Thus, senescent cell antigen is immunologically related to band 3 and may be derived from it.

Senescent cell antigen is mapped along the band 3 molecule using topographically defined fragments of band 3. Both binding of IgG eluted from senescent red blood cells ("senescent cell IgG") to defined proteolytic fragments of band 3 in immunoblots, and two-dimensional peptide mapping of senescent cell antigen, band 3, and defined proteolytic fragments of band 3 is used to localize senescent cell antigen along the band 3 molecule (Kay et al., 1983a). The data suggested that the antigenic determinants of senescent cell antigen that is recognized by physiologic IgG autoantibodies reside on an external portion of a naturally occurring transmembrane fragment of band 3 that has lost an Mr 40,000 cytoplasmic (NH2-terminal) segment and part of the anion transport region. A critical cell age specific cleavage of a band 3 appears to occur in the transmembrane, anion transport region of band 3.

BAND 3 AGING

The demise of band 3, which is synonymous with generation of senescent cell antigen, occurs in two distinct steps. Structurally, band 3 undergoes an as yet uncharacterized initial change during cellular aging that triggers a series of events terminating the life of the cell. We have recently developed antibodies against aged band 3 that recognize this change because they bind to a distinct region of band 3 in old but not middle-aged or young cells (Kay et al., 1989). Following the change in intact band 3 with aging, band 3 undergoes degradation, presumably catalyzed by an enzyme (Kay, 1985a). Preliminary experiments indicate that it is a calcium dependent membrane bound protease (Kay, 1985a). Cleavage of band 3 occurs in the transmembrane, anion transport region (Kay, 1984, 1985a, 1986; Kay et al., 1982, 1983a, 1986, 1989). Fragments of band 3 is detected in membranes of old but not young cells by immunoblotting with antibodies to normal band 3. Following degradation, band 3 undergoes a change in tertiary structure (Kay, 1985a) becoming senescent cell antigen (Kay, 1985a). A physiologic IgG autoantibody binds to senescent cell antigen and initiates cellular removal.

Since our previous studies indicated that senescent-cell antigen is derived from band 3 by cleavage in the transmembrane anion transport region (Kay, 1983, 1984, 1985a; Kay et al., 1982), we suspected that anion transport might be altered with cellular aging (Kay et al., 1986). If this suspicion proved to be correct, then we would have a functional assay for aging of band 3, the major anion transport protein of the erythrocyte membrane.

Transport studies on age-separated rat erythrocytes indicated that anion transport decreased with age (Kay *et al.*, 1986). The Michaelis constant (Km) increased, and the maximal velocity (V_{max}) decreased in old erythrocytes as compared to middle-aged erythrocytes in humans and rats (Table 2). These data provided us with another assay of cellular function to use to determine whether erythrocytes are "senescent". However, it is doubtful that the number of molecules of band 3 to which IgG is bound (100 per cell) is adequate to account to the magnitude of change in anion transport. Therefore, we suspect that another as yet unidentified change precedes events initiating IgG binding and is responsible for the observed changes in anion transport.

The following functional changes in band 3 occur as red cells age. These changes are: decreased anion transport activity (increased K_m; decreased V_{max}), decreased number of high affinity ankyrin binding sites, and binding of physiologic IgG autoantibodies *in situ* (Bosman and Kay, 1988). In addition, band 3 undergoes an as yet undefined change that results in binding of 980 antibodies to aged band 3 (Kay *et al.*, 1989). These 980 antibodies recognize band 3 that has aged prior to its formation of senescent cell antigen. Degradation of band 3 generates senescent cell antigen.

As part of our ongoing studies on mechanisms of cellular aging, we searched for models of accelerated and decelerated cellular aging. We anticipated that such models would allow us to dissect molecular aging and provide insight into mechanisms.

TABLE 1. Cellular Aging Changes

DECREASED ANION TRANSPORT

DECREASED GLUCOSE TRANSPORT

BAND 3 DEGRADATION

APPEARANCE OF SENESCENT CELL ANTIGEN

IgG BINDING

PHAGOCYTOSIS OF CELL

Models for Cellular Aging

Oxidation and Vitamin E. Generation of senescent cell antigen results from oxidation (Kay *et al.*, 1986). This was determined in studies of erythrocytes from vitamin E-deficient rats (Kay *et al.*, 1986). The importance of vitamin E as an antioxidant, providing protection against free radical-induced membrane damage, has been well documented (Kay *et al.*, 1986). Vitamin E is primarily localized in cellular membranes, and a major role of vitamin E is the termination of free-radical chain reactions propagated by the polyunsaturated fatty acids of membrane phospholipids. Vitamin E-deficient erythrocytes are defective in their ability to scavenge free radicals. It is interesting that there is a correlation between life span and natural antioxidant levels in a variety of species and that the level of such antioxidants appears to correlate with metabolic activity of individual species. Specific biochemical alterations in the membrane of erythrocytes from vitamin E-deficient rhesus monkeys have been described (Kay *et al.*, 1986). Furthermore, vitamin E deficiency represents a "physiological" method for rendering cells susceptible to free-radical damage and may simulate conditions encountered *in situ*. We used vitamin E deficiency as a model for studying oxidation because studies show that, in mammals, vitamin E functions as an antioxidant, and because vitamin E deficiency simulates conditions encountered *in situ* more closely than does chemical treatment of cells in vitro. Red cells from vitamin E deficient rats behaved like old erythrocytes in the phagocytosis assay, and in anion transport and glyceraldehyde-3-phosphate dehydrogenase activity. In addition, increased breakdown products of band 3 is observed in erythrocyte membranes from vitamin E-deficient rats. Vitamin E-deficient rats developed a compensated hemolytic anemia as is observed in vitamin E-deficient humans.

Middle-aged erythrocytes from vitamin E-deficient rats behaved like old cells based on the phagocytosis assay, anion transport studies, and immunoblotting studies (Kay *et al.*, 1986).

Immunoblotting studies revealed increased breakdown products of band 3 in cells from vitamin E-deficient rats as is observed in old cells (Kay *et al.*, 1986). Thus, vitamin E deficiency leads to accelerated red cell aging, presumably through oxidation.

Results of the experiments on vitamin E deficiency suggest that oxidation can cause aging of band 3. We suspect that this may be one of the mechanisms of cellular aging *in situ*. At this time, it appears that general cellular damage such as lysis (Kay, unpublished), and oxidation can result in the generation of senescent cell antigen. We suspect that many different cellular insults have a final common pathway that results in generation of senescent cell antigen.

Chemical Models

As part of our systematic ongoing studies of mechanisms of cellular and molecular aging, we developed a "biochemical profile" of senescent human red cells (Bosman and Kay, 1988). This "red cell aging" panel allows us to assess functional red cell age independent of chronological age. The panel used to obtain this profile includes IgG binding, phagocytosis, enzyme activity, anion transport, ankyrin binding and immunoblotting with antibodies to band 3.

As part of our ongoing studies on mechanisms of cellular aging, we searched for models of accelerated and decelerated cellular aging. We anticipated that such models would allow us to dissect molecular aging and provide insight into mechanisms. Initially, we investigated models for aging in vitro (Bosman and Kay, 1988).

Free Radical Generating Systems

We subjected intact human erythrocytes to treatments that have been reported to result in changes in band 3 and/or to mimic aging in vitro. The validity of these treatments as model systems for erythrocyte aging is evaluated using a "red cell aging panel" which provides a biochemical profile of a senescent red cell (Kay *et al.*, 1986, 1989; Bosman and Kay, 1988). Treatments are assessed for their ability to induce the following changes in vitro that is observed in normal erythrocytes aged *in vivo*: 1, increased breakdown of band 3 as detected by immunoblotting; 2, decrease in anion transport efficiency as detected with a sulfate self-exchange assay; 3, decrease in total glyceraldehyde 3-phosphate dehydrogenase (G3PDH) activity with an increase in membrane-bound activity; and 4, increase in the binding of autologous IgG as detected with a protein A-binding assay (Table 1). Neither incubation with the free radical-generating xanthine oxidase/xanthine system, nor treatment with malondialdehyde, an end product of free radical-initiated lipid (per)oxidation, results in age-specific changes (Bosman and Kay, 1988). Loading of the cells with calcium, and oxidation with iodate results in increased breakdown of band 3, but does not lead to increased binding of autologous IgG. Only erythrocytes that have been stored for 3-4 weeks show the same structural and functional changes as observed during aging *in vivo* (Bosman and Kay, 1988).

Hemoglobin Köln and G6PD Deficiency

We then began a search for "experiments of nature" that might provide insights into the process of normal cellular aging (Kay *et al*, 1988a). Initially, we studied glucose 6-phosphate dehydrogenase deficiency (G6PD) and hemoglobin Köln as potential models (Kay *et al.*, 1988a). Membranes from both the G6PD deficient and hemoglobin Köln cells which we studied have been reported to contain high molecular weight polymers (Kay *et al.*, 1988a). In addition, hemoglobin Köln cells contain hemoglobin precipitates. We used the red cell aging panel to compare the biochemical profile of glucose 6-phosphate dehydrogenase-deficient and hemoglobin Köln cells containing high molecular weight protein polymers or hemoglobin precipitates with that of normal senescent cells. However, accelerated cellular aging is not present as determined by a "red cell aging panel" including lack of phagocytosis and IgG binding to young and middle-aged erythrocytes, normal ankyrin binding, normal anion transport, normal G3PDH activity, and no increase in band 3 breakdown products (Kay *et al.*, 1988a; Table 1). We found no evidence in support of the concept that aggregation of band 3 plays a role in the mechanism for generating senescent cell antigen. Observations like these support the hypothesis that degradation of band 3, not aggregation is a critical event in IgG binding and normal erythrocyte aging.

TABLE 2. Comparison of Anion Transport by Middle-aged High Molecular Weight Band 3, "Fast-aging", and Neuro Band 3 Cells to Normal Middle-aged and Old Human Erythrocytes.

Cell Age	K_m[2]	V_{max}[3]
Middle-aged	0.9 ± 0.1	13.2 ± 1.4
Old	$1.6 \pm 0.1^*$	$7.7 \pm 1.1^*$
High MWBD3	0.5 ± 0.1	$22.3 \pm 1.0^*$
Fast Aging	$1.6 \pm 0.2^*$	$9.9 \pm 0.2^*$
Neuro BD3	0.9 ± 0.1	$29.0 \pm 1.0^*$

*$P < 0.01$ compared to control

1. Cells were separated according to age and their anion transport activity was measured with the sulfate self-exchange assay at 37oC and pH 7.2. All values are the mean $\pm$ one standard deviation of at least 6 determinations.
2. Km, exchange constant, i.e. the sulfate concentration (in mM) at which the transport rate is half the maximal value, as determined from a Lineweaver-Burk plot of the ascending branch of the rate curve.
3. V_{max}, the maximal velocity in moles x 10-8/108 cells/min.
4. Cells were separated according to age, and the middle-age red blood cells were stored in Alsevers for four weeks at 4oC.
5. High MWBD3, high molecular weight band 3 middle-aged cells.

Band 3 Mutations/alterations

We began a search for mutations and/or clinical alterations of erythrocyte band 3 (Table 2). Our search for band 3 protein alterations resulted in the discovery of the first band 3 alterations/mutations in any species. All three different alterations were discovered in humans (Kay *et al.*, 1988b, 1989, 1990a, 1990b). Anion and glucose transport of all three is summarized in Table 2. One mutation, high molecular weight band 3, results from an addition of in the transmembrane, anion transport region of band 3 (Kay *et al.*, 1988b). Cytochalasin B binding was slightly increased, but the increase was not statistically significant (Bosman and Kay, 1990). It appears to be an autosomal recessive. This mutation is associated with acanthocyte ("thorny" cell) formation. However, erythrocyte survival is normal *in situ* as determined by the reticulocyte count, and the erythrocytes do not exhibit accelerated aging as determined by the "red cell aging panel" (Bosman and Kay, 1988).

A second band 3 alteration also exhibits acanthocytosis (Kay *et al.*, 1990a, 1990b). This alteration is associated with ion and glucose transport abnormalities and neurologic disease. The neurological disease is an autosomal recessive.

The third alteration of band 3 alteration is characterized by accelerated cellular aging as determined by a "red cell aging panel", and cellular removal. The propositus' reticulocyte count is ~20% indicating the destruction and replacement *in situ* of 20% of circulating erythrocytes daily and there is increased IgG binding to middle-aged cells. Both peripheral blood findings (eg the presence of nucleated erythrocytes and precursors of monocytes and lymphocytes) and bone marrow biopsy are consistent with a hemolytic anemia (Kay *et al.*, 1989). The propositus had a low hematocrit and hemoglobin. The propositus was splenectomized to reduce red cell destruction. We gave this band 3 alteration the descriptive name "fast aging" band 3 because the propositus' young and middle-aged cells exhibit all the characteristics of old erythrocytes (e.g. increased IgG binding, decreased anion and altered glucose transport, and increased breakdown products of band 3 is observed on immunoblots). Aged band 3 antibodies, which do not bind to young or middle-aged cells, bind to a distinct region of band 3 in immunoblots of membranes of middle-aged red cells, and to intact middle-aged red cells as determined by immunoelectronmicroscopy. We suspect that "fast aging" band 3 is more susceptible to proteolysis than is normal band 3.

The data indicate that the band 3 alteration that results from an addition to band 3 does not alter red cell lifespan or produce clinical disease. In contrast, band 3 alterations that are associated with band 3 aging and/or degradation are characterized by shortened red cell lifespan and clinical diseases. We suspect that these latter alteration result from deletions or substitution in the band 3 gene.

The three band 3 mutations/alterations provide support for the hypothesis that the membrane spanning domain of the anion transporter and glucose transporter(s) is functionally and structurally related (Kay, 1985b). For example, high molecular weight band 3 containing cells exhibit changes in band 3 structure and function that is probably caused by a redundant segment in the anion transport region. Changes in other membrane proteins or lipids is not detected. The other parameter that is altered is the V_{max} of glucose efflux. External modifications of band 3, probably at its cytoplasmic domain, that occur in G6PD-deficient cells or in cells containing unstable hemoglobin Köln have no effect on anion transport and do not affect glucose transport. At the present time, we can not reconcile the sequence data of the available glucose transport proteins with the genetic, functional, cell biology, and immunological data which indicate a relationship between band 3 and glucose transport.

A band 3 mutation/alteration in Southeast Asian ovalocytosis has recently been reported (Jarolim et al., 1991). A deletion of band 3 amino acids 400-408 and a substitution of Glu-56 for Lys-56 was found in this alteration associated with rigid erythrocytes associated with resistance to invasion by several strains of malaria in vitro (Jarolim et al., 1991). Apparently, the Glu-56 for Lys-56 substitution is found in normal individuals and thus is a polymorphism. The ovaloytic individuals are reported to have reduced numbers of intracellular parasites in vivo (Jarolim et al., 1991). Since residues 400-408 are in the cytoplasmic segment of band 3, it would be interesting to know the mechanism by which a change in the cytoplasmic segment would interfere with invasion by a parasite at the cell surface.

MOLECULAR BIOLOGY OF SENESCENT CELL ANTIGEN AND BAND 3

Mapping of Transport Function and Expression of Senescent Cell Antigen onto a Structural Representation of the Band 3 Molecule

Based upon the translated sequence of cDNA, the human erythrocyte band 3 protein consists of 911 amino acids (Tanner et al., 1988). Protein sequences have also been derived for erythrocyte band 3 molecules of mouse and chicken as well as for molecules related to band 3 from lymphocytes, the cell line K562, and kidney. A schematic diagram of the band 3 model showing key peptides is shown in Figure 1. The N-terminal 393 amino acids are located within the cytoplasm of the cell where they associate with the cytoskeleton by binding to ankyrin. This so-called "cytoplasmic domain" can be isolated as a water-soluble fragment (NTr-41) by digestion of red cell ghosts with trypsin (Kellokumpu et al., 1988). The "membrane-associated domain" (TrC-55) consists of the C-terminal 518 amino acids. These separable fragments exhibit different behavior in solution: the NTr-41 is soluble in aqueous solution and the CTr-55 is soluble in diluted detergent. These properties of the protein fragments parallel the proposed states of aggregation of intact molecules within the cell plasma membrane where stable dimers and tetrameres are found.

Two crucial properties of the band 3 molecule namely the anion transport region formation of the senescent cell antigen (Kay, 1984) map to the membrane-associated "domain". Application of hydrophilicity hydropathy algorithms to the derived protein sequence indicates that the membrane associated region consists of 11 hydrophobic membrane spanning helical regions and four major external and three major internal hydrophilic loops. The C-terminal 40 residues are located within the cytoplasm.

The locations of these predicted structures and the residues implicated in anion transport are given in Figure 1. Peptide technology was used to map the regions reactive with autoantibody to senescent cell antigen as summarized in Figure 1 (Kay, et al. 1990c, 1991). Synthetic peptides were used to block either binding of senescent cell IgG autoantibody to intact red blood cells or the transport of sulfate. Peptides from the second (ANION 1) and fourth (COOH) major external loops blocked the antibody. Furthermore, these peptides shows a synergistic effect when admixed,thereby indicating that these regions which are distant in amino acid sequence are closely associated in the 3-dimensional structure of the cell-bound band 3 molecule (or molecular complex). It is of interest that the autoantibodies also bind other peptides that do not block in the competition assay. This reflects the fact that aged cells binding the anti-senescent cell antigen autoantibody are removed from the circulation by phagocytes (Kay, 1984) which presumably process the antigen into peptides presented to T cells.

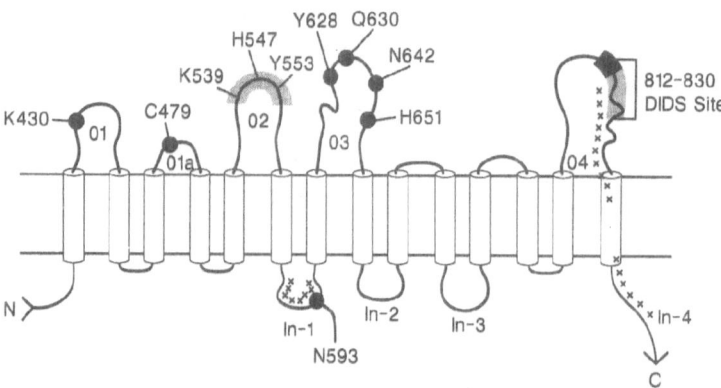

Figure 1.   Model of membrane associated and external regions of anion transport protein band 3, approximate residues (R) 400-900 showing aging antigenic sites (speckled area), minimal residues that function as aging epitopes ([ ]), anion transport sites (xxx), and DIDS binding site. K, lysine; C, cysteine; H, histidine; Y, tyrosine; N, asparagine; the number following a letter indicates residue number; In, inside the cell; O, outside of the cell; arrow C, carboxyl terminus; arrow N, amino terminus. The model was based upon application of the program PEPPLOT of the GCG package to identify membrane spanning nonpolar helices and intervening hydrophilic loops. The location of the hydrophilic loops as extracellular or intracellular was predicted on the basis of established chemical or biological markers, e.g., the demonstration that residues 814-829 contain a DIDS binding site accessible from the outside or external radioiodination of the tyrosine (Y) at position 553. Key residues are identified to facilitate their identification within the sequence. This is a two dimensional representation that does not reflect three dimensional associations of residues that are separated by long stretches of sequence. Results show, however, that close steric association must be maintained by external loops 02 and 04 and In-1 and In-4. The hydrophobicity plots indicate multiple turns/bends in the band 3 molecule in the region of the DIDS binding site. (From Kay, 1992.)

Peptides from external loop 04 ("COOH " region) and internal loop In-1 are the most effective inhibitors of anion transport (Kay, 1991b). The key residues tend to be positively changed lysyl (K) or arginyl (R) residues with some involvement of histidyl (H) and acidic glutamyl (E) or aspartyl amino acids. The involvement of lysines in external loop 04 is originally suggested by the capacity of these residues to bind the covalent inhibitor of anion-transport DIDS (Jennings, 1989). Comparison of band 3 sequences of different species strengthens the identification of residues Lys-539, 817, 829 and 851, Arg 589, 602,603, 808, 827, 832, 870 and 871 and His-703,

172

734, 819 and 834 as essential to anion transport because these are conserved among man, mouse and chicken (Figure 1). Comparisons among sequences of the N-terminal cytoplasmic regions do not provide clear indications of functional regions, however, because there are marked polymorphisms in length and no definite ankyrin-binding region shared with other ankyrin-binding proteins occurs. It is of further interest that searches of general computer databases by us and others showed considerable conservation of residues within the membrane-associated domains of band 3 of various species and band 3 related molecules but detected no strong homologies with other proteins.

Aging Antigenic Site

The aging antigenic site was identified using synthetic peptides and the IgG that binds to old cells (Kay, 1991a, 1991b; Kay et al., 1990c, 1990d; Kay and Lin, 1990; Kay and Marchalonis, 1991). Results indicate that crucial anion transport segments of the band 3 molecule carry the aging determinants and suggest that these may be the aging sites of the molecule.

Previous studies indicated that senescent cell antigen is a degradation product of band 3 that includes most of the ~35,000 Da carboxyl terminal segment and the ~17,000 Da anion transport region (Kay, 1984). Both immunoblotting studies with IgG isolated from senescent cells and peptide mapping studies of senescent cell antigen indicated that senescent cell antigen lacks a ~40,000 molecular weight cytoplasmic segment which contains the amino terminus and, possibly, additional peptides of band 3 (Kay, 1984; Kay et al., 1986, 1989). Peptide mapping studies and anion transport studies suggested that a cleavage of band 3 occurs in the anion transport region (Kay, 1984). Furthermore, breakdown products of band 3 are observed in the oldest cell fractions but not in young or middle-aged cell fractions, and anion transport is impaired in old cells (Kay, 1984; Kay et al., 1986, 1989).

Based on structural, biochemical, and immunological data (Kay, 1975, 1978, 1981, 1984; Kay et al., 1986, 1989), cleavage of old band 3 was hypothesized to occur approximately a third of the way into the transmembrane anion transport region from the carboxyl terminus end. Therefore, peptides were synthesized from the anion transport domain of erythrocyte band 3. "Walking" of the anion transport segment was performed (Kay, 1991a; Kay et al., 1990c, 1990d; Kay and Lin, 1990; Kay and Marchalonis, 1991). "Walking" means the antigenic analysis of a series of synthetic overlapping peptides that encompass the entire polypeptide chain of the anion transport domain. The synthetic peptides are of uniform size and overlap their adjacent neighboring peptides by a predetermined number of residues in the overlap regions in order to optimize the feasibility of synthesis and to expect reasonable resolution of individual antigenic sites.

These studies focused on the 511 amino acid anion transport domain. Results indicate that this domain carries both the SCA antigenic determinants (Kay, 1984, 1986, 1990; Kay et al., 1986, 1983a, 1988a, 1989, 1990b, 1991) and the aging vulnerable site (Kay, 1984; Kay et al., 1986, 1989); whereas, the 40,000 Da cytoplasmic segment does not. Therefore, investigation focused on segments within the transport domain that are most likely to be hydrophilic and therefore exposed based on hydrophobicity/hydrophilicity scales, amino acid composition, and location along the molecule based on our model described by Kay et al., (1990d). Peptides that corresponded to external regions predicted by the band 3 model presented earlier were synthesized. The peptides with residue number are: CYTO, 129-144: AGVANQLLDRFIFEDQ; 426-440, LLGEKTRNQMGVSEL; 515-531, FISRYTQEIFSFLISLI; 526-541, FLISLIFIYETFSKLI; 538-554, SKLIKIFQDHPLQKTYN; 549-566, LQKTYNYNVLMVPKPQGP; 561-578, PKPQGPLPNTALLSLVLM; 573-591, LSLVLMAGTFFFAMMLRKF; ANION 2, 588-602, LRKFKNSSYFPGKLR; 597-614, FPGKLRRVIGDFGVPISI; 609-626, GVPISILIMVLVDFFIQD; 620-637, VDFFIQDTYTQKLSVPD; GLYCOS, 630-648, QKLSVPDGFKVSNSSARGW; 645-659, ARGWVIHPLGLRSEF; 684-704,ITTLIVSKPERKMVKGSGFHL; 752-769, IQEVKEQRISGLLVAVL; 776-793, MEPILSRIPLAVLFGIFL; 788-805, FGIFLYMGVTSLSGIQL; 800-818, LSGIQLFDRILLLFKPPKY; 813-818, FKPPKY; COOH, 812-827, LFKPPKYHPDVPYVKR; 812-830, LFKPPKYHPDVPYVKRVKT; 818-827, YHPDVPYVKR; 822-839, VPYVKRVKTWRMHLFTGI; 869-879+881+883, LRRVLLPLIFRVL (The actual sequence in band 3 is: LRRVLLPLIFRNVEL, but the N and E were not included in the peptide used for these studies); 887-903, DADDAKATFDEEEGRDE; 902-910, DEYDEVAMP. As a control, a peptide from the cytoplasmic segment of band 3 within the region of the putative ankyrin binding site was used (Tanner *et al.*, 1988) (CYTO, 139-159). Peptides were synthesized based on the sequence data from the paper by Tanner *et al*,. (1988). Peptides were evaluated in a senescent cell IgG competitive inhibition assay. The specific physiologic autoantibody to the senescent cell antigen, senescent cell IgG (SCIgG) was isolated from old human red cells as described previously using affinity chromatography (Kay, 1978, 1981). IgG eluted from senescent cells, rather than serum IgG, was used because normal serum contains antibodies to spectrin, actin, 2.1, etc. (Kay, 1983, 1986; Kay *et al.*, 1982). Competitive inhibition studies were performed using synthetic peptides to absorb the IgG isolated from senescent erythrocytes. This is the same IgG that initiates phagocytosis *in situ*. The Fc portion IgG is required for binding and phagocytosis of cells by macrophages (Kay, 1975, 1981; Kay *et al.*, 1989) fragments were not used because we were simulating the physiological situation. Intact dimeric, senescent cell IgG containing the Fc portion binds to senescent cells *in situ* and initiates their removal (Kay, 1975, 1978, 1981, 1984; Kay *et al.*, 1986, 1989; Bennett and Kay, 1981). Only IgG isolated from aged erythrocytes binds specifically to senescent cells. For example, IgG eluted from young control erythrocytes did not bind to senescent cells (Kay, 1978). Moreover, the specific binding capacity of the autoantibody was eliminated by absorption with purified senescent cell antigen (Kay *et al.*, 1989). SCIgG ($3\mu g$) is absorbed with synthetic peptides at the concentrations indicated or purified SCA, as a control, for 60 min at room temperature, and incubated with stored red cells for 60 min at room temperature (Kay, 1975, 1978, 1981, 1990; Kay *et al.*, 1989, 1990c, 1990d, 1991; Kay and Lin, 1990). IgG binding and inhibition were determined with a protein A binding assay. This biological assay measures the fate of erythrocytes *in vitro* and *in vivo* (Kay, 1975; Bennett and Kay, 1981; Branch *et al.*, 1984). Storage mimics normal aging *in situ* immunologically and biochemically (Kay, 1975, 1978, 1981, 1983, 1984, 1986, 1988; Kay *et al.*, 1982, 1986, 1988a, 1989; Bennett and Kay, 1981; Bosman and Kay, 1988).

Results of these "walking" studies, summarized in Figure 1, indicate that senescent cell IgG recognizes antigenic determinants that lie within the anion transport region 538-554 and putative transport site containing a cluster of lysines toward the carboxyl terminus, 812-827. More recent studies indicate that that residues 812-830 is a more potent inhibitor of SCIgG binding than 812-827 (inhibition : 812-830, 45 ± 6 (P$\leq$0.001); 812-827, 32 ± 5 (P$\leq$0.01). The peptide "CYTO" from the cytoplasmic domain does not inhibit. Immunoblotting studies demonstrate binding of senescent cell IgG to peptides ANION 1 and COOH but not to CYTO, the peptide from the cytoplasmic segment of band 3 containing the putative ankyrin binding site (Kay, *et al.* 1990c, 1990d).

Results, to date, indicate that (a) an anion transport region of band 3 that appears to be extracellular carries active antigenic determinants of an aging antigen; (b) this transport site is located toward the carboxyl terminal and overlaps or contains a stilbene disulfonate binding site, (c) a putative ankyrin binding region peptide is not involved in senescent cell antigen activity; and (d) synthetic peptides alone, without carbohydrate attached, abolish binding of senescent cell IgG to red cells. Therefore, carbohydrate moieties are not required for the antigenicity or recognition of senescent cell antigen.

Synthetic Senescent Cell Antigen

A synthetic aging antigen was generated by mixing the two inhibitory peptides, pep-ANION 1 and pep-COOH. Although 3 μg of pep-ANION 1 alone only produced 16% inhibition, and pep-COOH alone produced 22% inhibition, the mixture containing 1.5μg of each produced 97% inhibition. This suggests that the two peptides interact together to form a three dimensional aging antigen. In other words, both peptides contribute amino acids to the configuration recognized by the antigen binding site on the IgG molecule. A mixture of 30μg pep-ANION 2 and pep-COOH did not produce synergy (18$\pm$4% inhibition). Synthetic peptides is usually not as effective as the native molecule from which they are derived in physiologic studies because short peptide segments do not assume the same tertiary configuration as that of the whole molecule. The synergism of peptides pep-ANION 1 and pep-COOH suggests that the conformation of the determinants of these two peptides interacting with each other is similar to that of the intact aging antigen and that these two determinants must be spatially close in the native molecule. Therefore, we refer to the synergistic mixture of pep-ANION 1 and pep-COOH as "synthetic senescent cell antigen." These results suggest that these peptides lie in close spatial proximity in the folded, tertiary structure of native band 3 even though they are separated by ~300 amino acids in the primary structure of the 911 amino acid band 3 molecule (Figure 1). An alternative explanation is that a change in aged band 3 alters its configuration and results in the three dimensional alignment of these two sites. We suspect that the three dimensional structure of band 3 is a ring in which the two external loops on which pep-ANION 1 and pep-COOH reside in close spacial proximity forming senescent cell antigen as cells age.

Active aging antigenic amino acids

In order to define the aging antigenic site along the band 3 molecule and to determine the active antigenic residues, a neutral or positively charged amino acid was substituted for the positively charged lysine in pep-COOH peptide (LFKPPKYHPDVPYVKR) during synthesis. Substitution of either neutral glycines (pep-COOH-G: LFGPPGYHPDVPYVGR) or positively charged arginines (pep-COOH-R: FRPPRYHPDVPYVRR) for lysine in pep-COOH reduced but did not abolish its activity in the competitive inhibition assay with senescent cell IgG (Kay and Lin, 1990). This suggests that (a) charge alone is not the critical determinant of antigenicity, and (b) lysines contribute to the antigenicity of the aging antigen. Since we changed all three of the lysines in the synthetic peptide pep-COOH, we can not determine at this time whether all lysines are critical or whether antigenicity depends on a specific lysine.

Results of studies using synthetic peptides to locate a crucial aging antigen on band 3 indicate that pep-ANION 1 (residues 538-554) and pep-COOH residues (812-827) may be the aging antigenic sites of the band 3 molecule, and that lysine(s) is required for antigenicity. Generation of senescent cell antigen initiates IgG binding and removal of cells *in situ*. These results are consistent with the physiological data demonstrating that old erythrocytes have impaired anion transport (Kay *et al.*, 1986, 1988a, 1989, 1990b, 1990c; Bosman and Kay, 1988), and the biochemical and immunological data indicating that band 3 undergoes degradation with loss of a cytoplasmic segment during the aging process (Kay, 1981, 1983, 1984, 1986, 1988; Kay *et al.*, 1982, 1983a, 1986, 1988a, 1988b, 1989; Bennett and Kay, 1981).

Mapping of Aging Antigenic Sites in Relation to Anion Transport and Anion Transport Inhibitor Binding Site(s)

Senescent cell antigen contains pep-ANION 1 and pep-COOH (Kay *et al.*, 1990c, 1990d; Kay and Lin, 1990). Data suggest that senescent cell antigen is related to the anion transport site because peptide mapping studies and anion transport studies indicate that a cleavage of band 3 occurs in the anion transport region (Kay, 1984). Furthermore, breakdown products of band 3 are observed in the oldest cell fractions, but not in young or middle-aged cell fractions (Kay, 1984; Kay *et al.*, 1989). Fragments of the cytoplasmic segment of band 3 are detected with antipeptide antibodies in the cytoplasm of old red cells and of red cells with a band 3 alteration associated with accelerated aging (Kay, 1984; Kay *et al.*, 1990d). Anion transport is impaired in old cells (Kay *et al.*, 1986, 1989, 1988a, 1990a; Bosman and Kay, 1988) and band 3 mutations/alterations with damage to the transmembrane transport region result in accelerated aging and increased IgG binding (Kay *et al.*, 1989). To clarify the relationship between senescent cell antigen and anion transport, we mapped the anion binding/transport properties of segments of band 3 using equimolar amounts of synthetic peptide and sulfate in a competitive inhibition assay (Figure 1). Studies showed that the peptide was competing for the sulfate because increased amounts of sulfate overcame the inhibition (Hughes, Haussler, and Kay, unpublished).

Anion Binding/transport Site

Peptides inhibited transport in a dose dependent manner (Kay, 1991b). Peptide residues 588-594 (a 7 amino acid peptide), 822-839, and 869-883 were the most active inhibitors of anion transport ($P \leq 0.001$ compared to control without peptide). The inhibitory activity of the last peptide, 869-883, could not be confirmed by testing adjacent peptides to the amino side since these regions are extremely hydrophobic. However, 6-7 amino acid peptides from this region produced inhibition of transport (Kay, 1991b), but to a lesser degree than 869-883. The component residues are probably additive. Peptide 869-879+881+883 seems to have special anion binding/transport properties although both it and 869-883 produced significant inhibition ($P \leq 0.001$). Synergy was not observed.

Anion transport has been attributed to residues 538-554 or, more recently, to a carboxyl segment of band 3 including pep-COOH, by investigators based on indirect evidence (Jennings *et al.*, 1985, 1986; Jennings, 1989). Residue 538-554 includes two important amino acids. The lysine at 538 (558 in the mouse) is a covalent binding site for the anion transport inhibitor, DIDS (Tanner *et al.*, 1988), and the tyrosine at residue 553 is radioiodinated by extracellular lactoperoxidase (Tanner *et al.*, 1988). However, our present results indicate that the residues 538-554 are not anion binding although segments carboxyl to them are. This agrees with the results of studies using mouse band 3 expressed in *Xenopus laevis* oocytes indicating that lysine 558 in the mouse sequence (539 on peptide 538-554 in the human sequence) is not involved in anion transport based on site directed mutagenesis (Bartel *et al.*, 1989; Garcia and Lodish, 1989). There are two other lysines on peptide 538-554. Since the peptide does not inhibit anion transport, the other two lysines on peptide 538-554 at positions 542 and 551 do not function as anion binding sites. Since peptide 538-554 does not inhibit anion transport even though it has 3 lysines, anion transport binding must involve more than a mere positive charge. The configuration of the peptide is probably important. Results of the amino acid substitution studies support this interpretation.

Modification of Anion Transport/binding Segments to Identify Active Amino Acids

Lysine has been implicated as an amino acid involved in anion transport based on DIDS inhibition studies (Jennings, 1989). Experiments were performed to test this hypothesis by substituting a neutral or positively charged amino acid for the positively charged lysine or arginine in pep-COOH peptide. pep-COOH peptide was selected because a) it is a highly conserved region of band 3 (Kay *et al.*, 1990d), b) it is a crucial sequence in senescent cell antigen, and c) it is an anion transport/binding peptide. Substitution of either neutral glycines (pep-COOH-G/K) or positively charged arginines (pep-COOH-R/K) for lysine in pep-COOH still resulted in significant inhibition, but the inhibition was significantly reduced compared to that of pep-COOH. Since we changed all three of the lysines in the synthetic peptide pep-COOH, we can not determine at this time whether all lysines are critical or whether antigenicity depends on a specific lysine.

pep-COOH related peptides were synthesized in which glycines were substituted for arginines (pep-COOH G/R) and glycines were substituted for arginines and lysines (pep-COOH G/KR). Significant inhibition occurred with both altered peptides, but only pep-COOH G/KR was significantly different than pep-COOH. No significant difference was observed between pep-COOH G/R and pep-COOH. Since all of the substituted peptides caused inhibition, it seems that the presence of lysine or arginine is not an absolute requirement for anion binding. Furthermore, it appears that another amino acid, perhaps glycine, can participate in anion binding. Because the amino acid sequence of human and chicken pep-COOH and pep-COOH-N6 differs in several key amino acids, peptides from the chicken sequence were synthesized. Chicken pep-COOH has an M instead of K at a position corresponding to residue 814 of the human sequence, a K and E instead of a D and V at residue 821 and 822, and a T instead of K at residue 826. For example, substitution of a methionine (M) for a lysine (K) is significant since the former is nonpolar and the latter is a positively charged amino acid. Likewise, substitution of a negatively charged glutamic acid (E) for a nonpolar valine (V) or a positively charged lysine for a negatively charged aspartic acid (D) would be expected to alter tertiary configuration and binding properties. Thus, nature has performed "site-specific" mutagenesis for us.

Chicken pep-COOH inhibited anion transport, but the inhibition was significantly less than that of human pep-COOH (P ≤ 0.05). Since chicken pep-COOH-N6 inhibited to the same degree as human pep-COOH-N6, the lysine at position 814 in the human sequence for which methionine is substituted in the chicken sequence is probably not critical for anion transport. This suggests that the change to a K and E instead of a D and V may be a significant change. A glutamic acid is implicated in transport (Jennings, 1989; Jennings and Anderson, 1987). Neither human nor chicken pep-ANION 1, residues 538-554, inhibited transport.

Some investigators have suggested that lysines are not themselves part of the transport mechanism but are close to the transport site, and that arginine is involved in anion transport (Bjerrum et al., 1983; Zaki, 1983). Residues 869-883 and 869-879+881+883, among the most potent inhibitors of anion transport, have arginines but no lysines. Other highly inhibitory peptides, residues 588-594 and 822-839, have lysines and arginines. Of the inhibitory peptides with $P \leq 0.01$, residues 804-811 and 830-835 have arginines and no lysines, and residues 549-566, 561-578, and 813-818 have lysines and no arginines. The substitution studies show that altered pep-COOH with glycines substituted for both lysine and arginine still inhibits anion transport although the percent inhibition is reduced. Histidine and glutamine have also been implicated in anion transport (Jennings, 1989). The peptide sequence recognized by SITS antiidiotypic antibodies overlaps a potent anion binding/transport peptide, 822-839, which is consistent with data indicating inhibition of transport by stilbene disulphonates. This site could also be adjacent to the other 2 potent inhibitory peptides in three dimensional structure.

Pep-COOH, a peptide from the carboxyl terminus region, contains both hydrophobic and hydrophylic regions. The lysines found in this region comprise another binding site for the stilbene disulphonates based on data presented here and that of Jennings (Jennings et al., 1986). Our data suggest that this region has anion binding/transport capability as well. Residues 812-827 (pep-COOH) and 813-818 (N6, the 6 amino acids on the amino side of pep-COOH) are inhibitors of anion transport. Pep-COOH (residues 812-827) is part of senescent cell antigen (Kay, 1991a; Kay et al., 1990b, 1990d; Kay and Lin, 1990; Kay and Marchalonis, 1991). N6 is both an inhibitor of anion transport and of senescent cell IgG binding (IgG binding inhibition: $48 \pm 1\%$ at $10\mu g$) even though it is only six amino acids long. However, COOH-N6 does contain a proline-proline bend which may contribute to an anion pocket. It probably forms a loop in the membrane. These experiments suggest that at least part of a transport site is located on the same region of band 3 that generates senescent cell antigen.

Localization of SITS Binding Site on Band 3 Membrane Protein

Antiidiotype antibodies recognize the receptor of the ligand against which the idiotype is prepared (Kay, 1985b, 1991b). This has provided an elegant method for preparing antibodies against membrane proteins without purifying them, and for localizing active ligand binding sites on membrane receptors and proteins (Kay, 1985b, 1991b).

SITS is an inhibitor of anion transport. Antiidiotypic antibodies to SITS react with band 3 and its breakdown products in erythrocyte membranes (Kay 1985b, 1991b) and with band 3 peptide residues 788-805, 800-818, and strongest with 812-830. Residues 812-827, 3 amino acids smaller that the peptide giving the strongest reaction, reacted as well as the other 2 peptides, 788-805 and 800-818, but weaker that 812-830. Antiidiotypic antibodies did not react with residues 813-818 which is to the amino end of the peptide which reacts the strongest. In contrast, the antibodies reacted with peptide 818-827 which is toward the carboxyl end of the peptide giving the strongest reaction. This is consistent with the band 3 model which we have presented predicting that these residues are on external loop 04 (Kay et al., 1990c, 1991b). Reaction of antiidiotypic antibodies with breakdown products of band 3 suggests that these breakdown products carry transport sites. This would indicate that these segments are not derived from the cytoplasmic segment of band 3.

In summary, residues 812-830 contain both a SITS binding site and an anion binding/transport site. Since the antiidiotypic antibody does not react with residue 813-818, gives a weak reaction with residues 812-827, and a strong reaction with residues 812-830 and reacts weakly with 818-827, a crucial epitope probably resides in the region of 828-830. band 3 residues 788-805 and 800-818 may also be part of a SITS binding site. Data from chicken pep-COOH-N6 suggest that lysine 814 is not critical for anion transport. Data of Jennings (Jennings et al., 1986) indicate that one end of the dihydro derivative of DIDS, H2DIDS, reacts covalently with a lysine that is between 70 and 168 residues from the C terminus of band 3. This would be between residues 772 and 840 in the human sequence.

Results of this study, summarized in Figure 1, indicate that: (a) regions with residues 588-594, 822-839, and 869-883 being the most active transport regions ($P \leq 0.001$); (b) residues 812-830, and, possibly, 788-805 and 800-818 are part of the stilbene disulphonate binding site; (c) residue 538-554, which has been reported to be a transport segment of band 3, does not bind anions; and (d) lysines themselves contribute to but are not required for anion binding and, thus, anion transport.

Results of these studies with synthetic peptides are consistent with the physiological data demonstrating that old erythrocytes have impaired anion transport (Kay et al., 1983b, 1986, 1989, 1988c), and the biochemical and immunological data indicating that band 3 undergoes degradation with loss of a cytoplasmic segment during the aging process (Kay, 1981, 1984, 1985a, 1986; Kay et al., 1983a, 1986, 1988a, 1989; Bosman and Kay, 1988). Localization of the active site of SCA will facilitate the next logical step, namely, definition of the molecular changes occurring during aging that initiate molecular as well as cellular degeneration. Peptides ANION 1 and COOH are in highly conserved regions of band 3 (Kay et al., 1990b).

BAND 3 AND SENESCENT CELL ANTIGEN IN BRAIN

The anion exchange family of proteins are ubiquitous (Kay, 1990; Kay et al., 1988c, 1991a; Dreckhahn et al., 1984, 1985; Hazen-Martin et al., 1986, 1987; Demuth et al., 1986; Alper et al., 1988; Vanderpuye et al., 1988; Verlander et al., 1988; Allen et al., 1987; Brosius et al., 1989; Wolpaw and Martin, 1986; Dreckhahn and Merte, 1987; Kudrycki and Shull, 1989). Band 3 is also present in nuclear (Kay et al., 1983b), Golgi (Kellokumpu et al., 1988), and mitochondrial membranes (Schuster et al., 1986) as well as in cell membranes. Band 3-like proteins in nucleated cells participate in band 3 antibody induced cell surface patching and capping (Kay et al., 1983b). Band 3 maintains acid-base balance by mediating the exchange of anions (eg. chloride, bicarbonate) (Steck, 1978; Goodman and Shiffer, 1983), and is the binding site for glycolytic enzymes (Jennings, 1989). Because of its central role in respiration of CO2, band 3 is the most heavily used ion transport system in vertebrate animals.

The presence of band 3 related molecules in nonerythroid tissues was first demonstrated in 1983 (Kay *et al.*, 1983b). A protein immunologically related to band 3 was demonstrated in such diverse cell types as isolated neurons, hepatocytes, squamous epithelial cells, alveolar (lung) cells, lymphocytes, neurons *in vivo*, and fibroblasts using an antibody to band 3 that reacts with the transmembrane, anion transport domain of band 3 (Kay *et al.*, 1983b). The band 3-like protein in many of these cell types appeared to be a truncated version of the erythroid protein based on its molecular weight of ~60,000 estimated from its migration in polyacrylamide gels. It was suggested that part of the cytoplasmic amino terminus segment was altered in the band 3-like protein in these cell types, and that band 3 protein was modified to perform functions in different environments (Kay *et al.*, 1983b). Since then, band 3 has been described in numerous cell types and tissues including fibroblasts, hepatoma cells, and lymphoid cells (Demuth *et al.*, 1986; Kopito and Lodish, 1985; Kudrycki and Shull, 1989; Alper *et al.*, 1988; Hazen-Martin *et al.*, 1986). A truncated version of band 3 which lacks the amino terminus has also been described in kidney (Kudrycki and Shull, 1989; Alper *et al.*, 1988). In erythrocytes, band 3 constitutes over 30% of the total membrane protein (Steck, 1978; Goodman and Shiffer, 1983). It maintains acid-base balance by mediating the exchange of anions (eg. chloride, bicarbonate) (Jennings, 1989). Its membrane domain catalyzes rapid Cl-HCO3- exchange which facilitates CO2 entry into the blood in the systemic capillaries and CO2 excretion in pulmonary capillaries. The anion exchanger allows HCO3- to leave the cell in exchange for Cl-. This allows organisms to maintain the observed venous-arterial CO2 differences that permit survival. Anion exchange stabilizes extracellular pH by allowing rapid pH equilibration between plasma and the heavily buffered intracellular space.

The anion exchange, band 3 related protein(s) in mammalian brain performs the same functions as that of erythroid band 3 (Kay 1991a; Kay *et al.*, 1991a, 1991b). These functions are anion transport, ankyrin binding, and generation of senescent cell antigen. Structural similarity of brain and erythroid band 3 is suggested by the reaction of antibodies to synthetic peptides of erythroid band 3 with brain band 3, the inhibition of anion transport by the same inhibitors, and an equal degree of inhibition of brain and erythrocyte anion transport by synthetic peptides of erythroid band 3. One of these segments, COOH, contains antigenic determinants of SCA. These findings suggest that the transport domain of erythroid and neural band 3 is similar functionally and structurally, and support the hypothesis that the immunological mechanism of maintaining homeostasis is a general physiologic process for removing senescent and damaged cells in mammals and other vertebrates. cDNA data indicate that the anion transport segment of band 3 (anion exchangers) from different tissue are similar.

Band 3 Location in Brain

Antibodies to synthetic peptides of human erythroid band 3 reacted with band 3 and breakdown products of band 3 in erythrocyte membranes and multiple polypeptides immunologically related to erythrocyte band 3 in membranes from perfused rat brains. Band 3 breakdown products have been found to migrate in ranges between ≈5,000 to 85,000 Daltons in erythrocytes (Kay, 1981, 1984; Kay *et al.*, 1991a, 1991b). Some of the band 3 antibody reactive bands migrating at different molecular weights in brain may be brain band 3 breakdown products. Binding of antibodies to synthetic peptides of erythroid band 3 to brain membranes suggests that the primary structure of these segments that is recognized by the antibodies is similar in brain and erythrocytes. Senescent cell antigen has been mapped to erythroid band 3 residues 538-554 and 812-830 (Kay *et al.*, 1990c, 1990d).

IgG eluted from senescent erythrocytes binds to band 3 in erythrocytes and band 3 related polypeptides in membranes of saline perfused rat brains as determined by immunoblotting. Absorption of senescent cell IgG with brain membranes resulted in inhibition of binding in the IgG binding/inhibition assay ($80\pm2\%$; Table 1). For comparison, absorption with purified erythroid senescent cell antigen itself resulted in a $74\pm6\%$ inhibition of binding. This suggests that senescent cell antigen, an aging antigen that terminates the life of cells, is generated on brain band 3.

TABLE 3. Inhibition of Senescent Cell IgG Binding by Senescent Cell Antigen from Erythrocytes or Perfused Rat Brain

SAMPLE	INHIBITION (%)
CONTROL CYTOPLASMIC PEPTIDE	0
SENESCENT CELL ANTIGEN	$74\pm6^*$
BRAIN (PERFUSED)	$80\pm2^*$

Data are presented as the mean ±1 SD of quadruplicate samples; $P\leq0.001$. Erythrocytes were separated into populations of different ages on Percoll gradients as previously described (Bennett and Kay, 1981). Middle-aged cells were stored at 4oC for 3 weeks in Alsever's solution. Storage mimics normal aging *in situ* immunologically and biochemically (Kay, 1975, 1978, 1981, 1984; Kay *et al.*, 1986, 1988a, 1989, 1990b, 1990d; Bennett and Kay, 1981; Bosman and Kay, 1988). Senescent cell IgG was isolated from senescent erythrocytes as described (Kay, 1981) and $3\mu g$ incubated with senescent cell antigen from erythrocytes, perfused rat brain, or buffer for 90 mins. at room temperature. Intact erythrocytes were added to the IgG and samples were incubated for 90 mins at room temperature. Cells were washed and the amount of IgG on cells was quantitated using 125I-labeled Protein A (Kay *et al.*, 1989, 1990b, 1990d; Yam *et al.*, 1982). Control cytoplasmic peptide is band 3 peptide pep-CYTO.

Antibodies to the whole band 3 molecule and to synthetic peptides stained Purkinje cell soma, axons, glomeruli (areas of synaptic contact) in the pia, internal granule layer, ependymal cells lining the ventricles, and the choroid plexus as determined by rhodamine and peroxidase staining (Kay *et al*, 1991). The presence of band 3 in the choroid plexus is not surprising given the transport properties of this epithelium which maintains the chemical stability of the cerebrospinal fluid.

Band 3 function in brain

Ankyrin Binding. Brain band 3 performs the same structural function of stabilizing the plasma membrane and linking it to the internal cytoskeleton as does erythrocyte band 3 (Goodman and Shiffer, 1983). The 40,000 Dalton cytoplasmic segment of erythroid band 3 competed with brain band 3 and reduced ankyrin binding to brain membranes by 58%. This indicates that $\approx58\%$ of the ankyrin binding is to band 3.

Anion Transport Studies

Anion transport studies performed on cerebral cortex demonstrated the presence of DIDS inhibitable anion transport systems in brain. The effect of the anion transport inhibitors DIDS, phenylglyoxal, and furosemide on brain and erythrocyte anion transport was compared. Sulfate exchange is inhibited in erythrocytes by (DIDS), phenylglyoxal, and furosemide. Stilbenedisulfonate derivatives such as DIDS inhibit anion transport by binding to at least two lysine residues in the membrane-spanning region of band 3. DIDS is a specific, irreversible, inhibitor of anion transport. It binds covalently. Phenylglyoxal modifies an arginine involved in anion transport that is located on the 35,000 carboxyl segment of band 3 (Bjerrum et al., 1983). The diuretic furosemide inhibits NaCl cotransport by acting at a chloride transport site. The compounds that inhibit anion transport in erythrocytes also inhibit transport in brain. Transport by brain vesicles was reduced to a level ~74% of normal by DIDS, ~65% by furosemide, and ~45% by phenylglyoxal. In contrast, transport in erythrocytes was reduced by 100% by DIDS, and ~95% by furosemide. Inhibition by phenylgloxal was the same for brain and erythrocytes. Since brain anion transport was significantly inhibited but not abolished by DIDS, the data suggest the presence in brain of more than one anion transport system (i.e. a transport system[s] that is DIDS inhibitable and another or others that are not) and/or ~20% of the sulfate influx is due to leakage.

Peptide Inhibition of Anion Transport

Peptides from putative human erythroid transport regions inhibit sulfate transport into erythrocytes and brain vesicles (Kay et al., 1991). Pep-COOH-6 is on the amino end of pep-COOH and contributes a significant amount of the antigenicity of senescent cell antigen (Kay et al., 1990c, 1990d). pep-ANION 1, a putative transport region, does not inhibit anion transport in erythrocytes (Kay et al., 1990b; Kay and Lin, 1990). Thus, it serves as a negative control for these experiments. Results of these experiments showed that the same peptides that inhibited anion transport in erythrocytes inhibit transport in brain (Kay et al., 1991). This suggests that the transport site(s) in erythroid and neural tissue are similar.

Analogues to erythrocyte membrane cytoskeletal proteins have been found in most if not all tissues throughout the body (Davis and Bennett, 1986; Goodman et al., 1981; Goodman and Zagon, 1986). Thus, the proteins with which band 3 interacts in erythrocytes to provide structural stability to the membrane are present in neural tissue. Band 3 proteins are present in mammalian brain that perform the same functions as that of erythroid band 3. These functions are anion transport, ankyrin binding, and generation of senescent cell antigen, an aging antigen that terminates the life of cells. The anion transport segments of erythroid and brain band 3 must be similar since synthetic peptides from transport regions of erythroid band 3 inhibit anion transport by brain vesicles as well as erythrocytes. In addition the inhibitors of anion transport in erythrocytes (DIDS, phenylgloxal, and furosemide) also inhibit anion transport by brain membranes. Since senescent cell antigen is derived from band 3 and (a) IgG specific for this antigen binds to brain membranes in immunoblots and a senescent cell IgG binding-inhibition assay, and (b) antibodies to the segments of band 3 on which senescent cell antigen resides react with brain in immunoblots and tissue sections, the data suggest that senescent cell antigen may be involved in the removal of neurons in aging and disease. This supports the hypothesis that the immunological mechanism of maintaining homeostasis is a general physiologic process for removing senescent and damaged cells in mammals and other vertebrates (Kay, 1981).

NATURALLY OCCURRING SERUM ANTIBODIES TO SYNTHETIC PEPTIDES OF BAND 3

We suspected that the naturally occurring autoantibodies to senescent cell antigen could be quantitated in human serum with ELISA, and that titers might change with diseases such as rheumatoid arthritis (RA) and systemic lupus erythematosus (SLE). We tested this by titering antibodies to senescent cell antigen peptides pep-ANION 1 and COOH, and an anion transport region peptide, ANION 3 (Marchalonis *et al.* 1992; Tables 4 and 5). Results showed a significantly increased titers to pep-ANION 1 in individuals with rheumatoid arthritis. This is consistent with tissue destruction in RA.

TABLE 4. Individual IgG Serum Antibody Titers to Synthetic Senescent Cell Antigen Peptides (pep-ANION 1 and COOH) and an ANION Transport Region Peptide pep-ANION 2)

SERUM	BLANK	ANION 1	COOH	ANION 2
YOUNG	≤ 50	77	123	≤ 50
NORMAL	82	83	54	≤ 50
	55	83	≤ 50	≤ 50
OLD	≤ 50	80	1,578	≤ 50
	≤ 50	81	172	≤ 50
RA	87	221	75	60
	112	548	327	94
SLE	≤ 50	≤ 50	79	≤ 50
	≤ 50	96	≤ 50	≤ 50

TABLE 5. IgG Serum Antibody Titers to Synthetic Senescent Cell Antigen Peptides (pep-ANION 1 and COOH) and an ANION Transport Region Peptide (pep-ANION 2)

SERUM	BLANK	ANION 1	COOH	ANION 2
NORMALS	0.11 ± 0.01	0.34 ± 0.03	0.31 ± 0.03	0.22 ± 0.02
ELDERLY	0.08 ± 0.01	0.23 ± 0.03	0.36 ± 0.10	0.21 ± 0.05
RA	0.15 ± 0.04	$0.71 \pm 0.15*$	0.37 ± 0.05	0.30 ± 0.03
SLE	0.18 ± 0.08	0.56 ± 0.17	0.54 ± 0.12	0.31 ± 0.04

* Significant when compared to normals. "Normals" are a healthy group of 34 women and 15 men ages 20-60 years. The elderly group consists of 29 healthy individuals (15 women and 14 men, greater that 70 years of age.

BAND 3 AND SENESCENT CELL ANTIGEN IN AGING AND ALZHEIMER'S DISEASE

A polyclonal antibody against the segment of band 3 containing senescent cell antigen was used to determine whether senescent cell antigen was altered by differentiation. In order to assess the amount of band 3 protein on living mitotic and differentiated cells, immunofluorescence was measured on a single-cell basis by flow cytometry. The neuroblastoma cell line N2AB-1 were used. Differentiation of

N2AB-1 resulted in an increase senescent cell antigen/band 3 protein on the cell surface on an individual cell basis. Since these cells do not change in size following differentiation, increased surface band 3 protein must be due to a change in density of this protein. These results were confirmed using immunoelectron microscopy. Preliminary studies on anion (sulfate) transport indicated that it increased following differentiation of N2AB-1.

Examination of frozen brain sections from 10 year old and 96 year old individuals revealed labeling of fibrillary structures and processes with senescent cell antigen-band 3 antibodies in sections from old but not young brains (Kay et al., 1988).

In normal brains from elderly individuals, band 3 antibodies react with cortex neurons in layers III and IV, Purkinje cells and their dendrites extending into the molecular layer, cerebellar dentate nucleus neurons. Aged band 3 (pre-senescent cell antigen) antibodies reacted with astrocytes in the white matter, a "mossy fiber" distribution in the cerebellum, and select Purkinje cells. Dentate neurons were strongly reactive, especially those containing lipofuscin, but the staining did not resemble that of lipofuscin. There was a moderately strong reaction with many, but not all, large neurons in the cerebrum. Aged band 3 antibodies recognize old band 3 before senescent cell antigen is formed. They bind to band 3 in old but not young or middle-aged red cells (Kay et al., 1989, 1991c).

In brains from Alzheimer's disease patients, aged band 3 antibodies labeled the amyloid core of classical plaques and the microglial cells located in the middle of the plaque. Adjacent neurons displayed a stronger and more widespread reaction than normals. In contrast, band 3 antibodies labeled the neuritic components of plaques with some reaction noted in microglial cells, adjacent astrocytes and neurons.

Antibodies to synthetic peptides of band 3 differentiate band 3 proteins in brain membranes from normal individuals from those from patients with Alzheimer's (Figures 6-8). One of these, an antibody to band 3 synthetic peptide pep-COOH detects a triplet in the 60-70,000 Dalton range in brains from patients with AD but not from age-matched controls (Saitoh et al., in press). Antibodies to pep-COOH were used for immunohistochemical analysis of Alzheimer's disease (AD) and normal frontal cortex. Pep-COOH antibodies were used because COOH is a component of senescent cell antigen. Pep-COOH antibodies only weakly stained neurons in normal cortex. Only a few large neurons were strongly reactive. In AD, the number of neurons labeling with antibodies to pep-COOH increased 260% ($P \leq 0.01$). Because of a 50% neuronal loss in AD compared to controls ($P \leq 0.002$), the proportion of neurons remaining that were pep-COOH positive increased 540%. In AD, antibodies to pep-COOH labeled glial cells found around plaques. This suggests that band 3 exists in normal neurons in, predominately, an uncleaved state, and that degradation of band 3 occurs in AD thus allowing binding of antibodies to pep-COOH.

CONCLUSIONS

Degradation of band 3 generates senescent cell antigen, an aging antigen that marks cells for removal by initiating the binding of IgG autoantibody and subsequent removal by phagocytes. It is generated on the transport domain of band 3. This appears to be a general physiologic process for removing senescent and damaged cells in mammals and other vertebrates. Although the initial studies were done using human erythrocytes as a model, senescent cell antigen has been found on all cells examined. The aging antigen itself is generated by the degradation of band 3.

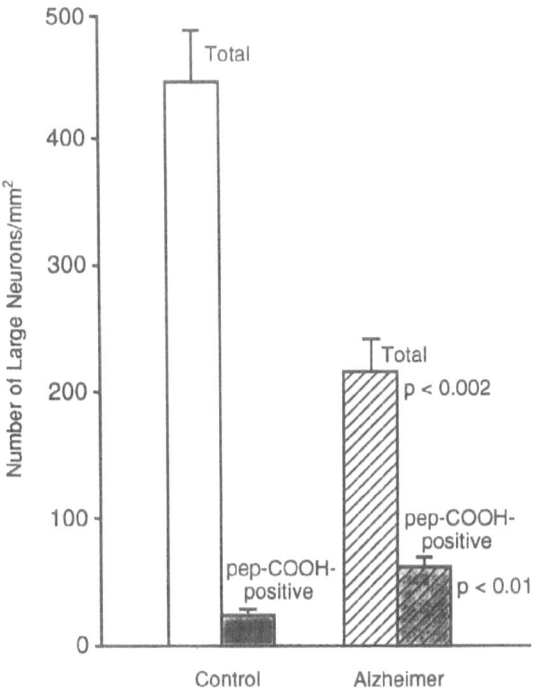

Figure 2. Computer -aided quantification of pyamidal neurons immunolabeled with antibodies to pep-COOH show that there is a 260% increase in the number of neurons labels with antibodies to pep-COOH in AD by immunohistochemical analysis. Because the pyramidal cell population is decreased in AD (Cresyl violet stain), the fraction of neurons that is labeled with antibodies to pep-COOH increased even more in AD. Quantification was performed with a Quantiment 970 (from Saitoh *et al.*, 1992).

Besides its role in the removal of senescent and damaged cells, senescent cell antigen also appears to be involved in the removal of erythrocytes in clinical hemolytic anemias, and the removal of malaria-infected erythrocytes. Oxidation generates senescent cell antigen *in situ*.

Band 3 is a crucial structural and functional protein and is intimately involved in cellular aging. Band 3 and senescent cell antigen are present in the central nervous system, and differences have been described in band 3 between young and aging brain tissue, and in Alzheimer's disease. This suggests that band 3 and senescent cell antigen may play a role in neurological health and disease.

ACKNOWLEDGEMENTS

This work was supported by NIH grants AG08444, AG08574, AG09258, a Veterans Administration Merit Review, the International Foundation for Biomedical Aging Research, and the Arizona Disease Control Commission.

REFERENCES

Allen, D.P., Low, P.S., Dola, A., and Maisel, H., 1987, Band 3 and ankyrin homologues are present in eye lens: evidence for all major erythrocyte membrane components in same non-erythroid cell, *Biochem. Biophys. Res. Commun.* 149:266.

Alper S.L., Kopito, R.R., Libresco, S.M., and Lodish, H.F., 1988, Cloning and characterization of a murine band 3-related cDNA from kidney and from a lymphoid cell line, *J. Biol. Chem.*, 263:17092.

Alper S.L., Natale, J., Gluck, S., Lodish, H.F., and Brown, D., 1989, Subtypes of intercalated cells in rat kidney collecting duct defined by antibodies against erythroid band 3 and renal vacuolar H+-ATPase, *Proc. Natl. Acad. Sci.* 86:5429-5433.

Bartel D., H. Hans, and H. Passow, 1989, Identification by site directed mutagenesis of lysine 558 as a covalent attachment site of DIDS in mouse erythroid band 3, *Biochem. Biophys. Acta*, 985:355.

Bartosz G., Sosynski, M., and Wasilewski, A., 1982a, Aging of the erythrocyte XVII. Binding of autologous immunoglobulin, *Mech. Ageing Dev.* 20:223.

Bartosz G., Sosynski, M., and Kredziona, J., 1982b, Aging of the erythrocyte. VI. Accelerated red cell membrane aging in Down's syndrome, *Cell Biol. Int. Rep.* 6:73.

Bennett G.D., and Kay, M.M.B., 1981, Homeostatic removal of senescent murine erythrocytes by splenic macrophages, *Exp. Hematol.* 9:297.

Bennett V., 1979, Immunoreactive forms of human erythrocyte ankyrin are present in diverse cells, *Nature, Lond.* 281:597.

Bjerrum, P.J., Wieth, J.O., and Minakami, S., 1983, Selective phenylglyoxalation of functionally essential arginyl residues in the erythrocyte anion transport protein, *J. Gen Physiol.* 81:453.

Bosman G., Bartholomeus,I., DeMan, C., VanKalmthout, P., and DeGrip, W., 1991, Alzheimer's disease: Indications for disturbed erythrocyte aging, *Neurobiol. Aging*, 12:13.

Bosman G.J.C.G.M., and Kay, M.M.B., 1988, Erythrocyte aging: A comparison of model systems for simulating cellular aging *in vitro*, *Blood Cells*, 14:19.

Bosman G.J.C.G.M., and Kay, M.M.B., 1990, Alterations of band 3 transport protein by cellular aging and disease: Erythrocyte band 3 and glucose transporter share a functional relationship, *Biochem. Cell Biol.* 68:1419.

Branch D.R., Gallagher, M.T., Mison, A.P., Sy Siok Hian, A.L., and Petz, L.D., 1984, *In vitro* determination of red cell alloantibody significance using an assay of monocyte-macrophage interaction with sensitized erythrocytes, *Br. J. Haematol.* 56:19.

Brosius F., Alpert, S., Garcia, A., and Lodish, H., 1989, The major kidney band 3 gene transcript predicts an amino-terminal truncated band 3 polypeptide, *J. Biol. Chem.*, 264:7784.

Davis J.Q., and Bennett, V., 1986, Association of brain ankyrin with brain membranes and isolation of brain ankyrin with brain membranes and isolation of active proteolytic fragments of membrane-associated ankyrin-binding protein(s), *J. Biol. Chem.* 261:16198.

Demuth D.R., Showe, L.C., Ballantine, M., Palumbo, A., Fraser, P.J., Cioe, L., Rovera, G., and Curtis, P.J., 1986, Cloning and structural characterization of a human non-erythroid band 3-like protein, *EMBO J.* 5:1205.

Drenckhahn D., and Merte, C., 1987, Restriction of the human kidney band 3-like anion exchanger to specialized subdomains of the basolateral plasma membrane of intercalated cells, *Eur. J. Cell. Biol.* 45:107.

Drenckhahn D., Schulter, K., Allen, D.P., and Bennett, V., 1985, Colocalization of band 3 with ankyrin and spectrin at the basal membrane of intercalated cells in the rat kidney, *Science*, 230:1287.

Drenckhahn D., Zinke, K., Schauer, U., Appell, K.C., and Low, P.S., 1984, Identification of immunoreactive forms of human erythrocyte band 3 in nonerythroid cells, *Eur. J. Cell Biol.* 34:144.

Friedman M.J., Fukuda, M., and Laine, R.A., 1985, Evidence for a malarial parasite interaction site on the major transmembrane protein of the human erythrocyte, *Science*, 228:75.

Garcia A.M., and Lodish, H.F., 1989, Lysine 539 of human band 3 is not essential for ion transport or inhibition of stilbene disulfonate, *J. Biol. Chem.*, 264:19607.

Glass G.A., Gershon, D., and Gershon. H., 1985, Some characteristics of the human erythrocyte as a function of donor and cell age, *Exp. Hematol.* 13:1122.

Goodman S.R., and Shiffer, K., 1983, The spectrin membrane skeleton of normal and abnormal human erythrocytes: a review, *Am. J. Physiol.* 244:C121.

Hazen-Martin D.J., Pasternack, G., Hennigar, R.A., Spicer, S.S., and Sens, D.A., 1987, Immunocytochemistry of band 3 protein in kidney and other tissues of control and cystic fibrosis patients, *Pediatr.Res.* 235-237.

Hazen-Martin, D.J., Pasternack, G., Spicer, S.S., and Sens, D.A., 1986, Immunolocalization of band 3 protein in normal and cystic fibrosis skin, *J. Histochem. Cytochem.* 34:823.

Hebbel R.P., and Miller, W.J., 1984, Phagocytosis of sickle erythrocytes. Immunologic and oxidative determinants of hemolytic anemia, *Blood,* 64:733.

Jarolim, P., Palek, J., Amato, D., Hassan, K., Sapak, P., Nurse, G.T., Rubin, H.L., Zhai, P., Sahr, K.E., and Liu, S.C., 1991, Deletion in erythrocyte band 3 gene in malaria-resistant Southeast Asian ovalocytosis, *Proc. Natl. Acad. Sci.* 88:11022.

Jennings, M.L., 1989, Structure and function of the red blood cell anion transport protein, *Annu. Rev. Biophys. Biophys. Chem.* 18:397.

Jennings, M.L., Anderson, M.P., and Monaghan, R., 1986, Monoclonal antibodies against human erythrocyte band 3 protein. Localization of proteolytic cleavage sites and stilbenedisulfonate-binding lysine residues, *J. Biol. Chem.* 261:9002.

Jennings, M.L., and Anderson, M.P., 1987, Chemical modification and labeling of glutamate residues at the stilbenedisulfonate site of human red blood cell band 3 protein, J. Biol. Chem. 262:1691.

Jennings, M.L., Monaghan, R., Douglas, S.M., and Nicknish, J.S., 1985, Functions of extracellular lysine residues in the human erythrocyte anion transport protein, *J. Gen. Physiol.* 86:653.

Kay, M.M.B., 1975, Mechanism of removal senescent cells by human macrophages *in situ. Proc. Natl. Acad. Sci.* 72:3521.

Kay, M.M.B., 1978, Role of physiologic autoantibody in the removal of senescent human red cells, *J. Supramol. Struct.* 9:555.

Kay, M.M.B., 1981, Isolation of the phagocytosis inducing IgG-binding antigen on senescent somatic cells, *Nature, Lond.* 289:491.

Kay, M.M.B., 1983, Appearance of a terminal differentiation antigen on senescent and damaged cells and its implications for physiologic autoantibodies, *Biomembranes,* 11:119.

Kay, M.M.B., 1984, Localization of senescent cell antigen on band 3, *Proc. Natl. Acad. Sci.* 81:5753.

Kay, M.M.B., 1985a, Aging of cell membrane molecules leads to appearance of an aging antigen and removal of senescent cells, *Gerontology,* 31:215.

Kay, M.M.B., 1985b, Glucose transport protein is structurally and immunologically related to band 3 and senescent cell antigen, *Proc. Natl. Acad. Sci.* 82:1731.

Kay, M.M.B., 1986. Senescent cell antigen: A red cell aging antigen, *in* Red Cell Antigens and Antibodies G. Garratty (ed.), American Association of Blood Banks, Arlington, VA. pp. 35-82.

Kay, M.M.B., 1988, Immunologic techniques for analyzing red cell membrane proteins, *in* Methods in Hematology: Red Cell Membranes, S. Shohet, and Mohandas (ed.), Churchill Livingston, Inc., New York, pp. 135-170.

Kay, M.M.B., 1991a, Drosophila to bacteriophage to erythrocyte: The erythrocyte as a model for molecular and membrane aging of terminally differentiated cells, *Gerontology*, 37:5.

Kay, M.M.B., 1991b, Molecular mapping of human band 3 anion transport regions using synthetic peptides, *Federation Proceedings*, 5:109.

Kay, M.M.B., Goodman, S., Sorensen, K., Whitfield, C., Wong, P., Zaki, L., and Rudoloff, V., 1983a, The senescent cell antigen is immunologically related to band 3, *Proc. Natl. Acad. Sci.* 80:1631.

Kay, M.M.B., Tracey, C.M., Goodman, J.R., Cone, J.C., and Bassel, P.S., 1983b, Polypeptides immunologically related to erythrocyte band 3 are present in nucleated somatic cells, *Proc. Natl. Acad. Sci.* 80:6882.

Kay, M.M.B., Bosman, G., Notter, M., and Coleman, P., 1988, Life and death of neurons: The role of senescent cell antigen, *Ann. N. Y. Acad. Sci.* 521:155.

Kay, M.M.B., and Bosman, G.J.C.G.M., 1985, Naturally occurring human "antigalactosyl" IgG antibodies are heterophile antibodies recognizing blood group related substances, *Exp. Hematol.* 13:1103.

Kay, M.M.B., Bosman, G.J.C.G.M., Johnson, G., and Beth, A., 1988, Band 3 polymers and aggregates, and hemoglobin precipitates in red cell aging, *Blood Cells*, 14:275.

Kay, M.M.B., Bosman, G.J.C.G.M., and Lawrence, C., 1988, Functional topography of band 3: A specific structural alteration linked to functional aberrations in human erythrocytes, *Proc. Natl. Acad. Sci.* 85:492.

Kay, M.M.B., Bosman, G.J.C.G.M., Shapiro, S.S., Bendich, A., and Bassel, P.S., 1986, Oxidation as a possible mechanism of cellular aging: Vitamin E deficiency causes premature aging and IgG binding to erythrocytes, *Proc. Natl. Acad. Sci.* 83:2463.

Kay, M.M.B., 1990, Aging of cell membrane molecules: Band 3 and senescent cell antigen in neural tissue, *in* Molecular Mechanisms of Aging, K. Beyreuther, and G. Schettler (eds.), Springer-Verlag, Berlin, Germany, pp. 110-123.

Kay, M.M.B., 1992, Molecular mapping of human band 3 aging antigenic sites and active amino acids using synthetic peptides, *J. Prot. Chem.* in press.

Kay, M.M.B., Flowers, N., Goodman, J., and Bosman, G.J.C.G.M., 1989, Alteration in membrane protein band 3 associated with accelerated erythrocyte aging, *Proc. Natl. Acad. Sci.* 86:5834.

Kay, M.M.B., Goodman, J., Goodman, S., and Lawrence, C., 1990a. Membrane protein band 3 alteration associated with neurologic disease and tissue reactive antibodies, *Clin. Exper. Immunogenet.* 7:181.

Kay, M.M.B., Goodman, J., Lawrence, C., and Bosman, G., 1990b, Membrane channel protein abnormalities and autoantibodies in neurological disease. *Brain Res. Bull.* 24:105.

Kay, M.M.B., Marchalonis, J.J., Hughes, J., Watanabe, K., and Schluter, S.F., 1990c, Definition of a physiologic aging auto-antigen using synthetic peptides of membrane protein band 3: Localization of the active antigenic sites, *Proc. Natl. Acad. Sci.* 87:5734.

Kay, M.M.B., Lin, F., Bosman, G., Marchalonis, J.J., and Schluter, S.F., 1990d, Human erythrocyte aging: Cellular and Mosicular biology, *Trans. Med. Revs.* 5:173.

Kay, M.M.B., and Lin. F., 1990, Molecular mapping of the active site of an aging antigen: Senescent cell antigen is located on an anion binding segment of band 3 membrane transport protein, *Gerontology*, 36:293.

Kay, M.M.B., Hughes, J., Zagon, I., and Lin, F., 1991, Brain membrane protein band 3 performs the same functions as erythrocyte band 3, *Proc. Natl. Acad. Sci.* 88:27780.

Kay, M.M.B., and Marchalonis, J.J., 1991, Synthetic aging antigen can be used to manipulate cellular lifespan, *Life Sci.* 48:1603.

Kay, M.M.B., Sorensen, K., Wong, P., and Bolton, P., 1982, Antigenicity, storage & aging: Physiologic autoantibodies to cell membrane and serum proteins and the senescent cell antigen, *Mol. Cell. Biochem.* 49:65.

Kellokumpu, S., Neff, L., Jamsa-Kellokumpu, S., Kopito, R., and Baron, R., 1988, A 115-kD polypeptide immunologically related to erythrocyte band 3 is present in Golgi membranes, *Science*, 242:1308.

Khansari, N., and Fudenberg, H.H., 1983, Immune elimination of autologous senescent erythrocytes by Kupffer cells *in vivo, Cell. Immunol.* 80:426.

Khansari, N., Springer, G.F., Merler, E., and Fudenberg, H.H., 1983, Mechanisms for the removal of senescent human erythrocytes from circulation: specificity of the membrane-bound immunoglobulin G, *J.Mech.Aging Dev.* 21:49.

Kopito, R.R., and Lodish, H.F., 1985, Structure of the murine anion exchange protein, *J. Cell. Biochem.* 29:1.

Kudrycki, K.E., and Shull, G.E., 1989, Primary structure of the rat kidney band 3 anion exchange protein deduced from a cDNA, *J. Biol. Chem.* 264:8185.

Lutz, H.U., Flepp, R., and Stringaro-Wipf, G., 1984, Naturally occurring autoantibodies to exoplasmic and cryptic regions of band 3 protein, the major integral membrane protein of human red blood cells, *J. Immunol.* 133:2160.

Marchalonis, J. J., Schluter, S.F., Wilson, L., Yocum, D., Boyer, J., and Kay, M. M.B., Natural antibodies to synthetic peptide autoantigens: correlations with age and autoimmune disease, *Gerontology*, 39:65.

Okoye, V.C., and Bennett, V., 1985, Plasmodium falciparum malaria: band 3 as a possible receptor during invasion of human erythrocytes, *Science*, 227:169.

Petz, L.D., Yam, P., Wilkinson, L., Garratty, G., Lubin, B., and Mentzer, W., 1984 Increased IgG molecules bound to the surface of red blood cells of patients with sickle cell anemia, *Blood*, 64:301.

Saitoh, T., Masliah, E., Baum, L., Sundsmo, M., Flanagan, L., Vikramkumar, R., and Kay, M.M.B., 1992, Degradation of proteins in the membrane-cytoskeleton complex in Alzheimer's disease: Might amyloidogenic APP processing be just the tip of the iceberg? *Ann. N.Y. Acad. Sci.* in press.

Scatchard, G., 1949, The attraction of proteins for small molecules and ions, *Ann. N. Y. Acad. Sci.*, 51:660-672.

Schuster, V.L., Bonsib, S.M., and Jennings, M.L., 1986, Two types of collecting duct mitochondria-rich (intercalated) cells: lectin and band 3 cytochemistry, *Am. J. Physiol.* 251:C347.

Singer, J.A., Jennings, J.K., Jackson, C., Doctker, M.E., Morrison, M., and Walker, W.S., 1986, Erythrocyte homeostasis: Antibody-mediated recognition of the senescent state by macrophages, *Proc. Natl. Acad. Sci.* 83:5498.

Steck, T.L., 1978, The band 3 protein of the human red cell membrane: A review, *J. Supramol. Struct.* 8:311.

Tanner, M.J.A., Martin, P.G., and High, S., 1988, The complete amino acid sequence of the human erythrocyte membrane anion-transport protein deduced from the cDNA sequence, *Biochem J.* 256:703.

Vanderpuye, O.A., Kelley, L.K., Morrison, M.M., and Smith, C.H., 1988, The apical and basel plasma membranes of the human placental syncytiotrophoblast contain different erythrocyte membrane protein isoforms. Evidence for placental forms of band 3 and spectrin. *Biochim. Biophys. Acta.* 943:277.

Verlander, J.W., Madsen, K.M., Low, P.S., Allen, D.P., and Tisher, C.C., 1988, Immunocytochemical localization of Band 3 protein in the rat collecting duct, *Am. J. Physiol.* 255:F115.

Walker, W.S., Singer, J.A., Morrison, M., and Jackson, C.W., 1984, Preferential phagocytosis of *in vivo* aged murine red blood cells by a macrophage-like cell line, *Br. J. Haemat.* 58:259.

Wolpaw, E.W., and Martin, D.W., 1986, A membrane protein in LRM55 glial cells cross-reacts with antibody to the anion exchange carrier of human erythrocytes, *Neurosci. Lett.* 67:42.

Zaki, L., 1983, Anion transport in red blood cells and arginine specific reagents. (1) Effect of chloride and sulfate ions on phenylglyoxal sensitive sites in the red blood cell membrane, *Biochem. Biophys. Res. Comm.* 110:616.

B-CELL ORIGIN OF COLD AGGLUTININS

Leslie E. Silberstein

Associate Professor of Pathology and Laboratory Medicine *and*
   Medicine, University of Pennsylvania School of Medicine
Department of Pathology and Laboratory Medicine
Hospital of the University of Pennsylvania
3400 Spruce Street
Philadelphia, PA 19104

INTRODUCTION

Cold agglutinins (CA) are RBC specific autoantibodies that agglutinate RBCs optimally at 4 - 22°C in vitro and may be of widely divergent clinical significance. Virtually all sera from healthy individuals contain low titered CA which cause no apparent immune hemolysis and may be defined as natural or benign RBC autoantibodies. In contrast, the pathologic (e.g. hemolytic) counterparts of these autoantibodies are generally derived from clonal B cell expansions that may progress to frank lymphoma (Table 1).

The occurrence of normal and pathogenic serum autoantibodies suggests that under certain conditions, autoreactive B cell clones escape from tolerance mechanisms. For example, in certain transgenic mice, it has been known that functionally inactivated (e.g. anergic) B-cells can be rescued and induced to produce self reactive antibody by polyclonal activation with bacterial polysaccharide antigens. These experiments may therefore explain why autoimmune hemoytic syndromes, especially in children, are often associated with viral and bacterial infections. Conceivably, the causative autoantibodies result from the dysregulation of certain autoreactive B-cell clones, and in some instances, these dysregulated B-cell clones may undergo neoplastic transformation (e.g. cold agglutinin disease).

Table 1. Classification of Cold Agglutinins

Monoclonal *	Polyclonal
idiopathic/chronic B-cell lymphoma	benign/natural post infectious (e.g. Mycoplasma pneumonia, Epstein-Barr virus (EBV); Human immunodeficiency virus (HIV); collagen vascular disorders.)

* Monoclonal cold agglutinins are derived from a spectrum of clonal B-cell expansion ranging from pre neoplastic (e.g. no evidence of malignancy) to frank lymphoma [1-3].

CLINICAL AND SEROLOGICAL ASPECTS

The majority of benign and pathologic CA are IgM autoantibodies. While the benign CA are considered to be polyclonal, the pathologic CA are invariably monoclonal and detectable as paraproteins in the serum. The pathologic CA usually occur spontaneously, during the course of a lymphoproliferative disorder, or rarely as a post-infectious complication of mycoplasma pneumonia or infectious mononucleosis [4,5]. Patients with cold agglutinin disease (i.e. CA induced autoimmune hemolytic anemia) may have CA`titers in the thousands or even millions compared to normal individuals who may have low titer (<32) IgM cold agglutinins. Because of their low thermal binding properties, these IgM autoantibodies appear to bind to red cells and fix complement in the peripheral circulation where temperatures fall below ~32°C. As the cells return to warmer parts of the circulation, the IgM dissociates leaving the cells coated with only complement. Complement activation may then lead to intravascular hemolysis.

Paroxysmal cold hemoglobinuria (PCH), like cold agglutinin disease, is caused by cold-reactive autoantibodies that react with red cells in cooler parts of the body, cause complement to irreversibly bind to cells, and then elute off of the erythrocyte surface when warmed. However, in PCH, the autoantibodies are IgG molecules usually directed at P blood group antigens [6] and are present in relatively low titers (<64). Because the presence in serum of these biphasic IgG antibodies may be difficult to detect by standard serological methods, specialized tests which use in vitro hemolysis as an indicator of IgG-induced complement sensitization can be employed (Donath-Landsteiner test, Donath and Landsteiner 1904). PCH may be idiopathic or secondary to syphilis or viral infection.

Although PCH is the rarest form of autoimmune hemolytic anemia (AIHA), historically it was the first type recognized probably because of its graphic clinical presentation, i.e. the sudden onset of shaking chills, back and leg pain, abdominal cramps, high fever, and the passage of black urine. Reports in the medical literature began to appear in the mid-1800's that described attacks of hemoglobinuria after exposure to cold [7,8]. In 1879, Rosenbach [9] reported how he had induced hemoglobinuria by immersing his patient's feet in ice water, and two years later Paul Erlich [10] reported his observation of hemolysis and erythrophagocytosis in the blood obtained from the chilled finger of a patient with the disease.

For the next fifty years, laboratory investigation into the biology of AIHA was limited to the study of rare, lytic antibodies, or to cold-reactive IgM autoantibodies which would directly agglutinate red cells in vitro due to their multivalent, red cell-bridging properties.

TARGETS FOR COLD-REACTIVE AUTOANTIBODIES

I/i-blood Group Specificity

Cold-reactive anti-red cell autoantibodies typically bind to carbohydrate structures on membrane glycolipids and/or glycoproteins. The majority of cold agglutinins are IgM immunoglobulins that are directed at I/i blood group antigens [11,12]. This antigen system comprises oligosaccharide chains composed of repeating N-acetyllactosamine $(Gal[\beta1\rightarrow4]GlcNAc[\beta1\rightarrow3])$ units (Feizi 1980) linked to ceramide or the membrane glycoproteins band 3 and band 4.5 [13,14] (Figure 1). The best available evidence indicates that the difference between I and i antigens relates to branching of the oligosaccharide chain; anti-i antibodies recognize a linear N-acetyllactosamine oligosaccharide while anti-I antibodies recognize a similar chain that is also branched [15-17]. There appears to be a developmentally-regulated transition in expression of the I/i antigens; fetal and newborn red cells express mostly i antigen while adult red cells demonstrate the opposite pattern [18] (Figure 1). This transition appears to involve the acquisition of a "branching enzyme" (a $[\beta1\rightarrow6]$-N-acetylglucaminyl transferase) [17]. It is of interest to note that the conversion of the

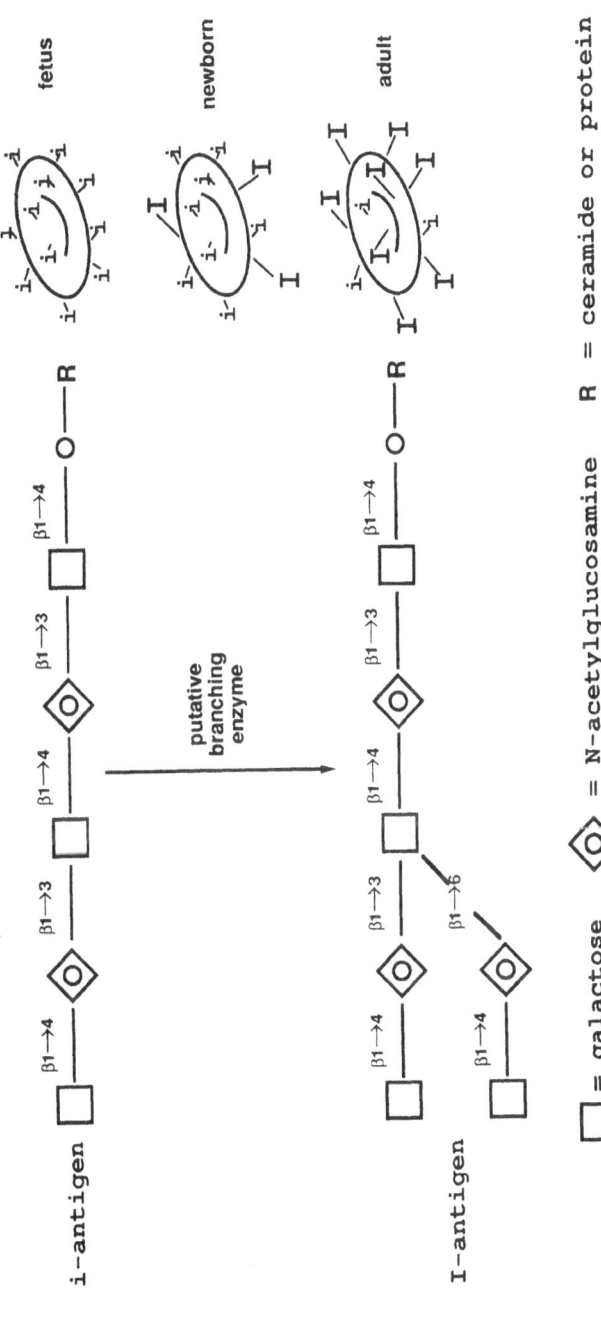

Figure 1. Examples of structures containing I and i antigens

unbranched i structure to a branched one coincides temporally with the switch from fetal to adult hemoglobin synthesis which has led some researchers to suggest that these processes may be coupled [19].

I and i antigens are also present on human granulocytes, macrophages, platelets and lymphocytes and antibodies to these antigens can be lymphocytotoxic [20, 21]. In addition, they are variably expressed on non-human red cells [22] and cultured cell lines of several animal species [23]. On human B-lymphocytes, the I/i antigens appear to be carried on a family of high molecular weight glycoproteins (~200 kD) known as the leukocyte common antigen (T200 or CD45) [24] while on normal T-cells, lower molecular weight (128-140kD) sialoglycoproteins express I/i-related structures [24-26]. Using multiparameter flow cytometry, we have recently demonstrated that anti i autoantibodies bind predominantly if not exclusively to a subset of B-cells in both mouse and man (Silberstein et al, manuscript in preparation;[27]). The presence of I/i-related antigens on CD45 is of particular interest since CD45, recently identified as a protein tyrosine phosphatase involved in regulating lymphocyte activation and proliferation (for review see [28]), can exist as isoforms abnormally distributed on cells in autoimmune disease [29, 30]. The finding that antibodies to CD45 on T- or B-cells can either inhibit or augment lymphocyte activity in vitro depending on what other membrane receptors are nearby [31] raises interesting questions regarding the effect(s) that autoantibodies to I/i-blood group antigens may have on lymphocyte function in vivo.

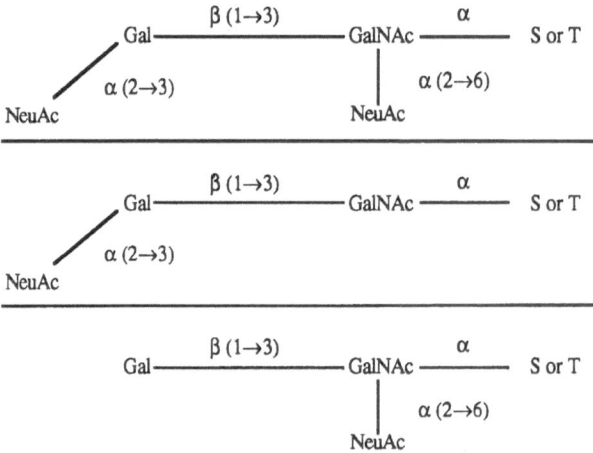

Figure 2. The structures of a tetrasaccharide and two trisaccharides bound to human glycophorins.

NeuAc = N-acetylneuraminic acid, Gal = D-galactose, GalNAc = N-acetyl-D-galactose, GalNAc = N-acetyl-D-galactosamine, S = serine, T = threonine

Pr-blood Group Specificity

The term Pr refers to a group of red cell antigens defined serologically to possess the following characteristics: inactivation by proteases, equivalent expression on adult and newborn red cells, and inactivation by neuraminidase (Marsh and Jenkins 1968, Roelcke 1969). Anti-Pr antibodies were originally divided into the subclasses Pr_1, Pr_2, Pr_3 and Pr_a based on their reactivity with various animal red cells and by hemaggulination inhibition experiments with chemically-modified human red cell sialoglycoprotein preparations

(Roelcke et al 1976). Although the biochemistry of this blood group system is still not completely understood, the best available evidence indicates that the Pr antigens comprise O-linked sialylated oligosaccharide structures that reside within the N-terminal 26 amino acid residues of glycophorins A and B [32-34]; see figure 2. This is based in part on the findings that anti-Pr antisera fail to agglutinate neuraminidase-treated human red cells and are inhibited by the O-linked sialotetrasaccharides isolated from red cell sialoglycoprotein preparations. That they are located on glycophorins A and B (and not glycophorin C) is based on the observation that anti-Pr reactivity is dramatically reduced with En(a-) erythrocytes (which lack glycophorin A) and totally eliminated with homozygous M^kM^k red cells (which lack glycophorins A and B but have normal amounts of glycophorin C) [35]. Like the I/i-blood group antigens, Pr antigens have been found on human lymphocytes, granulocytes, and monocytes [36] as well as on kidney, liver, stomach, pancreas, lung and brain tissue [37]. The significance of these findings has not been reported to date. Other targets for cold-reacting IgM autoantibodies include the Gd (for glycolipid-dependent) gangliosides [38] and other neuraminidase- and/or protease-sensitive antigens referred to as Sa, Lud, Fl, Vo, and Li [36, 38].

P-blood Group Specificity

As noted in the previous section, complement-fixing, cold-reactive, IgG autoantibodies characteristic of PCH, are directed at the P blood group system. The P antigens comprise 3 glycosphingolipids, P^k, P, and P_1, which are structurally and biosynthetically related [39] (Table 2). Gal($\beta1\rightarrow4$)Glc-ceramide serves as a common precursor with the P^k and P_1 antigens sharing the common dissaccharide Gal($\beta1\rightarrow4$)Gal($\beta1\rightarrow4$)-R at the non-reducing end of the glycosphingolipid. The presence of P blood group antigens on red cell membrane polypeptides has not been demonstrated.

Table 2. P Blood Group Antigens

pk	Gal($\alpha1\rightarrow4$)Gal($\beta1\rightarrow4$)Glc-ceramide
P	GalNAc($\beta1\rightarrow3$)Gal($\alpha1\rightarrow4$)Gal($\beta1\rightarrow4$)Glc-ceramide
P_1	Gal($\alpha1\rightarrow4$)Gal($\beta1\rightarrow4$)GlcNAc($\beta1\rightarrow3$)Gal($\beta1\rightarrow4$)Glc-ceramide

Gal = galactose
Glc = glucose
GalNac = N-acetylgalactosamine

MOLECULAR ANALYSES OF CA

The cellular origin of cold reactive autoantibodies has recently been assessed by model systems of EBV transformed cell lines secreting anti - i, anti - I and anti - Pr2 pathogenic autoantibodies. First, it was recognized that in approximately 40% of patients, a circulating B-cell clone could be readily found with a distinctive karyotypic marker, characterized by trisomies 3 and 12 (+3, +12) [1] (Table 3). This chromosomal aberrancy was associated with the chronic/idiopathic cold aggutinin syndrome (i.e. patients with a monoclonal serum CA but no associated B-cell x neoplasm) as well as with monoclonal cold agglutinins secondary to a B-cell neoplasm. This marker proved very useful for demonstrating the close relationship between the EBV transformed B-cell clones and the B-cell populations in the peripheral blood of the patients. Additionally, it could be demonstrated that both the in vitro and in vivo B-cells not only shared the same karyotypic marker and immunoglobulin restriction fragements but their secreted monoclonal antibodies had the same serologic specificity and isoelelctric focusing spectrotype. These studies therefore collectively showed that the monoclonal cold agglutinins are derived from the associated spectrum of (pre-) neoplastic B-cell populations found in patients with cold agglutinin disease [1,2].

Table 3. [1] Chromosomal Abnormalities in 4/9 Patients With Cold Agglutinin Disease

Patient*	Tissue	Abnormal Cells†/Total Cells	Karyotype of Abnormal Clone
1	Blood	28/56	51,XX,+3,+9,+12,+13,+18
	Spleen	12/23	51,XX,+3,+9,+12,+13,+18
	(bulk EBV culture)		
2	Blood	6/16	47,XY,+3
3	Blood	8/23	47,XX,+3
4	Blood	13/84	47,XX,+12
			48,XX,+3,+12

* Cytogenic studies on peripheral blood lymphocytes of patients 5 through 9 did not show evidence of a chromosomally abnormal clone.
† Abnormal cells = cells with karyotype of abnormal clone(s).

I/i-blood Group Specific Autoantibodies

Initial studies of the structure of these antibodies focussed on the constant regions of the immunoglobulin molecules. The composition of nearly all anti-i and anti-I autoantibodies are IgMκ although IgM cold agglutinins with associated λ light chains have been described [37]. More recently, attention has focussed on studying the structural diversity of the variable regions of these autoantibodies through the use of anti-idiotypic reagents and by direct nucleotide sequencing of the rearranged immunoglobulin variable region genes. Idiotypic variation results from differences in amino acid sequence in both heavy and light chain variable regions and anti-idiotypic antibodies can be used to recognize variable region determinants within or outside of the antigen combining site. Through the use of such antiidiotypic antisera, cross-reactive idiotypes were found among red cell autoantibodies [40]. This cross-reactivity appeared to be restricted to red cell autoantibodies with similar specificity. For example, anti-idiotypic antibodies raised against anti-Pr cold agglutinins generally did not cross-react with anti-I cold agglutinins, and vice versa. Subsequent studies by Stevenson et al [41] described a monoclonal anti-idiotypic antibody termed "9G4" which recognized an idiotypic determinant present on the heavy chain of both anti-I and anti-i cold agglutinins as well as on neoplastic B-cells secreting cold agglutinins. This suggested that the heavy chain variable regions of anti-I and anti-i may share some of the same structural elements. Studies from several laboratories [42,43], thave found that the 9G4 idiotype is present on all (at least 48/48) pathogenic anti-I/i cold agglutinins studied to date (Table 4). As can be seen in table 4, the 9G4 reagent, although highly associated with anti-i/I CA, also binds to rheumatoid factors (e.g. LES) encoded by V_H4 genes (non $V_H4.21$). A structural basis for this idiotypic cross-reactivity could not be fully appreciated at the time because of insufficient data on variable region gene usage which had been derived from amino acid sequences limited to the first framework region [44]. Using primary sequence-dependent polyclonal antibodies to heavy chain variable region determinants Silverman et al [45] first showed that both the anti-I and anti-i cold agglutinins are derived from a distinct subset of V_H4 family genes [42,43]. More recently, the establishment of EBV-transformed B-cell lines secreting either anti-I or anti-i autoantibodies has allowed for nucleotide sequence analyses of

198

```
                       FRAMEWORK 1                    CDR1       FRAMEWORK 2            CDR2                    FRAMEWORK 3
Germline VH4.21  QVQLQQWGAGLLKPSETLSLTCAVYGGSFS    GYYWS    WIRQPPGKGLEWIG    EINHSGSTNYNPSLKS    RVTISVDTSKNQFSLKLSSVTAADTAVYYCAR
VH anti-I        ------------------------------N-  -----    -------A------    --------------T-    --------------------------------
VH anti-i        ------------------------------    -----    --------------    ----------------    --------------------------------

                      FRAMEWORK 1              CDR1         FRAMEWORK 2             CDR2            FRAMEWORK 3                       CDR3
Germline Humigk328  EIVMTQSPATLSVSPGERATLSC  RASQSVSSNLA  WYQQKPGQAPRLLIY   GASTRAT   GIPARFSGSGSGTEFTLTISSLQSEDFAVYYC   QQYNNWP
VL-anti-I           -----------------------  -------LN--  -------------M-   DI-----   -----------------D-------------    ----T--

                      FRAMEWORK 1              CDR1         FRAMEWORK 2             CDR2            FRAMEWORK 3                       CDR3
Germline VK1        DIQMTQSPSSLSASVGDRVTITC  RASQSISSYLN  WYQQKPGKAPKLLIY   AASSLQS   GVPSRFSGSGSGTDFTLTISSLQPEDFATYYC   QQSYSTP
VL-anti-i           -----------------------  -----------  ---------------   -------   -------------------------------   -------
```

Figure 3. Comparison of VH and VL genes encoding anti-i and anti-I to their relevant germline precursors

199

the entire expressed variable region genes to further assess the molecular basis for the autoimmune response [42, 43]. Collectively, the work from two laboratories studying four anti-I and two anti-i cold agglutinins indicates that anti-I as well as anti-i cold agglutinins likely derive from the same $V_H4.21$ (or closely-related) gene segments (fig.3)[42]. Furthermore, protein sequencing of another anti-I cold agglutinin has also indicated $V_H4.21$ usage [46].

In contrast, the variable region genes used by the anti-I/i *light* chains do not demonstrate such restriction. While the anti-I cold agglutinin light chains appear to derive from the VkIII gene family, the anti-i cold agglutinins use light chains from a number of different Vk families including VkIII [42]. To ask whether the anti-I/i light chains which use VkIII might exhibit similar idiotypic cross-reactivity as observed for the heavy chains, a panel of eight VkIII light chain-dependent monoclonal anti-idiotypic antibodies were raised against an anti-I autoantibody expressing V_H4 and VkIII genes [47]. Significant idiotypic heterogeneity was observed among VkIII-expressing cold agglutinin light chains. In addition, the idiotypic heterogeneity could be ascribed to the use of at least three different VkIII gene segments, as well as to somatic diversification of the germline-encoded genes.

The remarkable restriction in the variable region genes used for the anti-I/i heavy chains and diversification in the associated light chain variable region gene usage suggests a model for the relative contributions of heavy and light chains to antigen binding. The V_H sequence would be required for the global interaction with the I/i antigen complex, while the V_L sequence would confer the fine specificity that distinguishes between these distinct yet related carbohydrate structures. This model for cold agglutinin binding to the I/i antigens is currently being tested through the use of bacterial expression systems that permit the mixing-and-matching of heavy and light chains from different immunoglobulin molecules. From a practical standpoint, the similarities in idiotypic structure among cold agglutinin heavy chain variable regions and the ability to generate anti-idiotypic antibodies specific for these structures suggest potential therapeutic applications for these reagents in down-regulating the production of autoantibodies or inhibiting the growth of an underlying malignancy [41].

In this same study by Silberstein et al described above, the expressed variable region genes were also examined for the number and pattern of somatic mutations in order to evaluate the potential role for antigen-mediated selection. It was determined that both the V_H and V_L genes encoding the anti-i antibody were identical to germline sequences, whereas numerous base pair differences were noted in the anti-I response (Figure 3). Compared to its most likely germline precursor, $V_H4.21$, the V_H gene encoding anti-I had only three amino acid differences, two located in framework regions and one in a complementarity determining region (CDR or region of antigen contact). In contrast, the V_L sequence of anti-I had a relatively high number of amino acid substitutions (relative to the total number of silent mutations) when compared to its likely precursor germline sequence. These amino acid changes resulted from a non-random distribution of replacement mutations in the CDR's. Taken together, these results provide evidence that positive selection by antigen led to the accumulation of amino acid substitutions in the light chain of the anti-I antibody studied. If this proves to be a universal feature of anti-I cold agglutinins, it may represent a consequence of differential regulation of the immune responses to the related I/i antigens. Perhaps the high expression of i antigen on fetal red cells mediates tolerance to i, either by clonal anergy or deletion of B cells with anti-i specificity. The expression of I antigen occurs much later in development so immunologic tolerance for I may differ from that for i.

With respect to B-cell function, it was recently demonstrated that IgM autoantibodies found in patients with Wiskott-Aldrich syndrome, a rare X-linked recessive immunodeficiency disorder, often have anti-i cold agglutinin activity and are encoded by the $V_H4.21$ gene in germline configuration. Additionally, it was shown that these anti-i CA recognize a subpopulation of human B-cells present early in B-cell ontogeny [27]. This same early B-cell population is also recognized by anti-i cold agglutinins isolated from patients with cold agglutinin disease [42]. These studies suggest the possibility that the gene product encoded by this highly-conserved germline $V_H4.21$ gene may play a physiologic role in B-cell development and/or differentiation.

Table 4. [42] Structural and idiotypic diversity of CA assessed with V_H4 sequence dependent anti-peptide antibodies and with the 9G4 reagent.

| Name | L Chain | | H Chain | | | |
	Sequence	V_L	V_H	V_H4-FR1a	VH4-HV2a	9GA
Anti-I CA						
1. JOH(AJ)	+	$V_\kappa III$	V_H4	NT	NT	NT
2. BAT	-	$V_\kappa III$	V_H4	-	+	+
3. BON	-	$V_\kappa III$	V_H4	+	+	+
4. HIG	-	$V_\kappa III$	V_H4	+	+	+
5. KAU	+	$V_\kappa III$	V_H4	NT	NT	NT
6. LEA	-	$V_\kappa III$	V_H4	+	+	+
7. NAD	-	$V_\kappa III$	V_H4	+	-	+
8. ODO	-	$V_\kappa III$	V_H4	+	+	+
9. PER	-	$V_\kappa III$	V_H4	-	+	+
10. POS	-	$V_\kappa III$	V_H4	NT	NT	NT
11. RIC	-	$V_\kappa III$	V_H4	+	+	+
12. SOC	-	$V_\kappa III$	V_H4	NT	NT	NT
13. TRI	-	$V_\kappa III$	V_H4	NT	NT	NT
14. VOG	+	$V_\kappa III$	V_H4	+	+	+
		($V_\kappa III$) 13/14	14/14	7/9	8/9	9/9
Anti-i CA						
1. CAP	+	$V_\kappa I$	V_H4	+	+	+
2. FA	-	$V_\kappa III$	V_H4	+	-	+
3. GRO	-	$V_\kappa III$	V_H4	+	+	+
4. LG	-	$V_\kappa I$	V_H4	+	+	+
5. PER*	-	V_κ†	V_H4	-	+	-
6. PRO	-	$V_\kappa III$	V_H4	+	-	+
7. RID	-	$V_\kappa III$	V_H4	+	+	+
8. TM	-	V_κ†/V_λ	V_H4	+	+	+
		($V_\kappa III$) 4/8	8/8	7/8	6/8	7/8
Non-CA						
1. LES	+	$V_\kappa III$	V_H4	+	-	+
2. SA	-	V_λ	V_H4	-	-	+
3. GRE	-	$V_\kappa III$	V_H4	-	-	+

V_H and V_L family was determined by immunoblot analyses with antipeptide antibodies. The presence of the V_H4-FR1a and V_H4-HV2a sequences was determined by immunoblot analyses after preincubation of antipeptide antibodies with synthetic peptides with related sequences to improve fine binding specificity.

* Indicates the presence of two different patients with the same initials PER.

† These $_\kappa$ light chains were unreactive with the V_κ family-specific reagents

Pr2-blood Group Specific Autoantibodies

Most anti-Pr antibodies are monoclonal IgM,κ, which induce autoimmune hemolysis. Monoclonal anti-Pr antibodies of the IgA isotype have been rarely reported; these antibodies do not fix complement, are not hemolytic and are not associated with lymphoproliferative disorders.

In contrast to the structural uniformity of anti-I cold agglutinins discussed above, substantial structural differences occur in the two anti-Pr_2 cold agglutinins that have been sequenced [3, 48]. In the case of well characterized anti-Pr2 specific lymphoma, the heavy chain variable region was 88% homologous to a V_H1 germline gene while the light chain variable region was 97% homologous to a VkIII germline gene[3 49]. Anti-idiotypic antibodies raised against the anti-Pr_2 cold agglutinin from this patient did not show cross-reactivities with red cell autoantibodies from other individuals having anti-Pr or different specificities and were thus unique to this patient's anti-Pr_2 autoantibody [50]. The structural basis for this "private" idiotype was recently investigated by the establishment of Id negative mutant clones, which were derived from the parental Id positive clone. Through sequence analyses of V_H and V_L regions, it was demonstrated that a single VH CDR3 mutation (cysteine $\rightarrow$ tyrosine) altered both idiotype and specificity (anti Pr_2) [51].

THE B-CELL ORIGIN OF NATURAL ANTI-I/i CA

The issue of what makes an autoantibody "clinically-significant", i.e. pathologic, is another area worthy of investigation. Despite attempts to correlate antibody subclass, complement-fixing ability, or antibody titer with the presence or absence of hemolysis, it remains unclear what structural properties distinguish "benign" autoantibodies from hemolytic ones [52-54]. Furthermore, although the spectrum of pathogenic anti-I/i is now known to comprise premalignant clonal B-cell expansions on one end and true malignant lymphomas on the other, it is not knowm if these clonal disorders have a similar B-cell origin as the natural/benign anti-I autoantibodies. In this regard, recently our laboratory has isolated EBV-transformed anti-I-producing B-cell clones from two normal individuals with a 1:4 CA titer in their serum. Interestingly, unlike all of the pathologic anti-I/i autoantibodies we have studied thus far which have shown V_H-gene restriction to a $V_H4.21$ variable gene segment, the heavy chains of these benign autoantibodies are encoded by different gene segments from the V_H3 family. In addition, these natural/benign CA do not express the 9G4 idiotope.

In summary, the present findings suggest a diverse B-cell origin of natural CA which infers that B cell clones secreting natural CA do not necessarily represent precursors of the monoclonal B-cell expansions secreting pathogenic CA. As a result, the divergent physiologic behavior between natural and pathogenic anti-i/I CA may not simply be ascribed to quantitative differences in serum concentration. Furthermore, the mechanisms by which natural and pathogenic CA are triggered, may also differ. In view of the demonstrated presence of the I antigen on L.Monocytogenes[55], we speculate that natural anti-I autoantibodies are stimulated by microbial antigens, similar to naturally occurring anti-A and anti-B isoagglutinins.

REFERENCES

1. L. E. Silberstein, G. A. Robertson, A. C. Hannam Harris, L. Moreau, E. Besa and P. C. Nowell, Etiologic Aspects of Cold Agglutinin Disease: Evidence for Cytogenetically Defined Clones of Lymphoid Cells and the Demonstration That an Anti-Pr Cold Autoantibody is Derived from a Chromosomally Aberrant B Cell Clone, *Blood,* 67:1705 (1986)
2. L. E. Silberstein, J. Goldman, J. A. Kant and S. L. Spitalnik, Comparative biochemical and genetic characterization of clonally related human B-cell lines secreting pathogenic anti-Pr_2 cold agglutinins., *Arch. Biochem. Biophys.,* 264:244 (1988)
3. L. E. Silberstein, S. Litwin and C. E. Carmack, Relationship of variable region genes expressed by a human B-cell lymphoma secreting pathologic anti-Pr_2 erythrocyte autoantibodies, *J. Exp. Med.,* 169:1631 (1989)

4. F. A. Janney, L. T. Lee and C. Howe, Cold hemagglutinin cross-reactivity with Mycoplasma pneumoniae, *Infect. Immunol.,* 22:29 (1978)
5. R. Rosenfield, R. Schmidt, R. Calvo and M. McGinniss, Anti-i, a frequent cold agglutinin in infectious monoucleosis, *Vox Sang,* 10:631 (1965)
6. S. Worlledge and C. Rousso, Studies on the serology of PCH with special reference to its relationship with the P blood group sytem., *Vox Sang,* 10:293 (1965)
7. G. Harley, Notes of two cases of intermittent haematuria with remarks upon their pathology and treatment, *Lancet,* 1:568 (1865)
8. A. H. Hassall, On intermittant or winter haematuria, *Lancet,* 11:368 (1865)
9. O. Rosenbach, Zur lehre von der peroidischen Hamoglobinurie, *Dtsch Med Worchenschr.,* 5:613 (1879)
10. P. Ehrlich, Uber paroxysmate Hamoglobinurie, *Z. Klin. Med,* 3:383 (1981)
11. A. Weiner, L. Unger and L. Cohen, Type-specific cold autoantibodies as a cause of acquired hemolytic anemia and hemolytic transfusion reactions. Biologic test with bovine red cells., *Ann. Intern. Med.,* 44:221 (1956)
12. J. Dacie, The Haemolytic Anaemias, Congential and Acquired., Churchill Livingston, London (1962)
13. R. A. Childs, T. Feizi and M. Fukuda, Blood group I activity associated with band 3, the major intrinsic membrane protein of human erythrocytes, *J. of Biol. Chem.,* 173:333 (1978)
14. M. Fukuda, M. N. Fukuda and S. Hakomori, The developmental change and genetic defect in carbohydrate structures of band 3 glycoprotein of human erythrocyte membranes., *J. Biol. Chem.,* 254:3700 (1979)
15. T. Feizi, K. Childs, K. Watanabe and S. Hakomori, Three types of blood group specificities among monclonal anti-autoantibodies revealed by analogues of a branched erythrocyte glycoliped., *J. Exp. Med.,* 149:975 (1979)
16. H. Niemann, K. Watanabe and S. Hakomori, Blood group i and I activities of "lacto-N-nor-hexaosylceramide" and its analogues;the structural requirements for i-specificities., *Biochem. Biophys. Res. Commun.,* 81:1286 (1976)
17. K. Watanabe and S. Hakamori, Status of blood group carbohydrate chains in autogenesis and oncogensis., *J. Biol. Chem.,* 254:3221 (1976)
18. W. L. Marsh, Anti II: a cold antibody defining the I's relationship in human red cells, *Br. J. Hematol.,* (1961)
19. S. Hakomori, Blood group ABH and Li antigens of human erythrocytes: Chemistry polymorphism, and their developmental change., *Semin. Hematol.,* 18:39 (1981)
20. W. Pruzanski and K. Shumak, Biologic activity of cold-reacting antibodies. Parts I and II, *N. Eng. J. Med,* 297:538 (1977)
21. R. A. Dunstan, M. B. Simpson and M. J. Borowitz, Heterogenous distribution of antigens on human platelets demonstrated by fluorescence flow cytometry, *Br. J. Haemat.,* 61:603 (1985)
22. A. Wiener, J. Moore-Janowski, E. Gordon and J. Davis, The blood factors I and i in primates including man in lower species., *Am. J. Phys. Anthropol.,* 23:389 (1965)
23. R. A. Childs A. Kapadia, T. Feizi, R. Schauer ed, Georg Thieme, Stuttgart (1979)
24. M. B. Omary, I. Trowbridge and H. Battifora, Human homologue of mouse T200 glycoprotein., *J. Exp. Med.,* 152:842 (1980)
25. R. A. Childs and T. Feizi, Differences in carbohydrate moieties of high molecular weight glycoproteins of human lymphocytes of T and B origins revealed by monoclonal autoantibodies with anti-I specificities, *Biochem Biophys. Res. Commun.,* 102:1158 (1981)
26. R. A. Childs, P. Dalchau and P. Seudder, Evidence for the occurence of O-glycosidically linked oligosaccharides of poly-N-acetyllactosamine type on the human leucocyte common antigen, *Biochem. Biophys. Res. Commun.,* 110:424 (1983)
27. C. Grillot-Courvalin, J. Brouet, F. Piller, L. Z. Rassenti, S. La Baume, G. J. Silverman, L. Silberstein and T. J. Kipps, The anti-B cell autoantibodies from Wiskott-Aldrich sysdrome recognizes i blood group specificity on normal human B cells., *Eur. J. Immunol.,* 22:1781 (1992)
28. R. Thomas, The leukocyte common antigen family, *Annual Review of Immunology.,* 7:339 (1989)

29. Y. Yamashita, I. Yasuyuki and O. Toshiaki, Poly[N-acetyllactosamine]-type sugar chain in CD45 antigens of adnormal T cells of Ipr mice are different from those of normal T cells and B cells., *Molec. Immunol.*, 26:905 (1989)
30. L. Rose, A. Ginsberg and T. Rothstein, Selective loss of a subset of T helper cells in active multiple sclerosis, *Proc. Nat. Acad. Sci*, 82:7389 (1985)
31. M. Sanders, M. Makgoba and S. Shaw, *Immunology Today*, 9:195 (1988)
32. W. Ebert, J. Fey and H. Gartner, Isolation and partial charactization of the Pr autoantigen determinants, *Molec. Immunol.*, 16:413 (1979)
33. D. J. Anstee, The blood group MNSs-active sialoglycoprotein., *Semin. Hematol.*, 18:13 (1981)
34. K. Uemura, D. Roelcke, N. Yoshitakia and T. Feizi, The reactivity of human erythrocyte antoantibodies anti-Pr_2, and Gd, Fl, and Sa with gangliosides in a chromatogram binding assay., *Biochem J.*, 219:865 (1984)
35. D. Anstee, Blood group MNSs-active sialoglycoproteins of the human erythrocyte membrane, *in* : "Immunobiology of the Erythrocyte"., S. Sandler, J. Nusbacher and M. Schanfield ed, Alan R. Liss, New York (1980)
36. W. Pruzanski, D. Roelcke, M. Armstrong and M. Manly, Pr and Gd antigens on human B and T lymphocytes and phagocytes, *Clin. Immunol. Immunopathol.*, 15:631 (1980)
37. P. Mollison, C. Engelfriet and M. Contreras, Blood Transfusion in Clinical Medicine., Blackwell Scientific, Oxford (1988)
38. D. Roelcke, W. Riesen, H. Geisen and W. Ebert, Serological identification of the new cold agglutnin specificity anti Gd., *Vox Sang*, 33:304 (1977)
39. M. Naiki and D. Marcus, An immunochemcial study of the human blood group P_1, P and P_k glycosphingolipid antigens, *Biochem.*, 14:4837 (1975)
40. R. Williams, H. Kunkel and J. Capra, Autigenic specificites related to the cold agglutinin activity of gamma M. globulins, *Science*, 161:379 (1968)
41. F. Stevenson, G. Smith, J. North, T. Hamlin and M. Glennie, Idenificaton of normal B-cell counterparts of neoplastic cells which secrete cold agglutinins of anti-I and anti-i specficity., *Br. J. Haematol.*, 72:9 (1989)
42. L. E. Silberstein, L. C. Jefferies, J. Goldman, D. F. Friedman, J. S. Moore, P. C. Nowell, D. Roelcke, W. Pruzanski, J. Roudier and G. J. Silverman, Variable region gene analysis of pathologic human autoantibodies to the related i and I red blood cell antigens, *Blood*, 73:2372 (1991)
43. V. Pascual, K. Vistor, D. Lelsz, M. B. Spellerberg, T. J. Hamblin, K. M. Thompson, I. Randem, J. B. Natvig and J. D. Capra, Nucleotide sequence analysis of the V regions of two IgM cold agglutinins: evidence that the V_H4-21 gen segment is responsiblile for the major cross-reactive idiotype, *J. of Immunol.*, 146:4385 (1991)
44. J. Gergely A.C. Wang and H.H. Fudenberg, Chemical Analyses of Variable Regions of Heavy and Light Chains of Cold Agglutinins, *Vox Sang*, 24432-440:(1973)
45. G. Silverman, F. Goni and J. Fernandez, Distinct patterns of heavy chain variable region subgroup use by human monoclonal antoantibodies of different specificity, *J. Exp. Med.*, 168:2361 (1988)
46. J. Leoni, J. Ghiso, F. Goni and B. Frangione, The primary structure of the Fab fragment of protein KAU, a monoclonal immunoglobulin M cold agglutinin, *J. Biol. Chem*, 266:2836 (1991)
47. L. Jefferies, G. Silverman and L. Silberstein, Idiotypic heterogeneity of V_kIII-restricted cold aggultin light chains., *Clin. Immun. Immunopath.*, in press:(1992)
48. A. C. Wang H.H. Fudenberg, J.V. Wells, D. Roelcke., A new subgroup of the kappa chain variable region associated with anti-Pr cold agglutinins., *Nature*, 243:126 (1973)
49. D. F. Friedman, E. A. Cho, J. Goldman, C. E. Carmack, E. C. Besa, R. R. Hardy and L. E. Silberstein, The role of clonal selection in the pathogenesis of an autoreactive B-cell lymphoma., *J. Exp. Med.*, 174:525 (1991)
50. L. C. Jefferies, F. K. Stevenson, J. Goldman, I. M. Bennett, S. L. Spitalnik and L. E. Silberstein, Anti-idiotypic antibodies specific for a pathological anti-Pr_2 cold agglutinin, *Transfusion*, 30:495 (1990)

51. L. S. Reidl, D. F. Friedman, J. Goldman, R. R. Hardy, L. C. Jefferies and L. E. Silberstein, Structural Basis of a Conserved Idiotope Expressed by an Autoreactive Human B cell Lymphoma: Evidence that a V_H CDR3 Mutation Alters Idiotypy and Specificity, *J. of Immunol.*, 147:3623 (1991)

52. W. Rosse and J. Adams, The variability of hemolysis in the cold agglutinin sysdrome., *Blood*, 56:(1980)

53. A. Schreiber, B. Herskovitz and M. Goldwein, Low-titer cold hemagglutin disease mechanism of hemolysis and reponse to corticosteroids., *N. Engl. J. Med.*, 296:1490 (1977)

54. G. Garratty, The significance of IgG on the red cell surface, *Transfus. Med. Rev.*, 1:47 (1987)

55. Costea N, Yakulis V and H. P., Experimental production of cold agglutinins in rabbits, *Blood*, 26:323 (1965)

INITIATION OF AUTOIMMUNE TYPE 1 DIABETES AND MOLECULAR CLONING OF A GENE ENCODING FOR ISLET CELL-SPECIFIC 37KD AUTOANTIGEN

Hee Sook Jun and Ji-Won Yoon

Julia McFarlane Diabetes Research Centre and Dept. of Microbiology
and Infectious Diseases, Faculty of Medicine, University
of Calgary, 3330 Hospital Drive NW, Calgary, Alberta, T2N 4N1

INTRODUCTION

Diabetes mellitus and its complications are now considered to be the third leading cause of death in North America (Canada and the United States), trailing only cancer and cardiovascular disease. There are two major forms of diabetes mellitus. One is type I diabetes, which is also known as insulin-dependent diabetes mellitus (IDDM), and the other is type II diabetes, which is also known as noninsulin-dependent diabetes mellitus (NIDDM). Most patients with IDDM have a common pathological picture: the nearly total disappearance of insulin-producing pancreatic beta cells, which results in hyperglycemia (Yoon, 1989; Yoon, 1990; Yoon, 1991).

Considerable evidence shows that most insulin-dependent diabetes mellitus (IDDM) is the consequence of progressive, autoimmune beta cell destruction during an asymptomatic period that often extends over many years. Humoral immunity is thought to be involved, as circulating islet cell autoantibodies (Yoon et al., 1988b) can be detected during the prediabetic period. These autoantibodies include: islet cell cytoplasmic antibodies, islet cell surface antibodies, autoantibodies against islet cell-specific 64 kd and 37 kd autoantigens (64 kd and 37 kd islet cell autoantibodies), and insulin autoantibodies (Baekkeskov et al., 1982; Rossini et al., 1985; Drell and Notkins, 1987; Ko et al., 1991). Cell-mediated immunity is also thought to be involved, as autopsy specimens of pancreata from acutely diabetic patients show lymphocytic infiltration (Gepts, 1984). The hypothesis that IDDM is an autoimmune disease has been considerably strengthened by studying diabetic animal models such as the BioBreeding (BB) rat (Marliss et al., 1982; Marliss, 1983) and the nonobese diabetic (NOD) mouse (Katoka et al., 1983). Both of these animals spontaneously develop IDDM, and their syndromes share many pathological features with IDDM in humans. Although there is evidence that T-cell mediated autoimmune mechanisms are involved in the development of IDDM in the BB rat and NOD mouse (Like et al., 1982; Laupacis et al., 1983; Dyrberg, 1986), the initial event that causes destruction of the beta cells remains unknown. Over the last several years, we have focused our research on identifying this initial event. In this brief review, we will summarize our results, as well as those of others, that contribute to our understanding of the initiation of autoimmune IDDM.

ROLE OF T CELLS IN THE PATHOGENESIS OF IDDM

In the BB rat, the diabetic syndrome results from the destruction of pancreatic beta cells by cell-mediated and/or humoral immune responses (Like et al., 1979; 1982; Laupacis et al., 1983; Koevary et al., 1983; Dyrberg, 1986). Several observations indicate that cell-mediated immunity plays a role in the pathogenesis of the disease: 1) the development of diabetes was prevented either by performing a neonatal thymectomy (Like et al., 1982) or by suppressing the immune system (Like et al., 1979; Laupacis et al., 1983); 2) BB rats treated with monoclonal antibodies (OX19) against T-cell antigens did not develop insulitis or thyroiditis (Like et al., 1986); and 3) diabetes was adoptively transferred when concanavalin-A-stimulated spleen cells, derived from acutely diabetic BB rats, were injected into nondiabetic BB or Wistar-Furth rats (Koevary et al., 1983). Similarly, substantial evidence indicates that a T cell-dependent autoimmune mechanism is responsible for the development of IDDM in NOD mice (Katoka et al., 1983; Bach, 1988). More specifically, $CD4^+$ T cells have been shown to be necessary both for the development of insulitis (Koike et al., 1987; Shizuru et al., 1988) and for the recurrence of the disease in islet allografts (Wang et al., 1987). Furthermore, both $CD4^+$ and $CD8^+$ T cell subsets are required for the transfer of diabetes in NOD mice both in vivo (Bendelic et al., 1987; Miller et al.,1988; Haskins et al., 1989; Haskins and McDuffie, 1990; Nakano et al., 1991) and in vitro (Nagata et al., 1989).

T cells not only induce diabetes in NOD mice, but certain T cell populations, especially $CD4^+$ T cell clones, can prevent the development of the disease. For example, a $CD4^+$ islet-specific T cell clone derived from NOD mice accelerated the autoimmune process, leading to diabetes in young NOD mice (Haskins and McDuffie, 1990). In contrast, the injection of autoreactive T cells, isolated as a T cell line from NOD islets, into young nondiabetic NOD mice profoundly inhibited the development of diabetes and nearly eliminated insulitis, indicating the presence of regulatory T cells in the islet tissue of the NOD mouse (Reich et al., 1989a). Therefore, the islets of the diabetic NOD mouse apparently contain both effector cells and cells capable of inhibiting these effector cells. Thus, it is believed that this immunoregulatory balance can be shifted in favour of either suppressing effector cells, which impairs beta cell destruction, or enhancing effector cells, which leads to beta cell destruction (Reich et al., 1989b).

ROLE OF MACROPHAGES IN THE INITIATION OF BETA CELL DESTRUCTION

Although it is clear that T cells are involved in the progression of autoimmune IDDM, it is not clear whether they are primary causative agents or are secondary to some other initial event. An examination of the time course of insulitis with respect to the types of infiltrating lymphocytes is likely to provide clues regarding the initial events of autoimmune destruction. As earlier studies of this type were equivocal (Like et al., 1983; Dean et al., 1985), we used various highly specific monoclonal antibodies to distinguish the types of immunocytes present in the islets of BB rats at different stages of insulitis. These antibodies included: MRC OX19 (a marker for pan-T lymphocytes), W3/25 (a marker for helper T cells and macrophages), MRC OX8 (a marker for cytotoxic/suppressor lymphocytes and natural killer cells), MRC OX12 (a marker for B lymphocytes), and MRC OX41 (a marker for macrophages, granulocytes, and dendritic cells). We found that the major population of infiltrating cells during the early stages of insulitis is macrophages/dendritic cells (Lee et al., 1988a). These cells precede invasion by T lymphocytes, natural killer cells, and B lymphocytes (Table 1). This finding was supported by electron microscopic observations (Kolb et al., 1986), and further studies have confirmed that macrophages play an important role in the pathogenesis of IDDM (Walker et al., 1988; Yoon et al., 1988a; Voorbij et al., 1989).

Intraperitoneal administration of silica, which is known to be toxic to macrophages, to young BB rats (Table 2) and NOD mice completely prevented the development of insulitis and diabetes (Oschilewshki et al., 1985; Lee et al., 1988b; Lee et al., 1988c). Further studies showed that this preventative effect was probably due to a decrease in macrophage-dependent T lymphocytes (Lee et al., 1988c; Amano and Yoon, 1990). However, the precise role of macrophages in the autoimmune destruction of beta cells in the BB rat remains unclear. Macrophages are antigen presenting cells (APCs), and the cytotoxic action of the activated macrophage is generally assumed not to be specific for any given antigen or cell. We hypothesize that prior to the development of insulitis, there might be antigenic changes on the beta cells that can be recognized as non-self and are processed and presented by APCs.

Table 1. Distribution of lymphoid cells at different stages of insulitis.

Islets observed[+]		Monoclonal Antibodies				
Insulitis[*] (n)		MRC OX41	W3/25	MRC OX19	MR COX8	MRC OX12
Early	25	+ + + +	+ + + +	0	0	0
Intermediate	18	+ + +	+ + +	+	+	+ +
Late	23	+ +	+ +	+ +	+ +	+ +

Stages of insulitis:[*] **Early:** defined as an accumulation of mononuclear cells at the periphery of or just within the islets; **Intermediate:** defined as an infiltration of mononuclear cells in the center of islets but with relatively good preservation of the islet architecture; **Late:** defined as an increased infiltration by mononuclear cells and distortion of islet architecture.

[+] **Serially sectioned islets were stained with different monoclonal antibodies for identification of specific cell types.** MRC OX41 reacts with macrophages, dendritic cells, and granulocytes but not with lymphocytes. W3/25 reacts with helper T-lymphocytes and macrophages. MRC OX19 reacts with all thymocytes, T-lymphocytes and <2% of B-lymphocytes. MRC OX8 reacts with cytotoxic or suppressor T-lymphocytes and majority of natural killer cells. MRC OX12 reacts with rat Ig kappa chain (B-lymphocytes). The percentage of stained cells was expressed on 0 to 4+ scale: 0 = no staining; 1+ = <15% of lymphoid cells stained; 2+ = 15-20% of lymphoid cells stained; 3+ = 50-90% of lymphoid cells stained; 4+ = almost all lymphoid cells stained. The total number of animals used for this experiment was 20.

DELAYED EXPRESSION OF BETA CELL-SPECIFIC AUTOANTIGEN

An antigenic change that could lead to autoimmune responses in BB rats is thought to be the delayed expression of a particular beta cell-specific antigen, resulting in a break in normal self-tolerance (Adams et al., 1987). To test this possibility, we transplanted islets from neonatal BB rats and islets from macrophage-depleted, insulitis-free adult BB rats into acutely diabetic BB rats (Ihm et al., 1991). Islet grafts from neonatal BB rats remained intact without insulitis, while islet grafts obtained from adult BB rats were destroyed in diabetic recipients (Table 3). These results suggest that neonatal BB rat beta cells are antigenically distinct from adult BB rat beta cells, with respect to their recognition by immunological effectors, and support the hypothesis that a beta cell-specific autoantigen is absent in the neonatal BB rat beta cells. The absence of such an antigen in neonates would result in the absence of self-tolerance to that antigen if it is expressed later in life. In transgenic mice, Adams et al. (1987) showed that the delayed expression of a transgene resulted in a failure to establish self-tolerance, and consequently produced autoimmune lesions in pancreatic islets. In contrast, mice that expressed the transgene early in life were tolerant. Considering our present results and previous information on the initial role of macrophages/dendritic cells in the development of insulitis in BB rats, we hypothesize that the delayed expression of a putative beta cell-specific autoantigen precedes insulitis, and that presentation of the autoantigen by APCs constitutes the initial step in the development of insulitis. However, the amplification of immune responses by both T lymphocytes and NK cells is required for the clinical expression of overt diabetes in BB rats.

More specifically, among the key roles played by macrophages in the immune process is the presentation of processed antigen to helper T cells in conjunction with the MHC Class II molecules present on the surface of macrophages (Unaue, 1984), which results in T helper cell activation (Unaue and Allen, 1987). Thus, we hypothesize that the presentation of a beta cell-specific autoantigen by macrophages and the recognition of the antigen by $CD4^+$ (helper) T cells might be the initial steps in the development of autoimmune diabetes (Lee et al., 1988c; Nagata and Yoon, 1992). Helper T lymphocytes in turn may secrete mediators to recruit more macrophages, which then activate inflammatory T lymphocytes and cytotoxic T lymphocytes (Kolb, 1990). The secretory products of the macrophages and T lymphocytes, such as interleukin-1, tumor necrosis factor-α, lymphotoxin, interferon-γ, may act synergistically in the destruction of beta cells, leading to the clinical expression of diabetes.

Table 2. The effect of silica administration on the incidence of diabetes, lymphocytic thyroiditis and insulitis in BB rats

	BB rats	
	Untreated	**Silica treated**
Diabetes	23/33 (70%)	1/32^a (3%)
Lymphocytic thyroiditis	13/33 (39%)	0/32^a (0%)
Insulitis	33/33 (100%)	1/32^a (3%)

^{a} $p<0.005$ compared with untreated animals.

210

IDENTIFICATION OF THE 37 KD BETA CELL-SPECIFIC AUTOANTIGEN

It is not known what particular beta cell-specific autoantigen(s) may be absent or insufficiently expressed in neonatal BB rats but is expressed later in life. The most often considered target antigens for T lymphocytes is a 64 kd islet cell antigen (Kolb, 1990). Over the past year, we have studied the expression of the islet cell-specific 64 kd autoantigen, and found that it is expressed on both neonatal and adult BB rat islets. In contrast, an islet cell-specific 37 kd autoantigen (which was previously identified as 38 kd autoantigen, Ko et al., 1991) was not expressed early in the life of the BB rat (Ko and Yoon, 1991), but was expressed at about 30 days of age. It is not yet known whether this 37 kd autoantigen is truly involved in the triggering of beta cell-specific autoimmune disease (IDDM), but it is the only delayed-expressed antigen found in BB rats to date, and is therefore a leading candidate antigen for an important role in the triggering of the autoimmune destruction of beta cells. Insufficient or delayed expression of this beta cell antigen may lead to the initial infiltration of macrophages into the islet. Interactions of macrophages with beta cells, e.g., beta cell breakdown, antigen uptake, and antigen processing and presentation would lead to the activation of T-helper cells, followed by the involvement of both cellular and humoral immune responses against pancreatic beta cells.

Table 3. Transplantation of neonatal and adult islets to diabetic BB rats

Number of Graft	Islets	Recipient	* Histological score
Neonatal Islets	300	1	0
		2	0
		3	1
		4	0
		5	0
		6	1
		7	1
Adult Islets	300	1	4
		2	3
		3	2
		4	1
		5	1
		6	3
		7	4
	1000	8	3
		9	3

* 0, intact islets with no mononuclear cell (MNC) infiltration; 1, intact islets with minor focal MNC infiltration; 2, intact islets with extensive MNC infiltration; 3, damaged islets with extensive MNC infiltration; 4, end stage of islets with few β-cells and some MNCs.

Recently we identified, for the first time, an anti-37 kd autoantibody against a rat islet cell 37 kd protein in the sera of BB rats by both immunoprecipitation and differential Western blotting. We found that the cumulative incidence of anti-37 kd antibody in the BB rats at the age of 90 days was approximately 92% (Fig. 1) and 11 of 12 BB rats in which the antibody was detected eventually developed IDDM (Ko et al., 1991).

The antibody was detectable at about 30 days of age, but was not detectable within a week after the onset of diabetes. In contrast, the antibody remained detectable in the sera of BB rats that had received silica treatment to prevent the destruction of beta cells (Fig. 2). In a previous study (Baekkeskov et al., 1982), autoantibodies in the sera from 4 of 4 tested new-onset IDDM patients immunoprecipitated an islet cell protein of approximately 38 kd prepared from HLA-DR3-positive donor islets. However, whether this 38 kd human antigen is homologous to the 37 kd antigen that we found in rat islet preparations has not been determined. In our recent study, a 37 kd protein band was detected when rat islet cell lysates were reacted with sera from newly diagnosed IDDM

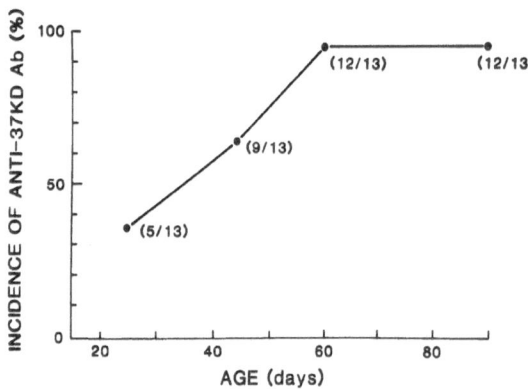

Fig. 1. **Cumulative incidence of autoantibodies to a 37 kilodalton (kd) islet cell protein in DP-BB rats of different ages.** Thirteen DP-BB rats were subjected to follow-up studies to determine the incidence of anti-37 kd islet cell autoantibodies by immunoprecipitation. At around 30, 45, and 60 days of age, 5, 9, and 12 rats respectively had anti-37 kd islet cell autoantibodies.

patients, suggesting that there may be a common antigenic determinant(s) between the rat and human 37 kd proteins. Recently, Roep et al. (1990) identified a 38 kd antigen from a rat beta cell line (RINm5F) that is recognized by a T cell clone established from a newly diagnosed IDDM patient. Subcellular fractionation studies using RINm5F cells indicated that the antigenic determinant recognized by the T cell clone is an integral membrane component of the insulin secretory granule. As the granular membrane of 38 kd proteins is transiently exposed on the cell surface during exocytosis, its accessibility to components of the immune system is thought to be a function of the secretory activity of pancreatic beta cells (Roep et al., 1990). The relationship between a 38 kd integral membrane component of the insulin secretory granule and the 37 kd islet cell-specific autoantigen localized in plasma membrane is not known.

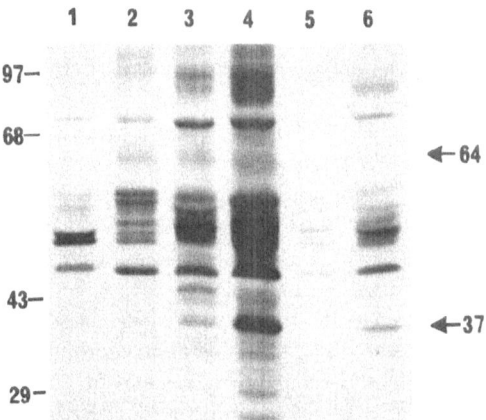

Fig. 2. Autoradiograph of sodium dodecyl sulphate-polyacrylamide gel electrophoresis of immunoprecipitates of RINm5F cell proteins with sera from control Wistar Furth (WF) rats and different ages of BB rats. Metabolically labelled RINm5F cell lysates were pre-cleared with pooled sera from ten WF rats (30-100 days old), and reacted with individual test sera. Equal amounts (20 μl) of control sera (lane 1) and test sera (lane 2 through lane 6) were used for immunoprecipitation. Similarly, an equal amount (80 μl) of bound immune complex was applied to each lane. Molecular weight markers are indicated at left. Antigens of 37 kilodalton (kd) and 64 kd are indicated by an arrow with appropriate number at right. Lane 1: control WF (110-day-old) sera failed to immunoprecipitate a 37 kd protein from RINm5F cell extract. Lane 2: neonate BB rat sera (at 5 days of age) also failed to immunoprecipitate a 37 kd protein. Lane 3: 37 kd protein recognized by sera from 30-day-old BB rats. Lane 4: 37 kd protein recognized by sera from 75-day-old BB rat. Lane 5: 110-day-old diabetic BB rat sera (7 days after the onset of diabetes) failed to immunoprecipitate a 37 kd protein. Lane 6: 37 kd protein recognized by sera from 140-day-old BB rat which was treated with silica for the preservation of beta cells. Control sera (lane 1) and 110-day-old diabetic BB rat sera (lane 5), which does not contain anti-37 kd islet cell antibody, appeared to react less strongly with RINm5F cell proteins. All sera (lane 1 through lane 6) failed to immunoprecipitate 64 kd protein, as the protein extracts were prepared from RINm5F cells.

PURIFICATION OF THE 37 KD AUTOANTIGEN PROTEIN AND SEQUENCING OF N-TERMINAL AMINO ACIDS

For years, several groups have been trying to identify the antigens on the pancreatic beta cells that stimulate the immune system to produce autoantibodies against them. Finding these target antigens, purifying them and making them in quantity would be a boon and a major step toward massive screening, therapeutic and preventive purposes. Practically, the preparation of a sufficient amount of islets from small animals for the purification of a putative autoantigen is difficult. Fortunately, the anti-37 kd antibody that reacted with a rat islet cell protein also reacted with a protein from a rat beta cell line (RINm5F). Using BB rat sera containing the anti-37 kd antibody, a 37 kd protein was purified from RINm5F cells by the preparative immunoprecipitation method. As our previous work had determined that 37 kd islet cell-specific autoantigen was located in the plasma membrane fraction (Ko, 1992), only the plasma membrane fraction was isolated from RINm5F cells by differential centrifugation. The solubilized plasma membrane fraction was immunoprecipitated with affinity purified BB rat sera containing the anti-37 kd antibody, and the samples from the preparative immunoprecipitation were run on 10% SDS-PAGE. After transblotting to a polyvinylidene difluoride membrane, the band containing the 37 kd protein was cut out and directly sequenced (Matsudaira, 1987). We obtained the N-terminal 41 amino acid sequence of the 37 kd autoantigen protein. This amino acid sequence was used for making a probe to clone the 37 kd gene in a RINm5F cell cDNA library.

CLONING AND SEQUENCING OF THE 37 KD AUTOANTIGEN GENE

To study the molecular nature of the 37 kd islet cell-specific autoantigen and to test the pathogenic role of this antigen, we tried to clone the 37 kd autoantigen gene from RINm5F cells (Fig. 3). Poly (A$^+$) mRNA from RINm5F cells, whose 37 kd gene expression was induced by incubation in 20mM glucose, was purified and double stranded cDNAs were prepared (Gubler and Hoffman, 1983). A RINm5F cell cDNA library was constructed in lambda ZAP II vector and screened with radiolabelled synthetic oligonucleotides deduced from the N-terminal amino acid sequence of the 37 kd protein. The selected positive clone was subcloned by an in vivo excision protocol as described elsewhere (Short et al., 1988) and sequenced by dideoxy sequencing method. The gene encoding the 37 kd autoantigen protein, which has a full-length cDNA of 1.3 kb, was cloned and the cDNA was shown to contain an open reading frame of 333 amino acids. The cloned 37 kd gene can be used as a probe to clone the 37kd gene from rat islet cells or human islet cells and can be expressed in procaryotes (E. coli, Bacillus, etc.) or eucaryotes (mammalian cells, baculovirus, vaccinia virus, yeast, etc.) for functional studies of the 37 kd autoantigen.

SUMMARY AND CONCLUSION

Insulin-dependent diabetes mellitus (IDDM) is believed to be an autoimmune disease, characterized by lymphocytic infiltration of the islets and the presence of islet cell autoantibodies. Autoimmunity may result from an intrinsically abnormal immune system, primary alterations of the target beta cell or both. However, the initial event that causes the beta cell-specific autoimmunity remains unknown.

Our recent experimental results showed that islet grafts from neonatal BB rats remained intact without insulitis when transplanted into the renal subcapsular space of

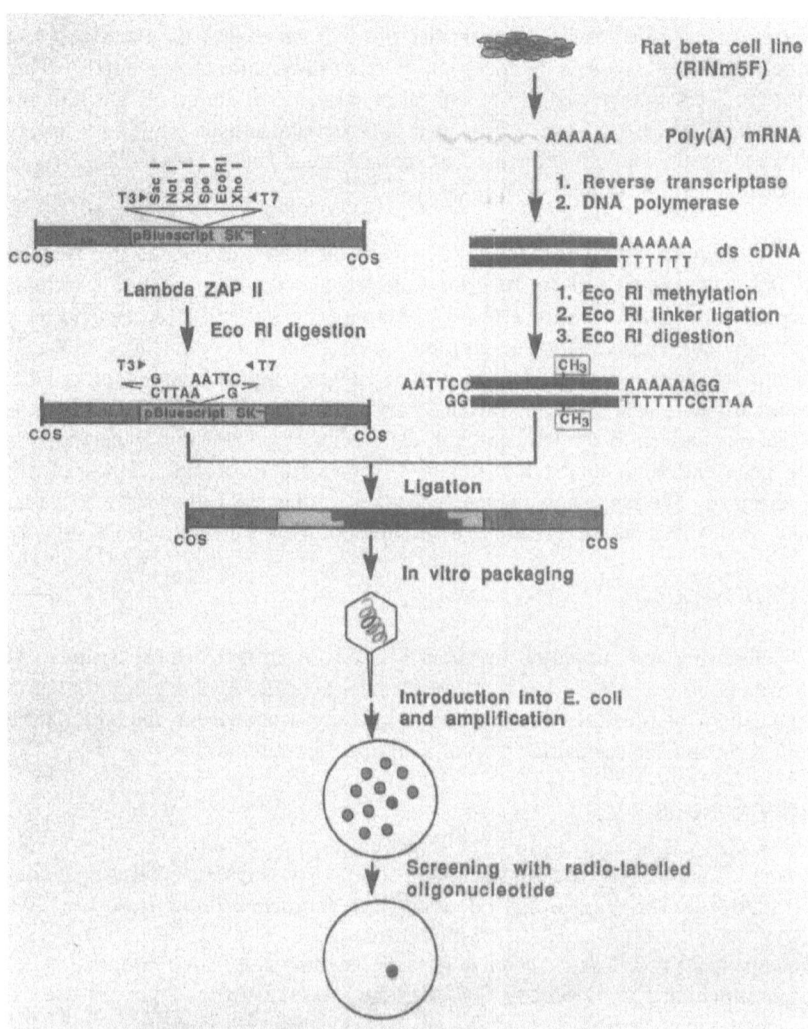

Fig. 3. **Cloning strategy of the 37 kd autoantigen gene.** Total RNA was isolated from RINm5F cells by the guanidium isothiocyanate method. Poly (A^+) RNA was purified by affinity chromatography on oligo-d(T) cellulose. The double-stranded cDNAs were synthesized with AMV reverse transcriptase and DNA polymerase using the modified method of Gubler and Hoffman (1983). After EcoRI methylation and EcoRI linker tailing, double-stranded cDNAs between 0.8 kb and 2.0 kb were purified by size fractionation on a Sephadex G-50 column. EcoRI-attached cDNAs were ligated into lambda Zap II vector (Stratagen, California) digested with Eco RI. Following in vitro packaging, the lambda ZAP II phages were plated on E. coli for amplification. The 33 oligonucleotides deduced from N-terminal amino acid sequence of the 37 kd autoantigen protein were synthesized and end-labelled with $r-^{32}P$ ATP. Phages from the plaques were transferred to nitrocellulose filters and screened with radio-labelled oligonucleotide.

215

acutely diabetic BB rats. In contrast, islet grafts from adult BB rats (which had been treated with silica for the prevention of insulitis) revealed severe insulitis and were rapidly destroyed. These results suggest that the delayed expression of a beta cell-specific autoantigen may result in the initiation of beta cell-specific autoimmunity. The islet cell-specific 37 kd autoantigen is not expressed early in the life of BB rats, but is expressed at around 30 days of age. This islet-specific autoantigen might be recognized and attacked by the immune effectors. In contrast, nondiabetic Wistar Furth rats express the autoantigen from birth.

We have attempted to clone and sequence the gene encoding for the islet cell-specific 37 kd autoantigen and to find the role of the autoantigen in the pathogenesis of autoimmune type I diabetes. The 37 kd protein, which cross-reacts with an anti-37 kd antibody (BB rat sera), was purified from a rat beta cell line (RINm5F) by the immunoprecipitation method. Using the synthetic oligonucleotide deduced from the N-terminal amino acid sequence of the 37 kd protein, we cloned the gene encoding for the 37 kd autoantigen in the RINm5F cell cDNA library. The entire nucleotide sequence of the gene encoding the 37 kd protein, which has a full length cDNA of 1.3 kb, was determined. We suggest that a gene encoding for the islet cell-specific 37 kd autoantigen may be involved in the initiation of autoimmune type I diabetes in BB rats.

ACKNOWLEDGEMENTS

This work was supported by Grant MA9584 from the Medical Research Council of Canada (J.W.Y.). J.W.Y. is a Heritage Medical Scientist Awardee of the Alberta Heritage Foundation for Medical Research. We gratefully acknowledge the editorial help of Dr. Ann Kyle and the secretarial help of Lorraine Pearson.

REFERENCES

Adams, T.E., Alpert, S., Hanahan, D., 1987, Non-tolerance and autoantibodies to a transgenic self antigen expressed in pancreatic beta cells. Nature 325: 223-228.

Amano, K., Yoon, J.W., 1990. Studies on autoimmunity for the initiation of beta cell destruction: V. Decrease of macrophage-dependent T-lymphocytes and NK cytotoxicity in silica-treated DP-BB rats. Diabetes 590-596.

Bach, J.F., 1988, Mechanisms of autoimmunity in insulin-dependent diabetes mellitus. Clin Exp Immunol 72: 1-8.

Baekkeskov, S., Nielsen, J.H., Marner, B., Bilde, T., Ludvigsson, J., Lernmark, A., 1982, Autoantibodies in newly diagnosed diabetic children immunoprecipitate human pancreatic islet cell proteins. Nature 298: 167-169.

Bendelic, A., Carnaud, C., Boiutard, C., Bach, J.F., 1987, Syngeneic transfer of autoimmune diabetes from diabetic NOD mice to healthy neonate: requirement for both L3T4$^+$ and Lyt2$^+$ T cells. J. Exp. Med. 166: 823-832.

Dean, B.M., Walker, R., Bone, A.J., Baird, J.D., Cooke, A., 1985, Pre-diabetes in the spontaneously diabetic BB/E rat: Lymphocyte subpopulations in the pancreatic infiltrate and expression of rat MHC Class II molecules in endocrine cells. Diabetologia 28: 464-466.

Drell, D.W., Notkins, A.L., 1987, Multiple immunological abnormalities in patients with Type 1 (insulin-dependent) diabetes mellitus. Diabetologia 30:132-143.

Dyrberg, T., 1986, Humoral autoimmunity in the pathogenesis of insulin-dependent diabetes mellitus. Studies in the spontaneously diabetic BB rat. Acta. Endocrinol. (Suppl.) (Copenh) 280: 1-29.

Gepts, W., 1984, Islet morphology in Type 1 diabetes. Behring Inst Mitt 75: 33-38.

Gubler, U. and Hoffman, B.J., 1983, A simple and very efficient method for generating cDNA libraries. Gene 25: 263-269.

Haskins, K., Portas, M., Bergman, B., Lafferty, K., Bradley, B., 1989, Pancreatic islet-specific T cell clones from nonobese diabetic mice. Proc. Natl. Acad. Sci. USA 86: 8000-8004.

Haskins, K., McDuffie, M., 1990, Acceleration of diabetes in young NOD mice with a $CD4^+$ islet-specific T cell clone. Science 249: 1433-1436.

Ihm, S.H., Lee, K.U., Yoon, J.W., 1991, Studies on autoimmunity for the initiation of beta cell destruction: VII. Evidence for antigenic changes on the beta cells leading to the autoimmune destruction of beta cells in BB rats. Diabetes 40: 269-274.

Katoka, S., Satoh, J., Fujiya, H., Toyota, T., Suzuki, R., Itoh, K., Kumagai, K., 1983, Immunological aspects of the nonobese diabetic (NOD) mouse: abnormalities of cellular immunity. Diabetes 32: 247-253.

Ko, I.Y., Yoon, J.W., 1991, Delayed expression of islet specific autoantigen in BB rats may be responsible for the initiation of autoimmune type I diabetes. Diabetes 40 (Suppl. 1): 54A.

Ko, I.Y., Ihm, S.H. and Yoon, J.W., 1991, Studies on autoimmunity for the initiation of beta cell destruction. VIII Pancreatic beta cell-dependent autoantibody to a 38 kd protein precedes the clinical onset of diabetes in BB rats. Diabetologia 34: 548-554.

Ko, I.Y., 1992, Molecular identification and characterization islet cell-specific 37 kd autoantigen in diabetes-prone BB rats, Ph.D. Thesis, University of Calgary.

Koevary, S., Rossini, A.A., Stoller, W., Chick, W., 1983, Passive transfer of diabetes in the BB/W rat. Science 220: 727-728.

Koike, T., Itoh, Y., Ishi, T., Ito, I., Takabayashi, K., Maruyama, T., Tomioka, H., Yoshida, S., 1987, Preventive effect of monoclonal anti-L3T4 antibody on development of diabetes in NOD mice. Diabetes 36: 539-541.

Kolb, H., Kantwert, G., Treichal, U., Kurner, T., Keisel, U., Hoppe, T., Kolb-Bachofen, V., 1986, Natural history of insulitis in BB rats. Pancreas 1: 370.

Kolb, H., 1990, Cellular immunity in Type 1 diabetes. IDF Bulletin 35: 37-39.

Laupacis, A., Stiller, C.R., Cardell, C., Keown, P., Dupre, J., Wallace, A.C., Thibert, P.,

1983, Cyclosporin prevents diabetes in BB Wistar rats. Lancet 1: 10-12.

Lee, K.U., Kim, M.K., Amano, K., Pak, C.Y., Jarworski, M.A., Mehta, J.G., Yoon, J.W., 1988a, Preferential infiltration of macrophages during early stages of insulitis in diabetes-prone BB rats. Diabetes 37: 1053-1058.

Lee, K.U., Amano, K., Yoon, J.W., 1988b, Evidence for initial involvement of macrophages in development of insulitis in NOD mice. Diabetes 37: 1989-1991.

Lee, K.U., Pak, C.Y., Amano, K., Yoon, J.W., 1988c, Prevention of lymphocytic thyroiditis and insulitis in diabetes-prone BB rats by the depletion of macrophages. Diabetologia 31: 400-402.

Like, A.A., Rossini, A.A., Appel, M.C., Guberski, D.L., Williams, R.M., 1979, Spontaneous diabetes mellitus: reversal and prevention in the BB/W rat with antiserum to rat lymphocytes. Science 206: 1421-1423.

Like, A.A., Kislauskis, E., Williams, R.M., Rossini, A.A., 1982, Neonatal thymectomy prevents spontaneous diabetes mellitus in the BB/W rat. Science 216: 644-646.

Like, A.A., Anthony, M., Guberski, D.L., Rossini, A.A., 1983, Spontaneous diabetes in the BB/W rat: effects of glucocorticoids, cyclosporin-A and antiserum to rat lymphocytes. Diabetes 32: 326-330.

Like, A.A., Biron, C.A., Weringer, E.J., Byman, K., Sroczynski, E., Guberski, D.L., 1986, Prevention of diabetes in Biobreeding/Worcester rats with monoclonal antibodies that recognize T-lymphocytes or natural killer cells. J. Exp. Med. 164: 1145-1159.

Marliss, E.B., Nakhooda, A.F., Poussier, P., Simma, A.A.F., 1982, The diabetic syndrome of the "BB" Wistar rat: possible relevance to Type 1 (insulin-dependent) diabetes in man. Diabetologia 22: 225-232.

Marliss, E.B., (Ed), 1983, The Juvenile Diabetes Foundation workshop on spontaneously diabetic BB rat as potential for insight into human juvenile diabetes. Metab Clin Exp 32 (Suppl 1) 1-166.

Matsudaira, P., 1987, Sequence from picomole quantities of proteins electroblotted onto polyvinylidene difluoride membranes. J. Biol. Chem. 262: 10035-38.

Miller, B.J., Appel, M.C., O'Neil, J.J., Wicker, L.S., 1988, Both the Lyt-2$^+$ and L3T4$^+$ T cell subsets are required for the transfer of diabetes in nonobese diabetic mice. J. Immunol. 140: 52-58.

Nagata, M., Yokono, K., Hayakawa, M., Kawase, Y., Hatamori, N., Ogawa, W., Yonezawa, K., Shii, K., Baba, S., 1989, Destruction of pancreatic islet cells by cytotoxic T lymphocytes in nonobese diabetic mice. J. Immunol. 143: 1155-1162.

Nagata, M. and Yoon, J.W., 1992, Studies on autoimmunity for T cell-mediated beta cell destruction: Distinct difference in beta cell destruction between CD4$^+$ and CD8$^+$ T cell clones derived from lymphocytes infiltrating the islets of NOD mice. Diabetes 41: 998-1008.

Nakano, N., Kikutani, H., Nishimoto, H., Kishimoto, T., 1991, T cell receptor V gene usage of islet β cell-reactive T cells is not restricted in non-obese diabetic mice. J. Exp. Med. 173: 1091-1097.

Oschilewski, U., Kiesel, U., Kolb, H., 1985, Administration of silica prevents diabetes in BB rats. Diabetes 34: 197-199.

Reich, E.P., Scaringe, D., Yagi, J., Sherwin, R.S., Janeway Jr. C.A., 1989a, Prevention of diabetes in NOD mice by injection of autoreactive T-lymphocytes. Diabetes 38: 1647-51.

Reich, E.P., Sherwin, R.S., Kanagawa, O., Janeway, C.A. Jr., 1989b, An explanation for the protective effect of the MHC Class II I-E molecule in murine diabetes. Nature 341: 326-329.

Roep, B.O., Arden, S.D., De Vries, R.R.P., Hutton, J.C., 1990, T-cell clones from a Type 1 diabetes patient respond to insulin secretory granule proteins. Nature 345: 632-634.

Rossini, A.A., Mordes, J.P., Like, A.A., 1985, Immunology of insulin-dependent diabetes mellitus. Ann Rev Immunol. 3: 289-320.

Shizuru, J.A., Taylor-Edwards, C., Banks, B.A., Gregory, A.E., Fathman, C.G., 1988, Immunotherapy of the nonobese diabetic mouse: Treatment with an antibody to T-helper lymphocytes. Science 240: 659-662.

Short, J.M., Fernandez, J.M., Sorge, J.A. and Huse, M.D., 1988, λZAP: A bacteriophage λ expression vector with in vivo excision properties. Nucleic Acids Res. 16: 7583-7600.

Unaue, E.R., 1984, Antigen-presenting function of the macrophage. Ann. Rev. Immunol. 2: 395-428.

Unaue, E.R., Allen, P.M., 1987, The basis for the immunoregulatory role of macrophages and other accessory cells. Science 236: 551-557.

Voorbij, H.A.M., Jeucken, P.H.M., Kabel, P.J., DeHaan, M., Drexhage, H.A., 1989, Dendritic cells and scavenger macrophages in pancreatic islets of prediabetic BB rats. Diabetes 38: 1623-1629.

Walker, R., Bone, A.J., Cooke, A., Baird, J.D., 1988, Distinct macrophage subpopulations in pancreas of prediabetic BB/E rats: possible role for macrophages in pathogenesis of IDDM. Diabetes 37: 1301-1304.

Wang, Y.I., Hao, L., Gill, R.G., Lafferty, K.J., 1987, Autoimmune diabetes in NOD mouse is L3T4 T-lymphocyte dependent. Diabetes 36: 535-538.

Yoon, J.W., Ihm, S.H., Lee, K.U., Amano, K., Pak, C.Y., 1988a, The initial step in the development of organ specific autoimmune disease in BB rats. Diabetologia 31: 779-780.

Yoon, J.W., Leiter, E.H., Coleman, D.L., Kim, M.K., Pak, C.Y., McArthur, R.G. and

Roncari, D.A.K., 1988b, Genetic control of organ-reactive autoantibody production in mice by obesity (ob)-Diabetes (db) genes. Diabetes 37: 1287-1293.

Yoon, J.W., 1989, Viral pathogenesis of insulin-dependent diabetes mellitus. Autoimmunity and the Pathogenesis of Diabetes (eds. Ginsberg-Fellner and R.C. McEvoy), Springer-Verlag, New York, pp. 206-257.

Yoon, J.W., 1990, Role of viruses and environmental factors in induction of diabetes. In: Human Diabetes: Genetic, Environmental and Autoimmune Etiology. (Eds. Baekkeskov, S. and Hansen, B.) Springer-Verlag, Current Topics in Microbiology and Immunology 164: 95-123.

Yoon, J.W., 1991, Role of viruses in the pathogenesis of IDDM. Annals of Medicine 23: 437-445.

MAPPING OF THE POLYPEPTIDE CHAIN ORGANIZATION OF THE MAIN EXTRACELLULAR DOMAIN OF THE α-SUBUNIT IN MEMBRANE-BOUND ACETYLCHOLINE RECEPTOR BY ANTI-PEPTIDE ANTIBODIES SPANNING THE ENTIRE DOMAIN

M. Zouhair Atassi[1]
Biserka Mulac-Jericevic[1] and
Tetsuo Ashizawa[2]

Departments of Biochemistry[1]
and Neurology[2], Baylor College
of Medicine, Houston, Texas

ABSTRACT

To study the organization of the polypeptide chain of the main extracellular domain of the nicotinic acetylcholine receptor (AChR) α-subunit, we examined the ability of the native membrane-bound AChR of *Torpedo californica* (T-AChR) to bind a panel of antibodies against overlapping synthetic peptides which collectively encompassed this entire domain. Antibodies against the α-chain peptides α1-16, α89-104 and α158-174 were able to bind to membrane-bound T-AChR. Other anti-peptide antibodies showed little or no binding to T-AChR in the membrane. It is concluded that regions α1-16, α89-104 and α158-174 are highly exposed on the surface of the α subunit of membrane-bound AChR.

INTRODUCTION

Nicotinic acetylcholine receptor (AChR) expressed on the postsynaptic membrane of the neuromuscular junction is a pentameric glycoprotein of about 250,000 dalton composed of two α subunits and one each of β, γ, and δ subunits (Reynolds and Karlin, 1978). The primary structure has been deduced from the cDNA sequence (Noda *et al.*, 1982; Noda *et al.*, 1983) and the secondary structure was estimated from its amino sequence. At least three models for intramembranous organization of each subunit have been developed (Claudio *et al.*, 1983; Devillers-Thiery *et al.*, 1983; Noda *et al.*, 1983; Guy, 1983; Finer-Moore and Stoud, 1984; Criado *et al.*, 1983). Immunochemical and α-neurotoxin binding studies (Atassi *et al.*, 1988) with AChR α-chain synthetic peptides firmly confirmed the subunit organization model depicting five trans-membrane regions (Guy, 1983; Finer-Moore and Stroud, 1984). As to the tertiary structure of the main extracellular domain (residues 1-210), a neurotoxin binding cavity has been defined (Ruan *et al.*, 1990), but the rest of the domain is yet to be characterized.

Previously, this laboratory introduced (Kazim and Atassi, 1980) a comprehensive synthetic approach that was specifically designed to localize the full profile of the continuous regions of antibody and T-cell recognition (as well as other recognition regions) on a protein molecule. This approach consists of the examination of the activities of consecutive synthetic overlapping peptides, of uniform size and overlaps, that encompass the entire protein chain (Kazim and Atassi, 1980, 1982). This strategy has been applied (Mulac-Jericevic et al., 1987) to the α-chain of Torpedo californica AChR (T-AChR) to localize the continuous regions recognized by mouse and rabbit anti-AChR antibodies, as well as regions concerned with other activities (Atassi et al., 1987). In the present work, we have employed the synthetic overlapping peptides encompassing the entire main extracellular part of the α-chain of T-AChR to prepare specific anti-peptide antibodies. These have been used to map the continuous surface regions of membrane-bound T-AChR.

MATERIALS AND METHODS

AChR synthetic peptide

Eighteen overlapping peptides representing the main extracellular part of T-AChR α subunit (residues α1-210, Figure 1) were synthesized, purified and characterized by the methods reported elsewhere (Mulac-Jericevic et al., 1987).

Antibody production

Fifty μg of each synthetic peptide in 50 μl of 0.15M NaCl in 0.01 M sodium phosphate buffer, pH 7.2 (PBS) was mixed with an equal volume of Freund's complete adjuvant and subcutaneously injected in the hind foot pads of ICR female mice. The immunizations were done on days 1, 15, 30 and 45. These immunizations were performed six years ago which was prior to the implementation of the NIH guidelines for immunization of animals. Two weeks after the fourth immunization (day 60), the blood was collected from the tail and the antibody titer was assessed by a solid-phase plate radioimmune assay. Rabbit immune IgG fractions with specificity against mouse IgG and IgM were obtained from Accurate Chemical Scientific Corp. (Westbury, NY).

Binding of anti-peptide antibodies to the peptides

The titers of the antisera were measured by binding to the immunizing peptide, using a solid phase radioimmunoassay, before they were used in experiments for binding to membrane-bound T-AChR. Each peptide was bound to the wells of a 96-well polyvinyl assay plate (Falcon, Lincoln Park, NJ) as described previously (Oshima and Atassi, 1989). Each antiserum was pre-diluted 1:500 - 1:1000 (vol/vol) in PBS/0.05% casein and added to the peptide-coated wells, including wells pre-coated with the peptide used to raise the particular antiserum. After incubation at $37°C$, for 3 hours, the wells were washed 5 times with PBS, then incubated with ^{125}I-labeled rabbit antibodies against mouse (IgG + IgM) washed and cut out to measure the bound anti-peptide antibody.

Binding of anti-peptide antibodies to membrane-bound T-AChR

Electric organ tissue from T. californica was a homogenized in 10 mM Tris-buffer containing 1 mM EDTA, 1 mM EGTA, 10 unit/ml aprotinin and 0.1 mM phenyl methyl sulfonyl fluoride at pH 7.4 and centrifuged at 27,000 x g for 45 min. The pellet was suspended in the same buffer and used as a T-AChR membrane fraction.

Peptide Number	Sequence Position	Structure
1	$\alpha1-16$	S E H E T R L V A N L **L E N Y N**
2	$\alpha12-27$	**L E N Y N** K V I R P V **E H H T H**
3	$\alpha23-38$	**E H H T H** F V D I T V **G L Q L I**
4	$\alpha34-49$	**G L Q L I** Q L I S V D **E V N Q I**
5	$\alpha45-60$	**E V N Q I** V E T N V R **L R Q Q W**
6	$\alpha56-71$	**L R Q Q W** I D V R L R **W N P A D**
7	$\alpha67-82$	**W N P A D** Y G G I K K **I R L P S**
8	$\alpha78-93$	**I R L P S** D D V W L P **D V V L Y**
9	$\alpha89-104$	**D V V L Y** N N A D G D **F A I V H**
10	$\alpha100-115$	**F A I V H** M T K L L L **D Y T G K**
11	$\alpha111-126$	**D Y T G K** I M W T P P **A I F K S**
12	$\alpha122-138$	**A I F K S** Y G E I I V T **H F P F D**
13	$\alpha134-150$	**H F P F D** Q Q N G T M K **L G I W T**
14	$\alpha146-162$	**L G I W T** Y D G T K V S **I S P E S**
15	$\alpha158-174$	**I S P E S** D R P D L S T **F M E S G**
16	$\alpha170-186$	**F M E S G** E W V M K D Y R G **W K H**
17	$\alpha182-198$	**R G W K H** W V Y Y T G G **P T T P Y**
18	$\alpha194-210$	**P T T P Y** L D I T Y H F I M Q R I

Figure 1. Covalent structure of the synthetic overlapping peptides of *T. californica* AChR α-subunit. Boldface letters represent the residues of overlap.

The T-AChR membrane fraction (50 μl containing 5 μg protein) was plated in a 96-well polyvinyl assay plate as an antigen and allowed to stand overnight at room temperature. The mouse antisera against the peptides were diluted with PBS containing 0.05% casein to give in some experiments, equal amounts of antibody binding (50,000 cpm) to each of the immunizing peptides, and to 1:500 - 1:1000 (vol/vol) in others. After washing 5 times with PBS containing 0.05% Tween-20, the wells were incubated with 100 μl of diluted antisera at room temperature overnight. The wells were washed with PBS (5 times) and then incubated with the second (rabbit) antibody against mouse IgG + IgM for 2 hours at room temperature. After washing 5 times with PBS, the wells were incubated with ^{125}I-protein A (2x10^5 cpm) for 2 hours at room temperature, then washed 5 times with PBS and bound radioactivity of each well was assayed.

All experiments for antibody titer and antisera binding to T-AChR membrane fraction were performed in triplicates. The experiments were repeated at least twice. Binding of preimmune sera of the same mice to membrane-bound T-AChR and binding of the immune sera to proteins (myoglobin, bovine serum albumin, casein) that are unrelated to AChR were used as negative controls.

RESULTS

Anti-peptide antibodies and their specificity

Immunization of mice with each peptide in its free form (i.e without conjugation to any carrier) elicited the formation of anti-peptide antibodies. It was first shown by Young and Atassi (1982) that small synthetic peptides, when used in their free state as immunogens, will stimulate the formation of antibodies against the immunizing peptide (for review, see Atassi, 1986). Antisera against each peptide bound to the immunizing peptide. Some anti-peptide antisera bound to one or the other (but not to both) of the two peptides that overlap the immunizing peptide and precede it or follow it in the sequence, indicating that the recognition of some antisera may have shifted to the left or to the right part of the immunizing peptide. Other than these occasional cross-reactions with an adjacent overlapping peptide, no other peptide cross-reactions were found.

Binding of anti-peptide antibodies to membrane-bound T-AChR

At 1:500 to 1:1000 dilution of each of the anti-peptide antisera, those against T-AChR peptides α1-16, α89-104 and α158-174 showed the highest binding to T-AChR membrane fraction. Antisera against peptides α146-162 and α170-186 gave particularly low binding even though these antisera possessed very high initial titers. When the titers were adjusted to give equal amounts (50,000 cpm) of peptide-bound antibodies in the solid phase radioimmunoassay, the binding of antisera against the peptides α1-16, α89-104 and α158-174 to the T-AChR membrane fraction remained greater than that exhibited by the other anti-peptide antisera (Figure 2).

DISCUSSION

In this study, we prepared antisera against a panel of synthetic overlapping peptides encompassing the entire main extracellular domain of T-AChR α-subunit (residues 1-210), and examined the binding of each antiserum to native T-AChR in the membrane fraction of *T. californica* electric organ. The antisera against peptides α1-16, α89-104 and α158-174 bound to T-AChR, indicating that these peptides contain well-exposed regions in the α-chain of the membrane-bound receptor.

It should be noted, however, that there are variables which could potentially influence the binding profile of these anti-peptide antibodies to the membrane-bound T-AChR. First, the titers of the anti-peptide sera may be critical. It may be expected that the higher the titers are, the greater the binding. In our experiments, however, the binding profile was not altered by the dilution of the antisera. Furthermore, adjusting the dilutions of the antisera to equal titers did not alter the fundamental binding profile, clearly indicating that accessibility of the regions to the antisera, rather than the variations of antiserum titers, is the major determinant of the antibody binding profile. Second, the anti-peptide antibodies may not be able to bind to regions that are actually exposed but may be partially protected by the surrounding membrane or other parts of the tertiary structure, allowing accessibility only for molecules that are smaller than antibodies.

To initiate the immunopathological events at the motor end plates in myasthenia gravis (Ashizawa and Appel, 1985), autoantibodies must first bind exposed epitopes of the extracellular regions of the intact, native AChR in the membrane. The synthetic peptides comprising residues α182-198 bind α-neurotoxin when T-AChR sequences are employed (Mulac-Jericevic and Atassi, 1987a,b) but the binding is much weaker when mammalian sequences are used (Mulac-Jericevic et al.,1988). Our data showed that the anti-peptide antibodies exhibited little or no binding to this region in the membrane-bound T-AChR, and would indicate that this region is not exposed in the

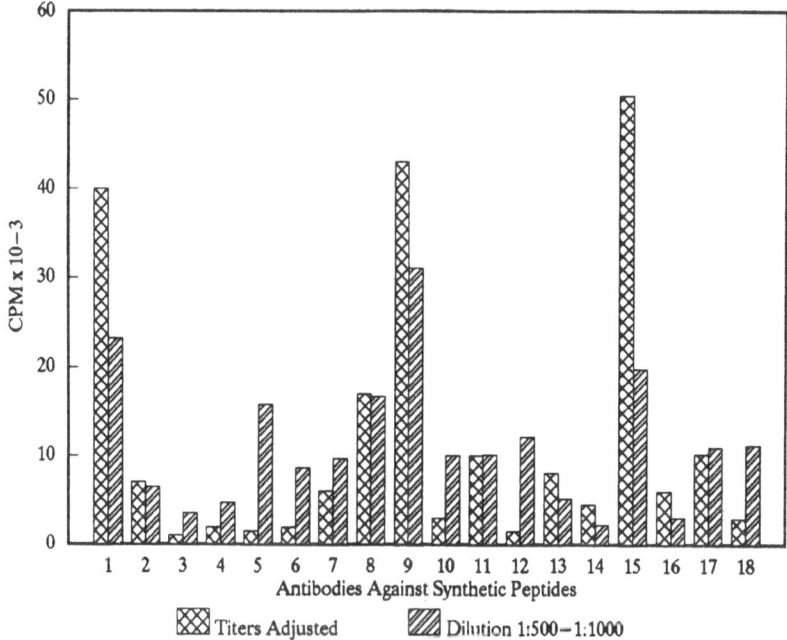

Figure 2. Binding profiles of antibodies against the synthetic overlapping peptides of the T-AChR α-subunit to the electric organ membrane of T. californica. The binding profile was obtained with anti-peptide antibodies at 1:500-1:1000 dilution and anti-peptide antibodies, the dilution of which were adjusted to equal titers (see the text for details).

native T-AChR. A T-AChR peptide (residues α183-200) induces EAMG in rats, and these EAMG animals had anti-peptide antibodies with rat AChR blocking activities (Takamori et al., 1988). This peptide, possessing the rat sequence α183-200, bound antibody in human myasthenia sera (Takamori et al., 1988). The human peptide corresponding to this region had neither of these two properties (Takamori et al., 1988; Ashizawa et al., 1992). Thus, antibodies against T-AChR peptide α183-200 may have antigenic cross-reactivity with a mammalian AChR region away from residues α183-200, and binding of the antibodies blocks the ligand binding either by inducing a conformational change or by steric hindrance. The primary cholinergic binding site is expected to be conserved across species. The region α183-200 shows poor ligand/α-bungarotoxin binding in mammals and is not exposed in T-AChR in spite of the high affinity binding to ligand and α-bungarotoxin. Furthermore, ^{3}H-acetylcholine mustard, which specifically binds in the cholinergic agonist binding site, failed to react with the tyrosyl or acidic side chains within T-AChR α180-200 (Cohen *et al.*, 1991).

It is evident from the foregoing that region α183-200 is not the primary cholinergic binding site of functional importance. On the other hand, T-AChR peptide α125-147 has been shown to bind acetylcholine (McCormick and Atassi, 1984), and immunization of rats with this peptide induces EAMG (Lennon *et al.*, 1985). This region is highly conserved in AChR of various species. Within this conserved part of AChR resides a universal binding region in *Torpedo* and mammalian AChRs for both long and short α-neurotoxin (Ruan *et al.*, 1991). Antibodies against this region did not bind to the native T-AChR bound in the membrane. One explanation for this apparent discrepancy is that the tertiary structure in this region of AChR may allow access of only α-neurotoxin but not anti-peptide antibodies. Indeed, modelling studies based on peptide-peptide interactions have shown this region to form a side of a deep cavity in which the toxin loops fit (Ruan *et al.*, 1990).

Previously, we have mapped the full profile of the binding regions for long (Mulac-Jericevic and Atassi, 1987a,b; Mulac-Jericevic *et al.*, 1988) and short neurotoxins (Ruan *et al.*, 1991) on *T. californica* and human AChR. Of the three regions (α1-16, α89-104 and α158-174) that were shown here to be exposed on the membrane-bound AChR by anti-peptide antibodies, only peptide α1-16 in T-AChR showed low binding to long neurotoxins (α-bungarotoxin and cobratoxin) and to the short neurotoxin, cobrotoxin. Peptides α89-104 and α158-174 did not bind any neurotoxin. This would indicate that peptides α89-104 and α158-174 are not in the toxin-binding cavity (Ruan *et al.*, 1990). Our mapping of the antigenic-profile with antibodies against isolated T-AChR or against membrane-bound AChR showed that only peptide α1-16 exhibited a small amount of antibody binding while the other two peptides showed no binding activity (Mulac-Jericevic *et al.*, 1987). Again, in the mapping of the full profile of the regions on the α-chain of human AChR that are recognized by human autoantibodies in MG patients, only peptide α1-16 showed low autoantibody binding with three out of 15 human MG autoantisera (Ashizawa *et al.*, 1992). This would suggest the peptide α1-16 might be exposed. The present studies clearly show that all three regions α1-16, α89-104 and α158-174 are well exposed on the surface of membrane-bound AChR. It is evident, therefore, that despite their full exposure, regions α89-104 and α158-174 do not appear to contribute to any of the known functional sites of AChR, while region α1-16 makes some slight contribution to antibody and α-neurotoxin recognition of the receptor. Clearly, surface exposure and participation in biological function do not necessarily coincide. Thus, not every exposed region participates in function and not every functional site needs to be exposed (e.g. a binding cavity).

Finally, it should be noted that the overlapping peptide strategy was designed (Kazim and Atassi, 1980, 1982) to map the continuous binding regions of a protein molecule. The strategy does not provide information about discontinuous binding sites. Accordingly, the present findings do not, in any way, preclude that the AChR molecule may have large surface areas that are assembled by a discontinuous architecture (i.e. surface areas formed by different regions of the molecule brought together by the folding of the polypeptide chain; for definitions, see Atassi and Smith, 1978). In the absence of a knowledge of the 3-D structure, it is not possible at the present time to map exposed regions of AChR that are assembled by a discontinuous architecture.

ACKNOWLEDGEMENTS

This work was supported by a grant (NS-26280) from the National Institutes of Health. The support of the Welch Foundation due to the award to MZA of the Robert A. Welch Chair of Chemistry is also gratefully acknowledged.

REFERENCES

Ashizawa, T., Ruan, K.H., Jinnai, K and Atassi, M.Z. (1992) Profile of the regions on the α-chain of human acetylcholine receptor recognized by autoantibodies in myasthenia gravis. *Mol. Immunol.* **29**, 1507-1514.

Ashizawa, T. and Appel, S.H. (1985) Immunopathological events at the end plate in myasthenia gravis. *Springer Sem. Immunopathol.* **8**, 177-196.

Atassi, M.Z. (1986) Preparation of monoclonal antibodies to preselected protein regions *Methods in Enzymol.* **12**, 69-95.

Atassi, M.Z. and Smith, J.A. (1978) A proposal for the nomenclature of antigen sites in peptides and proteins. *Immunochemistry* **15**, 609-610.

Atassi, M.Z., Mulac-Jericevic, B., Yokoi, T. and Manshouri, T. (1987) Localization of the functional sites on the α-chain of acetylcholine receptor. *Fed. Proc.* **46**, 2538-2547.

Claudio, T., Ballivet, M., Patrick, J. and Heinemann, S. (1983) Nucleotide and deduced amino acid sequences of *Torpedo californica* acetylcholine receptor subunit. *Proc. Natl. Acad. Sci. USA* **80**, 1111-1115.

Cohen, J.B., Sharp, S.D. and Liu, W.S. (1991) Structure of the agonist-binding site of the nicotinic acetylcholine receptor: 3H:acetylcholine mustard identified residues in the cation-binding subsite. *J. Biol. Chem.* **266**, 23354-23364.

Criado, M., Hochschwender, S., Sarin, V., Fox, J.L. and Lindstrom, J. (1985) Evidence for unpredicted transmembrane domains in acetylcholine receptor subunits *Proc. Natl. Acad. Sci. USA* **82**, 2004-2008.

Devillers-Thiery, A., Giraudat, J., Bentaboulet, M. and Changeux, J.-P. (1983) Complete mRNA coding sequence of the acetylcholine binding α-subunit of *Torpedo marmorata* acetylcholine receptor: a model for the transmembrane organization of the polypeptide chain. *Proc. Natl. Acad. Sci. USA* **80**, 2067-2071.

Finer-Moore, J. and Stroud, R.M. (1984) Amphipathic analysis and possible formation of the ion channel in an acetylcholine receptor. *Proc. Natl. Acad. Sci. USA* **81**, 155-159.

Guy, H.R. (1983) A structural model of the acetylcholine receptor channel based on partition energy and helix packing calculations. *Biophys. J.* **45**, 249-261.

Kazim, A.L. and Atassi, M.Z. (1980) A novel and comprehensive synthetic approach for the elucidation of protein antigenic structures. Determination of the full antigenic profile of the α chain of human haemoglobin. *Biochem. J.* **191**, 261-264.

Kazim, A.L. and Atassi, M.Z. (1982) Structurally inherent antigenic sites. Localization of the antigenic sites of the α-chain of human haemoglobin in three host species by a comprehensive synthetic approach. *Biochem. J.* **203**, 201-208.

Lennon, V.A., McCormick, D.J., Lambert, E.H., Griesmann, G.E. and Atassi, M.Z. (1985) Region of peptide 125-147 of acetylcholine receptor α-subunit is exposed at neuromuscular junction and induces experimental autoimmune myasthenia gravis, T-cell immunity and modulating autoantibodies. *Proc. Natl., Acad. Sci. USA* **82**, 8805-8809.

McCormick, D.J. and Atassi, M.Z. (1984) Localization and synthesis of the acetylcholine-binding site in the α-chain of the *Torpedo californica* acetylcholine receptor, *Biochem. J.* **244,** 995-1000.

Mulac-Jericevic, B. and Atassi, M.Z. (1987a) Profile of the α-bungarotoxin-binding regions on the extracellular part of the α-chain of *Torpedo californica* acetylcholine receptor. *Biochem. J.* **248**, 847-852.

Mulac-Jericevic, B. and Atassi, M.Z. (1987b) Neurotoxin binding to acetylcholine receptor: localization of the full profile of the cobra toxin-binding regions on the α-chain of *Torpedo californica* acetylcholine receptor by a comprehensive synthetic strategy. *J. Protein Chem.* **6**, 365-373.

Mulac-Jericevic, B., Kurisaki, J. and Atassi, M.Z. (1987) Profile of the continuous antigenic regions on the extracellular part of the α chain of an acetylcholine receptor. *Proc. Natl. Acad. Sci. USA,* **84**, 3633-3637.

Mulac-Jericevic, B., Manshouri, T., Yokoi, T.A and Atassi, M.Z. (1988) The regions of α-neurotoxin binding on the extracellular part of the α-subunit of human acetylcholine receptor. *J. Prot. Chem.* **7**, 173-177.

Noda, M., Takahashi, H., Tanabe, T., Toyosato, M., Furutani, Y., Hirose, T., Asai, M., Inayama, S., Miyata, T. and Numa, S. (1982) Primary structure of α-subunit precursor of *Torpedo californica* acetylcholine receptor deduced from cDNA sequence. *Nature (London)* **299**, 793-797.

Noda, M., Takahashi, H., Tanabe, T., Toyosato, M., Kikyotani, S., Furutani, Y., Hirose, T., Takashima, H., Inayama, S., Miyata, T. and Numa, S (1983) Structural homology of *Torpedo californica* acetylcholine receptor subunits. *Nature (London)* **302,** 528-532.

Oshima, M. and Atassi, M.Z. (1989) Comparison of peptide-coating conditions in solid phase plate assays for detection of anti-peptide antibodies. *Immunol. Invest.* **18,** 841-851.

Reynolds, J.A. and Karlin, A. (1978) Molecular weight in detergent solution of acetylcholine receptor from *Torpedo californica*. *Biochemistry* **17**, 2035-2038.

Ruan, K.-H., Spurlino, J., Quiocho, F.A. and Atassi, M.Z. (1990) Acetylcholine receptor α bungarotoxin interactions: determination of the region-to-region contacts by peptide-peptide interactions and molecular modeling of the receptor cavity. *Proc. Natl. Acad. Sci. USA* **87**, 6156-6160.

Ruan, K.-H., Stiles, B.G. and Atassi, M.Z. (1991) The short-neurotoxin binding regions on the α-chain of human and *Torpedo californica* acetylcholine receptors. *Biochem. J.* **274**, 849-854.

Takamori, M., Okumura, S., Nagat and Yoshikawa, H. (1988) Myasthenogenic significance of synthetic α-subunit peptide 183-200 of *Torpedo californica* and human acetylcholine receptor. *J. Neurol. Sci.* **85**, 121-129.

Young, C.R. and Atassi, M.Z. (1982) Antibodies to sperm whale myoglobin evoked by free synthetic peptides of an antigenic site. *Immunological Communications* **11**, 9-16.

INDEX

230